T0257529

Bacteriophages: Biological Aspects and Advances

Bacteriophages: Biological Aspects and Advances

Edited by **Dean Watson**

New York

Published by Callisto Reference,
106 Park Avenue, Suite 200,
New York, NY 10016, USA
www.callistoreference.com

Bacteriophages: Biological Aspects and Advances
Edited by Dean Watson

© 2015 Callisto Reference

International Standard Book Number: 978-1-63239-085-1 (Hardback)

Printed in the United States of America.

Contents

Preface

This book has been a concerted effort by a group of academicians, researchers and scientists, who have contributed their research works for the realization of the book. This book has materialized in the wake of emerging advancements and innovations in this field. Therefore, the need of the hour was to compile all the required researches and disseminate the knowledge to a broad spectrum of people comprising of students, researchers and specialists of the field.

Bacteriophages are considered as biological agents and their application as an appliance has been further enhanced in various areas of microbiology. Especially in drug design and advanced programs, phage and prophage genomics give an altogether new understanding in this field. This book provides knowledge on the organisms varying from their biology to their usage in agriculture and medicine. This book consists of a range of topics dealing with advancing technologies in this field, commencing with the biology and categorization of bacteriophages, discussing phage infections in industrial processes and their applications as therapeutic or biocontrol agents. This book will prove itself to be beneficial as a reference and resource for microbiologists, biotechnologists, agriculture, biomedical and sanitary engineers.

At the end of the preface, I would like to thank the authors for their brilliant chapters and the publisher for guiding us all-through the making of the book till its final stage. Also, I would like to thank my family for providing the support and encouragement throughout my academic career and research projects.

Editor

Part 1

Biology and Classification of Bacteriophages

Bacteriophages
and Their Structural Organisation

E.V. Orlova
Institute for Structural and Molecular Biology,
Department of Biological Sciences, Birkbeck College,
UK

1. Introduction

Viruses are extremely small infectious particles that are not visible in a light microscope, and are able to pass through fine porcelain filters. They exist in a huge variety of forms and infect practically all living systems: animals, plants, insects and bacteria. All viruses have a genome, typically only one type of nucleic acid, but it could be one or several molecules of DNA or RNA, which is surrounded by a protective stable coat (capsid) and sometimes by additional layers which may be very complex and contain carbohydrates, lipids, and additional proteins. The viruses that have only a protein coat are named "naked", or non-enveloped viruses. Many viruses have an envelope (enveloped viruses) that wraps around the protein capsid. This envelope is formed from a lipid membrane of the host cell during the release of a virus out of the cell.

Viruses interacting with different types of cells in living organisms produce different types of disease. Each virus infects a certain type of cell which is usually called "host" cell. The major feature of any viral disease is cell lysis, when a cell breaks open and subsequently dies. In multicellular organisms, if enough cells die, the entire organism will endure problems. Some viruses can cause life-long or chronic infections, where the viruses continue to replicate in the body despite the host's defence mechanisms. The other viruses cause lifelong infection because the virus remains within its host cell in a dormant (latent) state such as the herpes viruses, but the virus can reactivate and produce further attacks of disease at any time, if the host's defence system became weak for some reason (Shors, 2008).

Viruses have two phases in their life cycle: outside cells and within the cells they infect. Viral particles outside cells could survive for a long time in harsh conditions where they are inert entities called virions. Outside living cells viruses are not able to reproduce since they lack the machinery to replicate their own genome and produce the necessary proteins. Viruses can infect host cells, recognising their specific receptors on the cell surface. The viral receptors are normal surface host cell molecules involved in routine cellular functions, but since a portion of a molecular complex on the viral surface (typically spikes) has a shape complementary to the shape of the outer soluble part of the receptor, the virus is able to bind the receptor and be attached to the host cell's surface. After receptor-mediated attachment to its host the virus must find a way to enter the cell. Both enveloped and non-enveloped viruses use proteins present on their surfaces to bind to and enter the host cell employing the endocytosis mechanism (Lopez & Arias, 2010). The endocytic vesicles transport the viral particles to the perinuclear area of the host cell, where the conditions for viral replication are optimal. The other way of infection is to inject only the viral genome (sometimes accompanied by additional proteins) directly into the host cytoplasm.

The viruses are very economical: they carry only the genetic information needed for replication of their nucleic acid and synthesis of the proteins necessary for their reproduction (Alberts et al., 1989). Interestingly, the survival of viruses is totally dependent on the continued existence of their host, since after infection the viral genome switches the entire active host metabolism to synthesise the virion components. Without living host cells viruses will not be able to produce their progeny.

With the discovery of the electron microscope it became possible to study the morphology of viruses. The first studies immediately revealed that viruses could be distinguished by their size and shape, which became the important characteristics of their description. Viruses may be of a circular or oval shape, have the appearance of long thick or thin rods, which could be flexible or stiff. Some viruses have distinctive heads and a tail. The smallest viruses are around 20 nm in diameter and the largest around 500 nm.

The viruses that infect and use bacteria resources are classified as bacteriophages. Often we refer to them as "phages". The word "bacteriophage" means to eat bacteria, and is so called because virulent bacteriophages can cause the compete lysis of a susceptible bacterial culture. Bacteriophages, like bacteria, are very common in all natural environments and are directly related to the numbers of bacteria present. As a consequence they represent the most abundant 'life' forms on Earth, with an estimated 10^{32} bacteriophages on the planet (Wommack & Colwell, 2000). Phages can be readily isolated from faeces and sewage, thus very common in soil. Sequencing of bacterial genomes has revealed that phage genome elements are an important source of sequence diversity and can potentially influence pathogenicity and the evolution of bacteria. The number of phages that have been isolated and characterised so far corresponds to only a tiny fraction of the total phage population. Since bacteriophages and animal cell viruses have many similarities phages are used as model systems for animal cell viruses to study steps of the viral life cycle and to understand the mechanisms by which bacterial genes can be transferred from one bacterium to another.

1.1 Discovery of bacteriophages

Bacteriophages were discovered more than a century ago. In 1896, Ernest Hanbury Hankin, a British bacteriologist (1865 –1939), reported that something in the waters of rivers in India had unexpected antibacterial properties against cholera and this water could pass through a very fine porcelain filter and keep this distinctive feature (Hankin, 1896). However, Hankin did not pursue this finding. In 1915, the British bacteriologist Frederick Twort *(1877–1950)*, Superintendent of the Brown Institution of London, discovered a small agent that killed colonies of bacteria in growing cultures. He published the results but the subsequent work was interrupted by the beginning of World War I and shortage of funding. Felix d'Herelle (1873-1949) discovered the agent killing bacteria independently at the Pasteur Institute in France in 1917. He observed that cultures of the dysentery bacteria disappear with the addition of a bacteria-free filtrate obtained from sewage. D'Herelle has published his discovery of "an invisible, antagonistic microbe of the dysentery bacillus" (d'Herelle, 1917).

In 1923, the Eliava Institute was opened in Tbilisi, Georgia, to study bacteriophages and to develop phage therapy. Since then many scientists have been involved in developing techniques to study phages and using them for various purposes. In 1969 Max Delbrück, Alfred Hershey and Salvador Luria were awarded the prestigious Nobel Prize in Physiology and Medicine for their discoveries of the replication of viruses and their genetic structure.

1.2 Why do we need to study bacteriophages?

The first serious research of phages was done by d'Herelle which inspired him to do first experiments using phages in medicine. D'Herelle has used phages to treat a boy who had bad disentheria (d'Herelle, 1917). After administration of phages the boy successfully recovered. Later d'Herelle and scientists from Georgia (former USSR) have created an Institute to study the properties of bacteriophages and their use in treating bacterial infections a decade before the discovery of penicillin. Unfortunately a lack of knowledge on basic phage biology and their molecular organisation has led to some clinical failures. At the end of 1930s antibiotics were discovered; they were very effective, and nearly wiped out studies on the medical use of phages. However, a new problem of bacterial resistance to antibiotics has arisen after many years of using them. Bacteria adapted themselves to become resistant to the most potent drugs used in modern medicine. The emergence of modified pathogens such as *Mycobacterium tuberculosis*, *Enterococcus faecalis*, *Staphylococcus aureus*, *Acinetobacter baumannii* and *Pseudomonas aeruginosa*, and methicillin-resistant *S. aureus* (MRSA) has created massive problems in treating patients in hospitals (Coelho et al., 2004, Hanlon, 2007; Burrowes et al., 2011) and the time required to produce new antibiotics is much longer than the time of bacterial adaptation. Modern studies on the phage life cycle have revealed a way for their penetration through membrane barriers of cells. These results are important in the development of methods for using bacteriophages as a therapeutic option in the treatment of bacterial infections (Brussow & Kutter, 2005). Phages, like many other viruses, infect only a certain range of bacteria that have the appropriate receptors in the outer membrane. The antibiotic resistance of the bacteria does not affect the infectious activity of a phage. Knowledge of the phage structure, understanding the mechanism of phage-cell surface interaction, and revealing the process of switching the cell replication machinery for phage propagation would allow the design of phages specific for bacterial illnesses.

2. Classification of bacteriophages

All known phages can be divided in two groups according to the type of infection. One group is characterised by a lytic infection and the other is represented by a lysogenic, or temperate, type of infection (Figure 1). In the first form of infection the release of DNA induces switching of the protein machinery of the host bacterium for the benefit of infectious agents to produce 50-200 new phages. To make so many new phages requires nearly all the resources of the cell, which becomes weak and bursts. In other words, lysis takes place, causing death of the host bacterial cell. As result new phages are released into the extracellular space. The other mode of infection, lysogenic, is characterised by integration of the phage DNA into the host cell genome, although it may also exist as a plasmid. Incorporated phage DNA will be replicated along with the host bacteria genome and new bacteria will inherit the viral DNA. Such transition of viral DNA could take place through several generations of bacterium without major metabolic consequences for it. Eventually the phage genes, at certain conditions impeding the bacterium state, will revert to the lytic cycle, leading to release of fully assembled phages (Figure 1). Analysis of phages with lysogenic or lytic mode of infection has shown that there is a tremendous variety of bacteriophages with variations in properties for each type of infection. Moreover, under certain conditions, some species were able to change the mode of infection, especially if the number of host cells was falling down. Temperate phages are not suitable for the phage therapy.

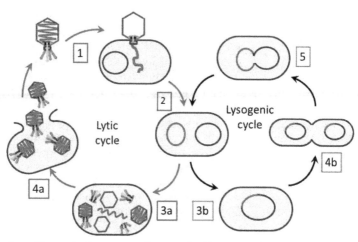

Fig. 1. Two cycles of bacteriophage reproduction. 1 - Phage attaches the host cell and injects DNA; 2 – Phage DNA enters lytic or lysogenic cycle; 3a – New phage DNA and proteins are synthesised and virions are assembled; 4a –Cell lyses releasing virions; 3b and 4b – steps of lysogenic cycle: integration of the phage genome within the bacterial chromosome (becomes prophage) with normal bacterial reproduction; 5- Under certain conditions the prophage excises from the bacterial chromosome and initiates the lytic cycle. (Copyright of E.V. Orlova)

Classification of viruses is based on several factors such as their host preference, viral morphology, genome type and auxiliary structures such as tails or envelopes. The most up-to-date classification of bacteriophages is given by Ackermann (2006). The key classification factors are phage morphology and nucleic acid properties. The genome can be represented by either DNA or RNA. The vast majority of phages contain double strand DNA (dsDNA), while there are small phage groups with ssRNA, dsRNA, or ssDNA (ss stands for single strand). There are a few morphological groups of phages: filamentous phages, isosahedral phages without tails, phages with tails, and even several phages with a lipid-containing envelope or contain lipids in the particle shell. This makes bacteriophages the largest viral group in nature. At present, more than 5500 bacterial viruses have been examined in the electron microscope (Ackermann, 2007) (Figure 2).

Pleomorphic and filamentous phages comprise ~190 known bacteriophages (3.6% of phages) and are classified into 10 small families (Ackermann, 2004). These phages differ significantly in their features and characteristics apparently representing different lines of origin. Pleomorphic phages are characterized by a small number of known members that are divided into three

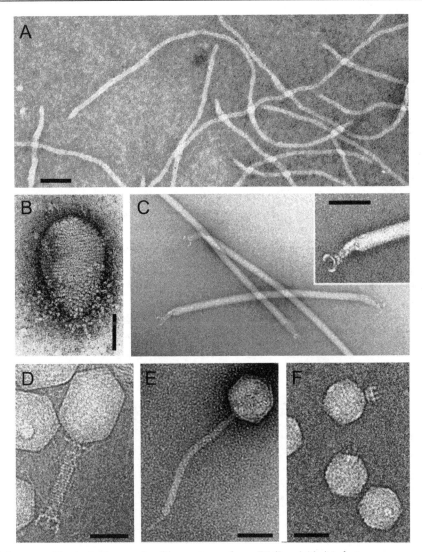

Fig. 2. Images of bacteriophages. A – filamentous phage B5 (*Inoviridae*) infects *Propionibacterium freudenreichi;* negatively stained with 2% uranyl acetate (UA) (Chopin et al., 2002, reproduced with permission of M.C. Chopin); B - *Sulfolobus neozealandicus* droplet-shaped virus (*Guttaviridae*) of the crenarchaeotal archaeon *Sulfolobus*, negatively stained by 2% UA (Arnold et al., 2000, reproduced with permission of W. Zillig); C - *Acidianus* filamentous virus 1 (*Lipothrixviridae*) with tail structures in their native conformation, negatively stained with 3% UA (Bettstetter et al., 2003, reproduced with permission of D. Prangishvili); D – Bacteriophage T4 (*Myoviridae*), in vitreous ice (Rossmann et al., 2004, reproduced with permission of M.G. Rossman); E – Bacteriophage SPP1 (*Siphoviridae*), negative stain 2% UA (Lurz et al., 2001, reproduced with permission of R. Lurz); F - Bacteriophage P22 (*Podoviridae*) in vitreous ice (Chang et al., 2006, reproduced with permission of W. Chiu). Bars are 50 nm.

families that need further characterization. *Plasmaviridae* (dsDNA) includes phages with dsDNA that are covered by a lipoprotein envelope and therefore can be called a nucleoprotein granule. Members of the *Fusseloviridae* family have dsDNA inside a lemon-shaped capsid with short spikes at one end; *Guttavirus* phage group (dsDNA) is represented by droplet-shaped virus-like particles (Figure 2B, Arnold et al., 2000).

There are phages with helical or filamentous organization. The *Inoviridae* (ssDNA) family includes phages that are long, rigid, or flexible filaments of variable length and have been classified by particle length, coat structure and DNA content. The *Lipothrixviridae* (dsDNA) phages are characterized by the combination of a lipoprotein envelope and rod-like shape (Figure 2C). The *Rudiviridae* (dsDNA) family represents phages that are straight rigid rods without envelopes and closely resemble the tobacco mosaic virus.

The next group of phages have capsids with an isosahedral shape. Phages from the *Leviviridae* family have ssRNA genome packed in small capsids and resemble enteroviruses. The known phages that form *Corticoviridae* family contain three molecules of dsRNA and, which is unusual, RNA polymerase. Phages with icosahedral symmetry for the capsids and a DNA genome compose the next three families *Microviridae*, *Cystoviridae* and *Tectiviridae*. The first includes small virions with a single circular ssDNA. The second family is currently represented only by a maritime phage, PM2, and has a capsid formed by the outer layer of proteins with an inner lipid bilayer (Huiskonen et al., 2004). The capsid contains a dsDNA genome. The last family, *Tectiviridae*, is characterised by presence of the lipoprotein vesicle that envelops the protein capsid with dsDNA genome. These phages have spikes on the apical parts of the envelope.

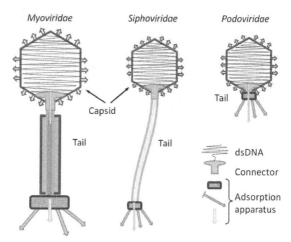

Fig. 3. Tailed phage families (copyright of E.V. Orlova).

The tailed phages were classified into the order *Caudavirales* (dsDNA) (Figure 2D,E,F) (Ackermann, 2006). Tailed phages can be found everywhere and represent 96% of known phages and are separated into three main phylogenetically related families. Tailed phages are divided into three families: A - *Myoviridae* with contractile tails consisting of a sheath and a central tube (25% of tailed phages); B - *Siphoviridae* , long, noncontractile tails (61%); C - *Podoviridae* , short tails (14 %). Since the tailed phages represent the biggest population of bacteriophages they are easy to find and purify; they are the most studied family both biochemically and structurally. For this reason the following part of the review will concentrate on results and analysis of the tailed bacteriophages.

3. Organisation of tailed bacteriophages

3.1 General architecture of bacteriophages

The basic structural features of bacteriophages are coats (or capsids) that protect the genome hidden inside a capsid and additional structures providing interface with a bacterium membrane for the genome release. The *Caudovirales* order of bacteriophages is characterised by dsDNA genomes and by the common overall organisation of the virus particles characterized by a capsid and a tail (Figure 3). Different phage species can vary both in size from 24-400 nm in length and genome length. Their DNA sequences differ significantly and can range in the size from 18 to 400 kb in length.

Structures obtained by electron microscopy (EM) do not typically provide detailed information on the atomic components owing to methods used for visualisation of particles. However, EM has allowed visualisation of these minuscule particles and morphological analysis. Each virion has a polyhedral, predominantly icosahedral, head (capsid) that covers the genome. The heads are composed of many copies of one or several different proteins and have a very stable organisation. A bacteriophage tail is attached to the capsid through a connector which serves as an adaptor between these two crucial components of the phage. The connector is a hetero-oligomer composed of several proteins (Lurz et al., 2001; Orlova et al., 2003). Connectors carry out several functions during the phage life cycle. They participate in the packaging of dsDNA into the capsid, and later they perform the function of a gatekeeper: locking the capsid exit of the phage, preventing leakage of DNA which is under high pressure and later, after a signal transmitted by the tail indicating that the phage is attached to the bacterium, the connector will be open allowing the release of DNA into the bacterium (Plisson et al., 2007). The tail and its related structures are indispensable phage elements securing the entry of the viral nucleic acid into the host bacterium during the infectivity process. The tail serves both as a signal transmitter and subsequently as a pipeline through which DNA is delivered into the host cell during infection. The tails may be short or long, the latter are divided into contractile and non-contractile tails. The long tails are typically composed of many copies of several proteins arranged with helical symmetry. All types of tails have outer appendages attached to the distant end of the tail and often include a baseplate with several fibres and a tip, or a needle that has specificity to the membrane receptors of the bacterium (Leiman et al., 2010). As soon the receptor has been found by the tail needle, which happens during multiple short living reversible attachments to the bacterium, the baseplate and tail fibres are involved in the binding of the phage to the bacterial outer membrane that makes the attachment irreversible (Christensen, 1965; São-José et al., 2006). The docking (irreversible attachment) of the phage induces opening of the phage connector and release of the genome through the tail tube into the bacterial cell.

3.2 DNA and its packaging

The virions of the bactriophage *Caudovirales* have a genome represented by linear molecules of dsDNA. The length of genome varies significantly between the phages . DNA is translocated through the central channel of the portal protein located at one vertex of the capsid. The portal complex provides a docking point for the viral ATPase complex (*terminase*). The terminase bound to the portal vertex forms the active packaging motor that moves the viral dsDNA inside the capsid. Encapsidation is normally initiated by an endonucleolytic cleavage at a defined sequence (*pac*) of the substrate DNA concatemer although some phages like T4 do not use a unique site for the initial cleavage. Packaging proceeds evenly until a threshold amount of DNA is reached inside the viral capsid. At the latter stages of packaging the increasingly dense arrangement of the DNA leads to a steep rise in pressure inside the capsid that can reach ~6 MPa (Smith et al., 2001). The headful cleavage of DNA is imprecise leading to variations in chromosome size of more than 1 kb (Casjens & Hayden, 1988; Tavares et al., 1996). The mechanism of packaging requires a sensor that measures the amount of DNA headfilling and a nuclease that will cleave DNA as soon the head is full. Termination of the DNA packaging is coordinated with closure of the portal system to avoid leakage of the viral genome. In tailed bacteriophages this is most frequently achieved through the binding of head completion proteins (or adaptor proteins). The complex of the portal dodecamer and these proteins composes the connector (Lurz et al., 2001; Orlova et al., 2003). After termination of the first packaging cycle initiated at *pac* (initiation cycle), a second packaging event is initiated at the non-encapsidated DNA end created by the headful cleavage and additional cycles of encapsidation follow. Some packaging series can yield 12 or more encapsidation events revealing the high processivity of the packaging machinery (Adams et al., 1983; Tavares et al., 1996).

4. Methods for study of bacteriophages

Microbiology and bacteriology were the first methods used to investigate viruses. Studies related to the life cycle of prokaryotic and eukaryotic microorganisms such as bacteria, viruses, and bacteriophages are combined into microbiology. This includes gene expression and regulation, genetic transfer, the synthesis of macromolecules, sub-cellular organization, cell to cell communication, and molecular aspects of pathogenicity and virulence. The earlier studies of phages were based on microbiological experiments including immunology. Nowadays the research of the biological processes is not limited to biochemical analysis and microbiology. To understand processes of virus/cell communication and interaction one often needs information on the molecular level and conformational changes of the components under different conditions. Gel filtration or Western blotting provides information for a protein on the macromolecular level such as size, molecular mass, binding to an antibody etc. These experiments will display how the proteins will change their characteristics with several chemical modifications and analysing what kind of change occurred, one could draw a conclusion for the structure. At the cellular level, optical microscopy can reveal the spatial distribution and dynamics of molecules tagged with fluorophores.

4.1 X-ray crystallography and NMR of phages

The methods of X-ray crystallography and NMR spectroscopy provide detailed information on molecular structure and dynamics. However, X-ray crystallography requires the growth of protein crystals up to 1 mm in size from a highly purified protein. Crystal growth is an experimental technique and there are no rules about the optimal conditions for a protein solution to result in a good protein crystal. It is extremely difficult to predict good conditions for nucleation or growth of well-ordered crystals of large molecular complexes. In practice, the best conditions are identified by screening multiple probes where a wide variety of crystallization solutions are tested. Structural analysis of viral proteins by crystallographic methods was very successful when separate proteins were studied. Protein crystals contain trillions of accurately packed identical protein molecules. When irradiated by X-rays, these crystals scatter X-rays in certain directions producing diffraction patterns. Computational analysis of that diffraction produces atomic models of the proteins. Viruses are much bigger than single proteins and may comprise thousands of components; it is difficult to pack them into crystals, and when successful, crystals have large unit cell dimensions (unit cell is an elementary part, from which the crystal is composed). Because of that the diffraction from virus crystals is far weaker than that of single proteins. It was an extremely challenging task to crystallise viruses for crystallographic studies although some icosahedral viruses were crystallised and the atomic structures have been obtained (Harrison, 1969; Grimes et al., 1998; Wikoff et al., 2000). Nowadays X-ray analysis has provided a wealth of information on atomic structures of many small protein components of large viruses including bacteriophages (Rossmann et al., 2005).

Nuclear Magnetic Resonance (NMR) is another very powerful method of structural analysis allowing studying dynamics of samples in solution. NMR methodology, combined with the availability of molecular biology and biochemical methods for preparation and isotope labelling of recombinant proteins has dramatically increased its usage for the characterization of structure and dynamics of biological molecules in solution. In NMR, a strong, high frequency magnetic field stimulates atomic nuclei of the isotopes H^1, D^2, C^{13}, or N^{15} and measures the frequency of the magnetic field of the atomic nuclei during its oscillation period before returning back to the initial state. NMR is able to obtain the same high resolution using different properties of the samples. NMR measures the distances between atomic nuclei, rather than the electron density in a molecule. Protein folding studies can be done by monitoring NMR spectra upon folding or denaturing of a protein in real time. However, NMR cannot deal with macromolecules in the mega-Dalton range, the upper weight limit for NMR structure determination is ~ 50 kDa.

4.2 Electron microscopy of tailed bacteriophages

For microbiological research, light microscopy is a tool of great importance in studies of the biology of microorganisms. However, light microscopy is not able to provide a high enough magnification to see viruses. The modern development and use of synchrotrons has revealed the structures of spherical viruses, nonetheless obtaining virus crystals remained problematic, especially for bacteriophages. EM has become a major tool for structural biology over the molecular to cellular size range. Bacteriophages do not have exact icosahedral symmetry since they have different appendages facilitating interactions and infection of the host cells, a fact that makes them very challenging objects for crystallography and their size makes them unsuitable for NMR. Members of the *Caudovirales* phage family with dsDNA genome are especially difficult to crystallise because they have tails. Here EM has become a tool of choice for structural analysis of these samples.

The simplest method for examining isolated viral particles is negative staining, in which a droplet of the suspension is spread on an EM support film and then embedded in a heavy metal salt solution, typically uranyl acetate (Harris, 1997). The method is called negative staining because the macromolecular shape is seen by its exclusion of stain rather than by binding of stain. During the last two decades other methods became widely used and demonstrated their efficiency when samples where fixed in the native, hydrated state by rapid freezing of thin layers of aqueous sample solutions at liquid nitrogen temperatures (Dubochet et al., 1988). Such rapid cooling traps the biological molecule in its native, hydrated state but embedded in glass-like, solid water – vitrified ice. This procedure prevents the formation of ice crystals, which would be very damaging to the specimen. EM images of particles are used to calculate their three-dimensional structures (Jensen, 2010).

EM was a major tool used in analysis of phage morphology and initiated a process of classification of viruses. The development of cryogenic methods has enabled EM imaging to provide snapshots of biological molecules and cells trapped in a close to native, hydrated state. High symmetry of the complexes is an advantage, but single particles of molecular mass –0.5-100 MDa with or without symmetry (e.g. viruses, ribosomes) can now be studied with confidence and can often reveal fine details of the 3D structure. The resulting images allow information not only on quaternary structure arrangements of macromolecular complexes but the positions of their secondary structural elements like helices and beta-sheets (Rossmann et al., 2005).

4.3 Hybrid methods

The components of bacteriophages and their interactions have to be identified and analysed. This can be done by localisation of known NMR or X-ray structures of individual viral proteins and nucleic acids combined with biochemical information to identify them in the EM structures. Electron cryo-microscopy and three-dimensional image reconstruction provide a powerful means to study the structure, complexity, and dynamics of a wide range of macromolecular complexes. One has to use different approaches for several reasons: there are limitations of the individual methods; some complexes do not crystallise; phages, being multi-protein complexes, have different conformational organisation at different conditions. Therefore all known structural and biochemical methods have to complement each other to generate structural information. When atomic models of components or subassemblies are accessible, they can be fitted into reconstructed density maps to produce informative pseudoatomic models. If atomic structures of the components are not known, it is helpful to perform homology modelling so that the generated models could be fitted into the EM maps. Fitting atomic structures and models into EM maps allows researchers to test different hypotheses, verify variations in structures of viruses and effectively increase the EM map resolution creating pseudo-atomic viral models (Lindert et al., 2009).

5. Examples of bacteriophage structures

In spite of the great abundance of the tailed phages, details of their organisation have emerged only during the last decade. The progress in structural studies of phages as a whole entity was slow because of their flexibility and complex organisation. The additional hindrance arises from intricate combination of different oligomerisation levels of the phage elements. Fully assembled capsids have at least 5-fold symmetry or more often, icosahedral symmetry where multiple structural units form a regular lattice with 2, 3, and 5 rotational symmetries. All known portal proteins were found to be dodecameric oligomers; tails have overall 6- or 3-fold rotational symmetry, the multiple repeats of major proteins have helical arrangement. The proteins related to the receptor sensor system at the far end of the tail could be in 6, 3, or only one copy. The information on the relative amount of different protein components has been revealed by biochemical and structural methods such as X-ray analysis of separate components. Development of hard and software has led to new imaging systems of better quality, new programs allowing processing of bigger data sets comprising hundreds of thousand images. The modern strategy is based on hybrid methods where structure determination at high resolution of isolated phage components is combined in three-dimensional maps of lower resolution obtained by electron microscopy. Electron microscopy by itself has reached such level of quality that for the complexes with icosahedral symmetry it has became nearly routine to obtain structures at 4-5 Å resolution (Hryc et al., 2011; Zhou, 2011; Grigorieff & Harrison, 2011)

5.1 Phage T4

The T4 phage of the *Myoviridae* family infects *E. coli* bacteria and is one of the largest phages; it is approximately 200 nm long and 80-100 nm wide with the capsid in a shape of an elongated icosahedron. The phage has a rigid tail composed of two main layers: the inner tail tube is surrounded by a contractile sheath which contracts during infection of the bacterium. The tail sheath is separated from the head by a neck. Phages of *Myoviridae* family have a massive baseplate at the end of the tail with fibres attached to it. The tail fibres help to find receptors of a host cell and provide the initial contact; during infection the tail tube penetrates an outer bacterial membrane to secure the pathway for genome to be injected into the cell.

The capsid of the T4 phage is built with three essential proteins: gp23* (48.7 kDa), which forms the hexagonal capsid lattice; gp24* (the * designates the cleaved form of the protein when the prohead matures to infectious virus) forms pentamers at eleven of the twelve vertices, and gp20, which forms the unique dodecameric portal vertex through which DNA enters during packaging and exits during infection. 3D-reconstruction has been determined at 22 Å resolution by cryo-EM for the wild-type phage T4 capsid forming a prolate icosahedron (Figure 4, Fokine et al., 2004). The major capsid protein gp23* forms a hexagonal lattice with a separation of ~140 Å between hexamer centres. The atomic structure of gp24* has been determined by X-ray crystallography and an atomic model for gp23* was built using its similarity to gp24* (Fokine et al., 2005). The capsid also contains two non-essential outer capsid proteins, Hoc and Soc, which decorate the capsid surface. The structure of Soc has been determined by X-ray crystallography and shows that Soc has two capsid binding sites which, through binding to adjacent gp23* subunits, reinforce the capsid structure (Qin et al., 2010). The failure of gp24* to bind Soc provides a possible explanation for the property of osmotic shock resistance of the phage (Leibo et al., 1979). The 3D maps of the empty capsids with and without Soc (Iwasaki et al., 2000) have been determined at 27 and at 15 Å resolution, respectively.

Single molecule optical tweezers and fluorescence studies showed that the T4 motor packages DNA at a rate of up to 2000 bp/sec, the fastest reported to date of any packaging motor (Fuller et al., 2007). FRET-FCS studies indicate that the DNA gets compressed during the translocation process (Ray et al., 2010).

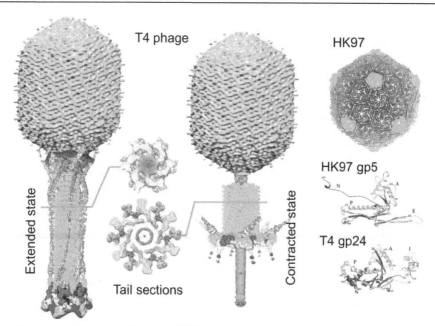

Fig. 4. Structures of T4 (cryo-EM) and HK97 (X-ray analysis) phages (reproduced with permission of M.G. Rossmann and J. E. Johnson). Ribbon diagrams compare the structure of HK97 (gp5) with the structure of the T4 gp24 capsid protein. Phages and sections are on a different scale.

Tails of *Myoviridae* phages have a long, non-contractible tube surrounded by a contractile sheath. Bacteriophage T4 has a tail sheath that is composed of 138 copies of gp18 (Leiman et al., 2004). The tail tube inside the sheath is estimated to be assembled from as many gp19 subunits as there are gp18 subunits in the sheath (Moody & Makowski, 1981). The tail sheath has helical symmetry with a pitch of 40.6 Å and a twist of 17.2° (Kostyuchenko et al., 2005). The tail sheath contraction can be divided into several steps. Previous studies of partially contracted sheath showed that conformational changes of the sheath are propagated 'upwards' starting from the disk of the gp18 subunits closest to the baseplate (Moody 1973). The cryo-EM reconstructions showed that during contraction, the tail sheath pitch decreases from 40.6 Å to 16.4 Å and its diameter increases from 24 nm to 33 nm (Figure 4, Leiman et al., 2004; Kostyuchenko et al., 2005). The combination of X-ray model and EM structures show that gp18 monomers remain rigid during contraction and move about 50 Å radially outwards while tilting 45° clockwise, viewed from outside the tail. During contraction of the tail the interactions between neighbouring subunits within a disk are broken so that the subunits from the disk above get inserted into the gaps formed in the disk below (Aksyuk et al., 2009).

The baseplate with the cell-puncturing device of the T4 phage is an ultimate element of the phage. This is an extremely complex multiprotein structure on the far end of the tail and represents multifunctional machinery that anchor the phage on the bacteria surface and provide formation of the DNA entrance into the bacteria. This important part of the phage structure is of ~27 nm in height and 52 nm in diameter at its widest part. The baseplate conformation is coupled to that of the sheath: the dome shape conformation is associated with the extended sheath, whereas the flat "star" conformation is associated with the contracted sheath that occurs in the T4 particle after attachment to the host cell. Short treatment of bacteriophage T4 with 3 M urea resulted in the transformation of the baseplate to a star-shape and subsequent tail sheath contraction (Kostyuchenko et al., 2003). During that switch the baseplate diameter increases to 61 nm and the height decreases to 12 nm although the protein composition of the baseplate does not change. It is composed of ~150 subunits of a dozen different gene products (Leiman et al., 2010). Proteins gp11, gp10, gp7, gp8, gp6, gp53, and gp25 form one sector of 6-fold structure. The central hub of the baseplate is formed by gp5, gp27, and gp29 and probably includes gp26 and gp28. Assembly of the baseplate is completed by attaching gp9 and gp12 to form the short tail fibres, and also gp48 and gp54 that are required to initiate polymerization of the tail tube, a channel for DNA (Leiman et al., 2010).

T4 tail has three types of fibrous proteins: the long tail fibres, the short tail fibres, and whiskers. Long tail fibres and short tail fibres are attached to the baseplate and whiskers extending outwardly in the region of the tail connection to the capsid. The long tail fibres, which are ~145 nm long and only ~4 nm in diameter, are primary reversible adsorption devices (Figure 4, Kostyuchenko et al., 2003). Each fibre consists of the rigid proximal halves, formed by gp34, and the distal ones composed by gp36 and gp37. The distal part of the fibre has a rod-like shape about 40 nm long that is connected to the first half of the fibre through the globular hinge. Gp35 forms a hinge region and interacts with gp34 and gp36. The N-terminal globular domain of gp34 interacts with the baseplate. Short tail fibres are attached to the baseplate by the N-terminal thin part, while the globular C-terminus binds to the host cell receptors (Boudko et al., 2002). The structure of this domain of the short tail fibres was determined by X-ray crystallography (Tao et al., 1997)

5.2 HK97

HK97 is a temperate phage from *Escherichia coli* which was isolated in Hong Kong by Dhillon (Dhillon & Dhillon, 1976). It shares a host range with the Lambda phage (Dhillon et al., 1976). HK97 has an isometric head and a long, flexible, non-contractile tail representing *Siphoviridae* family (Dhillon et al., 1976). The HK97 phage has multi step pathway of self-assembly revealing two forms of procapsids of ~470 Å in diameter. Capsid protein gp5 (42 kDa) forms capsids, with icosahedral symmetry characterised by $T = 7$ (Hendrix, 2005). A part of the gp5 (102 amino acids from the N terminus) plays the role of a scaffold, which is cleaved by gp4 (the phage protease) at maturation of the capsid (Conway at al., 1995). The first low resolution structures have shown conformation changes reflecting transition of the HK97 procapsids into expanded capsids (Conway at al., 1995). The diameter of procapsids during transition into the heads increases from 470 Å to 550 Å while the thickness of the capsid shell changes from 50 Å to ~ 25 Å.

The first atomic structure of a capsid for the tailed phage was only published in 2000. Gp5, if expressed alone, assembles into a portal-deficient version of prohead I. Co-expressing gp5 with the gp4 protease, which cleaves gp5 scaffolding domain, produces Prohead II that expands into the icosahedral head II (the diameter is ~650 Å) without DNA and portal complex; and it was used for the crystallisation. The crystal structure of the dsDNA bacteriophage HK97 mature empty capsid was determined at 3.6 Å resolution using icosahedral symmetry (Wikoff et al., 2000). The capsid crystal structure shows how an isopeptide bond is formed between subunits, arranged in topologically linked, covalent circular rings (Figure 4). The structure of the HK97 gp5 coat protein has revealed a new category of virus fold: it is mixture alpha-helices and beta-sheets organised into three domains that are not sequence contiguous (Figure 4). Domain A is located close to the centre of the hexamers and pentamers of the capsid. Domain P (peripheral) provides contacts between adjacent molecules within pentamers and hexamers. The third domain, represented by the E-loop, is an extension through which each subunit of the HK97 capsid is covalently linked to two neighbouring subunits. The bond organization explains why the mature HK97 particles are extraordinarily stable and cannot be disassembled on an SDS gel without protease treatment (Popa et al., 1991; Duda, 1998).

5.3 SPP1

SPP1 is a virulent *Bacillus subtilis* dsDNA phage and belongs to the *Siphoviridae* family. The virion is composed of an icosahedral, isometric capsid (~60 nm diameter) and a long, flexible, non-contractile tail. The SPP1 genome length is 45.9 kb. The procapsid (or prohead) of SPP1 consists of four proteins: the scaffold protein gp11, the major capsid protein gp13, the portal protein gp6, and a minor component gp7. The inside of the capsid is filled with gp11 which exits the procapsid during DNA packaging. Gp13 and the decoration protein gp12 form the head shell of the mature SPP1 capsid.

The portal protein is located at a 5-fold vertex of the icosahedral phage head and serves as the entrance for DNA during packaging. The structure of gp6 as a 13-subunit assembly was determined by EM and X-ray at 10 and 3.4 Å resolution correspondingly (Orlova et al., 2003; Lebedev et al., 2007). The 13–mer portal complex has a circular arrangement with an overall diameter of ~165 Å and a height of ~110 Å. A central tunnel pierces the assembly through the whole height. The portal protein monomer has four main domains: crown, wing, stem, and clip. The crown domain consists of three alpha-helices connected by short loops and has 40 additional C-terminal residues that are disordered in the X-ray structure. Mutations in the crown indicate the importance of this area for DNA translocation (Isidro et al., 2004a, b).The wing region is formed by alpha-helices flanked on the outer side by a beta-sheet. The stem domain connects the wing to the clip domain. It consists of two alpha-helices that are conserved in phi29 and SPP1 phages; a similar arrangement of helices was found in the P22 portal protein (Simpson et al., 2000; Guasch et al., 2002; Lebedev et al., 2007; Olia et al., 2011). The clip domain forms the base of the portal protein and is expected to be exposed to the outside of the capsid during viral particle assembly. The three-dimensional structures of the portal proteins of SPP1, phi29, and P22 phages demonstrate a strikingly similar fold. Although there is no detectable amino-acid sequence similarity between proteins, they have a nearly identical arrangement of two helices forming stem domains and in the clip domain which form a tightly packed ring of three stranded beta-sheets each made up of two strands from one subunit and one strand from an adjacent subunit.

After termination of DNA incorporation the portal pore needs to be rapidly closed to prevent leakage of the viral chromosome. In SPP1 this role is played by the head completion proteins gp15 and gp16 that bind sequentially to the portal vertex forming the connector (Lurz et al., 2001). Disruption of the capsids yielded connectors composed of gp6, gp15 and gp16. The connector is an active element of the phage that is involved into packaging the viral genome, serves as an interface for attachment of the tail, and controls DNA release from the capsid. The connector of *Bacillus subtilis* bacteriophage SPP1 was found to be a 12-fold cyclical oligomer (Lurz et al., 2001), though isolated gp6 is a cyclical 13mer. The structure of the connector was determined at 10 Å resolution, using cryo-EM (Figure 5, Orlova et al., 2003). Both the isolated portal protein and the gp6 oligomer in the connector reveal a similar arrangement of four main domains, the major changes take place in the clip domain through which gp15 contacts gp6. The connector structure shows that gp15 serves as an extension of the portal protein channel where gp16 binds. The central channel is closed by gp16 physically blocking the exit from the DNA-filled capsid (Orlova et al., 2003). Structures of SPP1 gp15 and gp16 monomers were determined by NMR and together with gp6 were docked into the EM map of the connector (Figure 5, Lhuillier at al., 2009). The channel of the connector will be opened when the virus infects a host cell. Comparison of the structures before and after assembly, provides details on the major structural rearrangements (gp15) and folding events (gp15 and gp16) that accompany connector formation.

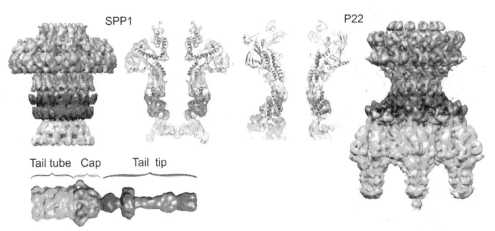

Fig. 5. Surface representation of SPP1phage connector (top left, Orlova et al., 2003), tail tip (bottom left, Plisson et al., 2007) and P22 phage tail machine (top right, Lander et al., 2009). The fit of the atomic coordinates into each connector is shown as a cut open view adjacent to its corresponding surface view. The portal proteins are shown in blue, the adaptor proteins in crimson, gp16 of SPP1 and the tail spikes of P22 are shown in green. (Copyright of E.V. Orlova)

The 160-nm-long tail of the SPP1 phage is composed of two major tail proteins (MTPs), gp17.1 and gp17.1*, in a ratio of about 3:1. They share a common amino-terminus, but the latter species is ~10 kDa more than gp17.1. The polypeptide sequence, identical in the two proteins is responsible for assembly of the tail tube while the additional module of gp17.1* shields the structure exterior exposed to the environment. The carboxyl-terminus domain of MTPs shares homology to motifs of cellular proteins (Fraser et al., 2006) or to phage components (Fortier et al., 2006) involved in binding to cell surfaces. Structures of the bacteriophage SPP1 tail before and after DNA ejection were determined by negative stain electron microscopy. The results reveal extensive structural rearrangements in the internal wall of the tail tube. It has been proposed that the adsorption device–receptor interaction triggers a conformational switch that is propagated as a domino-like cascade along the 160 nm -long helical tail structure to reach the head-to-tail connector. This leads to opening of the connector, culminating in DNA exit from the head into the host cell through the tail tube (Plisson et al., 2007).

The tail tip is attached to the cap structure that closes the tail tube (Figure 5). The absence of a channel for DNA traffic in the tip implies that it must dissociate from the cap for DNA passage to the cytoplasm during infection. The structural data show that the tail tip does not have a channel for DNA egress and that the signal initiated by interaction of the tip with the bacterial receptor causes release of the tip from the tail cap. Reconstructions were performed for two states of the tail: before and after DNA ejection. The cap structure was reconstructed separately from the tip and the main area of the tail. The reconstructions of the cap together with the first four rings of the tail tube demonstrate that the tail external diameter (before DNA ejection) tapers from ~110 to ~40Å at the capped extremity and changes symmetry from six-fold to three-fold. This arrangement provides a sturdy interface between the tail tube and the three-fold symmetric tip. Opening of the dome-shaped cap involves loss of the tip and movement of the cap subunits outwards from the tail axis, creating a channel with the same diameter as the inner tail tube (Plisson et al., 2007).

5.4 Phi29

The *Bacillus subtilis* bacteriophage phi29 (*Podoviridae* family) is one of the smallest and simplest known dsDNA phages. The bacteriophage phi29 (Figure 6) is a 19-kilobase (19-kb) dsDNA virus with a prolate head and complex structure. Proheads consist of the major capsid protein gp8, scaffolding protein gp7, head fibre protein gp8.5, head–tail connector gp10, and a pRNA oligomer. Mature phi 29 heads are 530Å long and 430Å wide, and the tail is 380Å long. The packaging of DNA into the head involves, besides the portal protein, other essential components such as an RNA called pRNA and the ATPase p16, required to provide energy to the translocation machinery. Once the DNA has been packaged, pRNA and p16 are released from the portal protein. In the mature phi29 virion, the narrow end of the portal protrudes out of the capsid and attaches to a toroidal collar (gp11). The collar has a diameter of about 130Å and is surrounded by 12 appendages that function to absorb the virion on host cells (Anderson et al., 1966). A thin, 160Å -long tube, with an outer diameter of 60Å, leads away from the centre of the collar (Hagen et al., 1976). The outer end of the tail (gp9) has a cylindrical shape and bigger diameter of ~ 80Å.

The three-dimensional structure of a fibreless variant has been determined to 7.9 Å resolution allowing the identification of helices and beta-sheets (Figure 6, Morais et al., 2005, Tang et al., 2008). For the prolate capsid phi29 there was not the advantage of using icosahedral symmetry for structural analysis, its cryo-EM three-dimensional reconstructions have been made of mature and of emptied bacteriophage phi29 particles without making symmetry assumptions (Xiang et al., 2006). Possible positions of secondary elements for gp8 indicate that the folds of the phi29 and bacteriophage HK97 capsid proteins are similar except for an additional immunoglobulin-like domain of the phi29 protein: the gp8 residues 348–429 are 32% identical to the group 2 bacterial immunoglobulin domain (BIG2) consensus sequence (Morais et al., 2005; Xiang et al., 2006). The BIG2 domain is found in many bacterial and phage surface proteins related to cell adhesion complexes (Luo et al., 2000; Fraser et al., 2006). The asymmetrical reconstruction of the complete phi29 has revealed new details of the asymmetric interactions and conformational dynamics of the phi29 protein and DNA components (Tang et al., 2008).

The DNA packaging motor is located at a unique portal vertex of the prohead and contains: the head-tail connector (a dodecamer of gp10); the portion of the prohead shell that surrounds the connector, a ring of 174-base prohead RNAs (pRNA), and a multimer of gp16, an ATPase that first binds DNA-gp3 and then assembles onto the connector/pRNA complex prior to packaging. The wide end of the portal protein contacts the inside of the head, whereas the narrow end protrudes from the capsid where it is encircled by the pentameric pRNA. The structure of the isolated phi29 portal complex has been studied by atomic force microscopy and electron microscopy (EM) of two-dimensional arrays (Carazo et al., 1985) and X-ray crystallography (Simpson et al., 2000; Guasch et al., 2002). X-ray crystallographic studies of the phi29 portal showed that it is a cone-shaped dodecamer with a central channel (Simpson et al., 2000). The three-dimensional crystal structure of the bacteriophage phi29 portal has been refined to 2.1Å resolution (Guasch et al., 2002). This 422 kDa oligomeric protein is part of the DNA packaging motor and connects the head of the phage to its tail. Each monomer of the portal dodecamer has an elongated shape and is composed of a central, mainly alpha-helical domain (stem domain) that includes a three-helix bundle, a distal a/b domain and a proximal six-stranded SH3-like domain (Simpson et al., 2000). The portal dodecamer has a 35 Å wide central channel, the surface of which is mainly electronegative. The narrow end of the head–tail portal protein is expanded in the mature virus. Gene product 3, bound to the 5′ ends of the genome, appears to be positioned within the portal, which may potentiate the release of DNA-packaging machine components, creating a binding site for attachment of the tail (Tang et al., 2008).

The process of DNA packaging is an extremely energy consuming act because electrostatic and bending repulsion forces of the DNA must be overpowered to package the DNA to near-crystalline density. Force-measuring laser tweezers were used to measure packaging activity of a single portal complex in real time where one microsphere has been used to hold on to a single DNA molecule as they are packaged, and the other was bound to the phage and fixed (Smith et al., 2001). These experiments have demonstrated that the portal complex is a force-generating motor which can work against loads of up to 57 pN, making it one of the strongest molecular motors reported to date. Notably, the packaging rate decreases as the prohead is filled, indicating that an internal force builds up to 50 pN owing to DNA confinement. These results suggest that the internal pressure provides the driving force for DNA injection into the host cell for the first half of the injection process.

The structure of the phi29 tail has revealed that 12 appendages protruding from the collar like umbrella with 12 ribs that end in 'tassels' (Xiang et al., 2006). Two of the 12 appendages are extended radially outwards (the 'up' position), whereas the other 10 have their tassels 'hanging' roughly parallel to the virus major axis. The adsorption capable 'appendages' were found to have a structure homologous to the bacteriophage P22 tail spikes. Two of the appendages are extended radially outwards away from the long axis of the virus, whereas the others are around and parallel to the phage axis. The appendage orientations are correlated with the symmetry mismatched positions of the five-fold related head fibres. The tail in the mature capsids, that have lost their genome have an empty central channel (Xiang et al., 2006). Comparisons of these structures with each other and with the phi29 prohead indicate how conformational changes might initiate successive steps of assembly and infection.

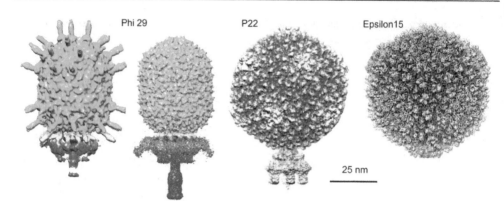

Fig. 6. Surface representation of EM structures. The Phi29 capsid is in green, the tassels are shown in magenta and blue, the tail is also in blue. The complete phage is shown at 16 Å (left) and a mutant phage without spikes at 8 Å (right, Tang et al., 2008) (Image copyright of E.V. Orlova). P22 phage is shown at 7 Å resolution (reproduced with permission of J. E. Johnson) and Epsilon15 is at 4.5 Å (Tang et al. 2011; Jiang et al., 2008, reproduced with permission of W. Chiu)

5.5 P22

Bacteriophage P22 infects *Salmonella enterica serovar Typhimurium* and is a prototypical representative of the *Podoviridae* family. The mature P22 virion presents an icosahedral $T = 7l$ capsid about 650 Å in diameter. The bacteriophage P22 procapsid comprises hundreds of copies of the gp5 coat and gp8 scaffolding proteins, multiple copies of three ejection proteins (gp7, gp16, gp20, also known as pilot proteins), and a unique multi-subunit gene 1 (gp1) portal (Prevelige, 2006).

Single-particle cryo-EM has been used to determine the P22 procapsid structure initially at low resolution then improved from 9 Å to 3.8 Å resolution (Figure 6, Jiang et al., 2006; Jiang et al., 2008; Chen et al., 2011). The procapsids were isolated from cells infected with mutants defective in DNA packaging and representing the physiological precursor prior to DNA packaging and capsid maturation. Coat protein gp5 is organized as pentamers and skewed hexamers as previously reported for the GuHCl treated procapsid (Thuman-Commike et al., 1999; Parent et al., 2010). The high resolution structure allowed Cα backbone models for each of the seven structurally similar but not identical copies of the gp5 protein in the asymmetric unit to be built. The analysis has shown that gp5 has fold similar to the HK97 coat protein (Jiang et al., 2008).

The first structures of the P22 assembly-naive portal formed from expressed subunits (gp1) were obtained at ~ 9 Å resolutions by cryo-EM (Lander et al., 2009). Later two atomic structures were obtained for the P22 portal protein: one is for a fragment of the portal, 1–602 aa (referred to as the 'portal-protein core'), bound to 12 copies of tail adaptor factor gp4 (Olia et al., 2006; Lorenzen et al., 2008). The second was the full-length P22 portal protein (725aa) at 7.5 Å resolution. To solve three independent crystal forms of the complex gp1/gp4 to a resolution of 9.5 Å, the EM structure of P22 tail at 9.4 Å resolution has been extracted computationally from the P22 tail complex and used as molecular replacement model. The high resolution atomic structure of the P22 portal protein has been obtained using a combination of multi- and intra crystal non-crystallographic-symmetry averaging, and by extension of EM phases to the resolution of the best diffracting crystal form (3.25 Å). The P22 portal complex is a ~0.96 MDa ring of 12 identical subunits, symmetrically arranged around a central channel of variable diameter, with an overall height of ~350 Å (Lander et al., 2009; Olia et al., 2011). A lower-resolution structure of the full-length portal protein unveils the unique topology of the C-terminal domain, which forms a ~200 Å long alpha-helical barrel. This domain inserts deeply into the virion and is highly conserved in the *Podoviridae* family. The quaternary structure of the P22 portal protein can be described as a funnel-shaped core ~170 Å in diameter, connected to an ~200 Å long, mostly α-helical tube formed by the C-terminal residues 603–725, which resembles a rifle barrel (Olia et al., 2011). The portal core is similar in topology to other portal proteins from phage SPP1 (Lebedev et al., 2007) and phi29 (Simpson et al., 2000; Guasch et al., 2002), but presence of the helical barrel is the first example of a dodecameric tube in a portal protein. Gp4 binds to the bottom of the portal protein, forming a second dodecameric ring ~75 Å in height (Olia et al., 2011).

In *Podoviridae*, the mechanisms of bacteria cell penetration and genome delivery are not well understood. P22 uses short, non-contractile tails to adsorb to the host cell surface. The tail machine comprises the tail spike, gp9; the tail needle, gp26; and the tail factors gp4 and gp10 (Tang et al., 2005). Protein gp4 serves as an adaptor between portal protein and tail elements. The tail has a special fibre known as the "tail needle" that likely functions as a cell membrane piercing device to initiate ejection of viral DNA inside the host. The structure of the intact tail machine purified from infectious virions has been obtained by cryo-EM at ~ 9 Å resolution (Figure 5, Lander et al., 2009). The structure demonstrated that the protein components are organized with a combination of 6-fold (gp10, trimers of gp9), and 3-fold (gp26, gp9) symmetry (Lander et al., 2009). The combined action of an adhesion protein (tailspike) and a tail needle (gp26) is responsible for binding and penetration of the phage into the host cell membrane (Bhardwaj et al., 2011a). Gp26 probably plays the dual role of portal-protein plug and cell wall–penetrating needle, thereby controlling the opening of the portal channel and the ejection of the viral genome into the host. In Sf6, a P22-like phage that infects *Shigella flexneri*, the tail needle presents a C-terminal globular knob (Bhardwaj et al., 2011b). This knob, absent in phage P22 but shared in other members of the P22-like genus, represents the outermost exposed tip of the virion that contacts the host cell surface. In analogy to P22 gp26, it was suggested that the tail needle of phage Sf6 was ejected through the bacterial cell membrane during infection and its C-terminal knob is threaded through peptidoglycan pores formed by glycan strands (Bhardwaj et al., 2011a; 2011b).

5.6 Epsilon 15

The Gram-negative *Salmonella anatum* is the host cell for bacteriophage Epsilon15 (ε15, *Podoviridae* family). The ~40kb Epsilon15 dsDNA is packed within the isometric icosahedral capsid with a diameter of ~680 Å. The virion capsid contains 11 pentons and 60 hexons made from the major capsid protein gp7 and a small decoration protein gp10 (12-kDa). Single-particle cryo-EM was used about ten years ago to determine the first structures of icosahedral viruses to subnanometre resolutions (Jiang et al., 2006). A 9.5 Å density map was generated from EM data using icosahedral symmetry. In the average subunit map, the locations of three helices were identified. Now the structure of the epsilon15 capsid has been refined to a 4.5 Å resolution (Figure 6, Jiang et al., 2008). The quality of the map allowed tracing the backbone chain of gp7. Comparison of the models has shown local discrepancies between subunits at the N- terminus and the E-loop in different subunits of gp7 within the hexamers of the capsid. Interestingly, a connection between E-loops of neighbouring subunits possibly exists, but the resolution was not sufficient to reveal it. Moreover, additional density was located between the gp7 monomers. This density has been assigned to the gp10 decoration protein that consists mainly of beta-sheets and two short alpha-helices. A back-to-back dimer of gp10 is positioned at the two-fold axes and makes contact with six gp7 subunits through the N-termini and the E-loops. It was suggested that gp10 'staples' the underlying gp7 capsomeres to cement the gp5 cage so that it withstands the pressure from packed dsDNA (Jiang et al., 2008).

The Epsilon15 capsid volume can accommodate up to 90kb dsDNA. Since the Epsilon15 genome is only ~40kb, there is ample space for a protein core of this size in the capsid chamber. The core has a cylindrical shape with a length of ~200 Å and diameter of ~180 Å. The protein core may facilitate the topological ordering of the dsDNA genome during packaging and/or release as suggested for T7 core. At the virion's tail vertex, six tails pikes attach to a central 6-fold-symmetric tail hub of the length ~170 Å. This hub may be equivalent to *Salmonella typhimurium* bacteriophage P22's hub. The hub is connected to the portal ring inside the capsid. The Epsilon 15 genome winds around the core, with a short segment of terminal DNA passing through the axis of the core and portal (Jiang et al., 2006)

6. Conclusions

Bacteriophages represent an example of amazing molecular machines with powerful motors energised by ATP hydrolysis and puncturing devices allowing to inject viral genome into the host cells. As more and more phage structures been studied a general theme emerges pointing to a common bacteriophage ancestor from which they all inherited essentially the same capsid protein fold and other elements of their organisation: capsids, tails, portal complexes, tail fibres, and other components. The number of phages that were discovered, purified, and studied by biochemical, and biophysical methods increased tremendously during the last decade. New technologies used for their studies both on the microbiological and molecular levels made it possible to analyse their evolutionary relationship and origins of the host range specificity. One of the powerful techniques in the structural biology of phages is the modern cryo-EM that recently allowed to reach close to atomic resolution level of details in the EM reconstructions (Hryc et al., 2011; Zhou, 2011; Grigorieff & Harrison, 2011). Understanding of the mechanisms which determine the host-range is required to solve many practical questions related to infectious human and animal diseases caused by bacteria, and quality food and its production (e.g. dairy products). A study conducted in Japan has demonstrated the efficiency of phages against bacterial infections of cultured fish (Nakai & Park, 2002). The use of bacteriophage as antimicrobial agents is based on the lytic phages that kill bacteria via lysis, which destroys the bacterium and makes its adaptation nearly impossible. High bacteriophage resistance for external factors is important for the stability of phage preparations. However, this stability is disadvantageous for industry when maintenance of the active bacterial strains is important.

Comparative studies demonstrate that bacteriophages have many common features on the molecular level and common principle of interaction with a bacterium cell, although components that trigger adsorption of phages to the host cell and the genome release are host dependent. Phage infection also depends on the availability of specific receptors on the cell surface, and investigation of the structure and biosynthesis of the bacterial cell membrane may be undertaken using phage-resistant mutants. Therefore there is a need to carry out further studies on phages, identifying receptors of targeted bacteria and environmental features that affect phage activity (Jończyk et al., 2011). The growing interest of the pharmaceutical and agricultural industries in phages requires more information on phage interactions, survivability and methods of their preservation. Structural studies revealed many similarities between bacteriophages and animal cell viruses. The chances of success in using bacteriophages as model systems for animal cell viruses and eventually as medical therapy are much better given our current extensive knowledge of bacteriophage biology following the advances in their molecular structural biology.

7. Acknowledgments

The author is grateful to Dr Helen White for the helpful comments during the preparation of this manuscript. EVO was also supported by BBSRC grant BB/F012705/1

8. References

Ackermann, H.W. (2006). Classification of bacteriophages. In *The Bacteriophages*, Ed. Calendar R, Oxford University Press, ISBN 0-19-514850-9, New York, USA, pp. 8–16

Ackermann, H.W. (2007). 5500 Phages examined in the electron microscope. *Archives of Virology* Vol.152, No.2, pp. 227-243. PMID 17051420

Ackermann, H.W. (2004). *Bacteriophage classification*. In *Bacteriophages. Biology and Applications. Eds Kutter E, Sulakvelidze A*, CRC Press ISBN 978-0-8493-1336-3, Boca Raton, USA, pp. 67–89

Adams, M.B.; Hayden, M. & Casjens, S. (1983). On the sequential packaging of bacteriophage P22 DNA. *The Journal of Virology*, Vol.46, No.2, pp. 673-677.

Aksyuk, A.A.; Leiman, P.G.; Kurochkina, L.P.; Shneider, M.M.; Kostyuchenko, V.A.; Mesyanzhinov, V.V. & Rossmann, M.G. (2009). The tail sheath structure of bacteriophage T4: a molecular machine for infecting bacteria. *The EMBO Journal*, Vol.28, pp 821–829

Alberts, B.; Bray, D.; Lewis, J.; Raff, M.; Roberts, K. & Watson, J. D. (1989). *Molecular Biology of the Cell*, 2nd ed. New York: Garland Publishing,. ISBN 0824036956

Anderson, D.L.; Hickman, D.D. & Reilly, B.E. (1966). Structure of Bacillus subtilis bacteriophage phi 29 and the length of phi 29 deoxyribonucleic acid. *The Journal of Bacteriology*, Vol.91, No.5, pp. 2081-9.

Arnold, H.P.; Ziese, U. & Zillig, W. (2000). SNDV, a novel virus of the extremely thermophilic and acidophilic archaeon Sulfolobus. *Virology*, Jul 5; Vol.272, No.2, pp. 409-416.

Bettstetter, M.; Peng, X.; Garrett, R.A. & Prangishvili, D. (2003). AFV1, a novel virus infecting hyperthermophilic archaea of the genus acidianus. *Virology*, Oct 10; Vol.315, No.1, pp. 68-79.

Bhardwaj, A.; Molineux, I.J.; Casjens, S.R. & Cingolani, G. (2011). Atomic structure of bacteriophage sf6 tail needle knob. *The Journal of Biological Chemistry*, Vol.286, No.35, pp. 30867-77.

Bhardwaj, A.; Walker-Kopp, N.; Casjens, S.R. & Cingolani, G. (2009). An evolutionarily conserved family of virion tail needles related to bacteriophage P22 gp26: correlation between structural stability and length of the alpha-helical trimeric coiled coil. *Journal of Molecular Biology*, Vol.391, No.1, pp. 227-245.

Boudko, S.P.; Londer, Y.Y.; Letarov, A.V.; Sernova, N.V.; Engel, J. & Mesyanzhinov, V.V. (2002). Domain organization, folding and stability of bacteriophage T4 fibritin, a segmented coiled-coil protein. *European Journal of Biochemistry*, Vol.269, pp. 833-841.

Brussow, H. & Kutter, E. (2005). Phage ecology. In: Kutter E, Sulakvelidze A, editors. *Bacteriophages: biology and applications*. Boca Raton, FL:CRC Press;. pp. 129–63.

Burrowes, B.; Harper, D.R.; Anderson, J.; McConville, M. & Enright, M.C. (2011). Bacteriophage therapy: potential uses in the control of antibiotic-resistant pathogens. *Expert Review of Anti-Infective Therapy*, Vol.9, No.9, pp. 775-85.

Carazo, J.M.; Santisteban, A. & Carrascosa, J.L. (1985). Three-dimensional reconstruction of bacteriophage phi 29 neck particles at 2 X 2 nm resolution. *Journal of Molecular Biology*, Vol.183, No.1, pp. 79-88.

Casjens, S. & Hayden, M. (1988). Analysis in vivo of the bacteriophage P22 headful nuclease. *Journal of Molecular Biology*, Vol.199, No.3, pp. 467-474.

Chang, J.; Weigele, P.; King. J.; Chiu, W. & Jiang, W. (2006). Cryo-EM asymmetric reconstruction of bacteriophage P22 reveals organization of its DNA packaging and infecting machinery. *Structure*, Vol.14, No.6, pp. 1073-1082.

Chen, D.H.; Baker, M.L.; Hryc, C.F.; DiMaio, F.; Jakana, J.; Wu, W.; Dougherty, M.; Haase-Pettingell, C.; Schmid, M.F.; Jiang, W.; Baker, D.; King, J.A. & Chiu, W. (2011). Structural basis for scaffolding-mediated assembly and maturation of a dsDNA virus. *Proceedings of the National Academy of Sciences of the United States of America*, Vol.108, No.4, pp. 1355-1360.

Chopin, M.C.; Rouault, A.; Ehrlich, S.D. & Gautier, M. (2002). Filamentous phage active on the gram-positive bacterium Propionibacterium freudenreichii. *The Journal of Bacteriology*, Vol.184, No.7, pp. 2030-2033.

Christensen, J.R. (1965). The kinetics of reversible and irreversible attachment of bacteriophage T 1, *Virology*, Vol.26, No.4, pp. 727-737.

Coelho, J.; Woodford, N.; Turton, J. & Livermore, D.M. (2004). Multiresistant *Acinetobacter* in the UK, pp. how big a threat? *Journal of Hospital Infection*, Vol.58, pp. 167–169.

Conway, J.F.; Duda, R.L.; Cheng, N.; Hendrix, R.W. & Steven, A.C. (1995). Proteolytic and conformational control of virus capsid maturation, pp. the bacteriophage HK97 system. *Journal of Molecular Biology*, Vol.253, No.1, pp. 86-99.

D'Hérelle, F. (1917). Sur un microbe invisible antagoniste des bacilles dysentériques. *Comptes rendus Acad Sci Paris*. 165, pp. 373–5. "On an invisible microbe antagonistic toward dysenteric bacilli, pp. brief note by Mr. F. D'Herelle. 2007 ". *Research in microbiology* Vol.158, No.7, pp. 553–4. Félix d'Hérelles (1917). Archived from the original on 2010-12-04. http, pp.//www.webcitation.org/5uicsPk41. Retrieved 2010-09-05)

Dhillon, T.S.; Dhillon, E.K.; Chau, H.C.; Li, W.K. & Tsang, A.H. (1976). Studies on bacteriophage distribution, pp. virulent and temperate bacteriophage content of mammalian feces. *Applied and Environmental Microbiology*, Vol.32, No.1, pp. 68-74.

Dhillon, T.S. & Dhillon, E.K. (1976). Temperate coliphage HK022. Clear plaque mutants and preliminary vegetative map. *Japanese Journal of Microbiology*, Vol.20, No.5, pp. 385-96.

Dubochet, J.; Adrian, M.; Chang, J.J.; Homo, J.C.; Lepault, J.; McDowall, A.W. & Schultz, P. (1988). Cryo-electron microscopy of vitrified specimens. *Quarterly Reviews of Biophysics*, Vol.21, No.2, pp. 129-228.

Duda, R.L. (1998). Protein chainmail: catenated protein in viral capsids. *Cell*, Vol. 94, No.1 pp. 55-60

Fokine, A.; Chipman, P.R.; Leiman, P.G.; Mesyanzhinov, V.V.; Rao, V.B. & Rossmann, M.G., (2004.) Molecular architecture of the prolate head of bacteriophage T4. *Proceedings of the National Academy of Sciences of the United States of America*, Vol.101, No.16, pp. 6003-6008.

Fokine, A.; Leiman, P.G.; Shneider, M.M.; Ahvazi, B.; Boeshans, K.M.; Steven, A.C.; Black, L.W.; Mesyanzhinov, V.V. & Rossmann, M.G. (2005). Structural and functional similarities between the capsid proteins of bacteriophages T4 and HK97 point to a common ancestry. *Proceedings of the National Academy of Sciences of the United States of America*, Vol.102, No. 20, pp. 7163-7168.

Fortier, L.C.; Bransi, A., & Moineau, S. (2006). Genome sequence and global gene expression of Q54, a new phage species linking the 936 and c2 phage species of *Lactococcus lactis*. *The Journal of Bacteriology*, Vol.188, No.17, pp. 6101-6114.

Fraser, J.S.; Yu, Z.; Maxwell, K.L. & Davidson, A.R. (2006). Ig-like domains on bacteriophages: a tale of promiscuity and deceit. *Journal of Molecular Biology*, Vol. 359, pp. 496-507.

Fuller, D.N.; Raymer, D.M.; Kottadiel, V.I.; Rao, V.B. & Smith, D. E. (2007). Single phage T4 DNA packaging motors exhibit large force generation, high velocity, and dynamic variability. *Proceedings of the National Academy of Sciences of the United States of America*, Vol. 104, No.43, pp. 16868-16873.

Grigorieff, N. & Harrison, S.C. (2011). Near-atomic resolution reconstructions of icosahedral viruses from electron cryo-microscopy. *Current Opinion in Structural Biology*, Vol. 21, No.2, pp. 265-73.

Grimes, J.M.; Burroughs, J.N.; Gouet, P.; Diprose, J.M.; Malby, R.; Ziéntara, S.; Mertens, P.P. & Stuart, D.I. (1998). The atomic structure of the bluetongue virus core. *Nature*, Vol. 395, No.6701, pp. 470-478.

Guasch, A.; Pous, J.; Ibarra, B.; Gomis-Rüth, F.X.,; Valpuesta, J.M.; Sousa, N.; Carrascosa, J.L .& Coll, M. (2002). Detailed architecture of a DNA translocating machine, pp. the high-resolution structure of the bacteriophage phi29 connector particle. *Journal of Molecular Biology*, Vol. 315, No.4, pp. 663-676.

Hagen, E.W.; Reilly, B.E.; Tosi, M.E. & Anderson, D.L. (1976). Analysis of gene function of bacteriophage phi29 of *Bacillus subtilis*: identification of cistrons essential for viral assembly. *The Journal of Virology*, Vol. 19, No.2, pp. 501-517.

Hankin, E H. (1896). "L'action bactericide des eaux de la Jumna et du Gange sur le vibrion du cholera" (in French). *Annales de l'Institut Pasteur* Vol. 10, pp. 511–523.

Hanlon, G.W. (2007). Bacteriophages, pp. an appraisal of their role in the treatment of bacterial infections. *International Journal of Antimicrobial Agents*, Vol. 30, No.2, pp. 118-28.

Harris, J. R. (1997). *Negative Staining and Cryoelectron Microscopy, The Thin Film Techniques;* BIOS Scientific Publishers, Oxford, UK,. ISBN 1859961207

Harrison, S.C. (1969) Structure of tomato bushy stunt virus. I. The spherically averaged electron density. *Journal of Molecular Biology*, Vol. 42, No.3, pp.457-83.

Hendrix, R.W. (2005). Bacteriophage HK97: Assembly of th capsid and evolutionary connections, in *Virus Structure and Assembly*, Vol. 64, pp. 1-14.

Hryc, C.F.; Chen, D.H. & Chiu, W. (2011). Near-Atomic-Resolution Cryo-EM for Molecular Virology. *Current Opinion in Virology*, Vol. 1, No.2, pp. 110-117.

Huiskonen, J.T.; Kivelä, H.M.; Bamford, D.H. & Butcher, S.J. (2004). The PM2 virion has a novel organization with an internal membrane and pentameric receptor binding spikes. *Nature Structural & Molecular Biology*, Vol. 11, No.9, pp. 850-856.

Isidro, A.; Henriques, A.O. & Tavares, P. (2004a). The portal protein plays essential roles at different steps of the SPP1 DNA packaging process. *Virology*, Vol. 322, No.2, pp. 253–263.

Isidro, A.; Santos, M.A.; Henriques, A.O. & Tavares, P. (2004b). The high resolution functional map of bacteriophage SPP1 portal protein. *Molecular Microbiology*, Vol. 51, No.4, pp. 949–962.

Iwasaki, K.; Trus, B. L.; Wingfield, P. T.; Cheng, N.;Campusano, G.; Rao, V. B. & Steven, A. C. (2000). Molecular architecture of bacteriophage T4 capsid, pp. vertex structure and bimodal binding of the stabilizing accessory protein, Soc. *Virology*, Vol. 271, No.2, pp. 321-333.

Jensen, G. (2010). *Cryo-EM* Part B: *3-D Reconstruction*, Vol. 482 (Methods in Enzymology) Elsevier Inc, ACADEMIC PRESS, San Diego ISBN 13:978-0-12-384991-5

Jiang, W.; Chang, J.; Jakana, J.; Weigele, P.; King, J. & Chiu, W. (2006). Structure of epsilon15 bacteriophage reveals genome organization and DNA packaging/injection apparatus. *Nature*, Vol. 439, No.7076, pp. 612–616.

Jiang, W.; Baker, M.L.; Jakana, J.; Weigele, P.R.; King, J. & Chiu, W. (2008). Backbone structure of the infectious epsilon15 virus capsid revealed by electron cryomicroscopy. *Nature*, Vol. 451, No.7182, pp. 1130-1134.

Jończyk, E.; Kłak, M.; Międzybrodzki, R. & Górski ,A. (2011). The influence of external factors on bacteriophages. *Folia Microbiologica (Praha).*, Vol. 56, No.3, pp. 191-200.

Kostyuchenko, V.A.; Chipman, P.R.; Leiman, P.G.; Arisaka, F.; Mesyanzhinov, V.V. & Rossmann, M.G. (2005). The tail structure of bacteriophage T4 and its mechanism of contraction. *Nature Structural & Molecular Biology*, Vol.12, No.9, pp. 810–813.

Kostyuchenko, V.A.; Leiman, P.G.; Chipman, P.R.; Kanamaru, S.; van Raaij, M.J.; Arisaka, F.; Mesyanzhinov, V.V. & Rossmann, M.G. (2003). Three-dimensional structure of bacteriophage T4 baseplate. *Nature Structural & Molecular Biology*, Vol. 10, No.9, pp. 688-693.

Kostyuchenko, V.A.; Navruzbekov, G.A.; Kurochkina, L.P.; Strelkov, S.V.; Mesyanzhinov, V.V. & Rossmann, M.G. (1999). The structure of bacteriophage T4 gene product 9, pp. the trigger for tail contraction. *Structure: Folding and Design*, Vol. 7, No.10, pp.1213-1222.

Lander, G.C.; Evilevitch, A.; Jeembaeva, M.; Potter, C.S.; Carragher, B. & Johnson, J.E. (2008). Bacteriophage lambda stabilization by auxiliary protein gpD, pp. timing, location, and mechanism of attachment determined by cryo-EM. *Structure*, Vol. 16, No.9, pp. 1399–1406.

Lander, G.C.; Khayat, R.; Li, R.; Prevelige, P.E.; Potter, C.S.; Carragher, B. & Johnson, J.E. (2009). The P22 tail machine at subnanometer resolution reveals the architecture of an infection conduit. *Structure*, Vol. 17, No.6, pp. 789-99.

Lebedev, A.A.; Krause, M.H.; Isidro, A.L.; Vagin, A.A.; Orlova, E.V.; Turner, J.; Dodson, E.J.; Tavares, P. & Antson, A.A. (2007). Structural framework for DNA translocation via the viral portal protein. *The EMBO Journal*, Vol. 26, No.7, pp. 1984-94.

Leibo, S.P.; Kellenberger, E.; Kellenberger-van der Kamp, C.; Frey, T.G. & Steinberg, C.M. (1979). Gene 24-controlled osmotic shock resistance in bacteriophage T4: probable multiple gene functions. *The Journal of Virology*, Vol. 30, No.1, pp. 327-338.

Leiman, P.G.; Arisaka, F.; van Raaij, M.J.; Kostyuchenko, V.A.; Aksyuk, A.A.; Kanamaru, S. & Rossmann, M.G. (2010). Morphogenesis of the T4 tail and tail fibers. *Virology Journal*, Vol. 7, pp. 355.

Leiman, P.G.; Chipman, P.R.; Kostyuchenko, V.A.; Mesyanzhinov, V.V. & Rossmann, M.G. (2004). Three-dimensional rearrangement of proteins in the tail of bacteriophage T4 on infection of its host. *Cell*, Vol. 118, pp. 419–429.

Leiman, P.G.; Kanamaru, S.; Mesyanzhinov, V.V.; Arisaka, F. & Rossmann, M.G. (2003). Structure and morphogenesis of bacteriophage T4. *Cellular and Molecular Life Sciences*, Vol. 60, pp. 235.

Lhuillier, S.; Gallopin, M.; Gilquin, B.; Brasilès, S.; Lancelot, N.; Letellier, G.; Gilles, M.; Dethan, G.; Orlova, E.V.; Couprie, J.; Tavares, P. & Zinn-Justin, S. (2009). Structure of bacteriophage SPP1 head-to-tail connection reveals mechanism for viral DNA gating. *Proceedings of the National Academy of Sciences of the United States of America*, Vol. 106, No.21, pp. 8507-12.

Lindert, S.; Stewart, P.L. & Meiler, J. (2009). Hybrid approaches, pp. applying computational methods in cryo-electron microscopy. *Current Opinion in Structural Biology*, Vol. 19, No.2, pp. 218-25.

Lopez, S. & Arias, C. (2010) How Viruses Hijack Endocytic Machinery. *Nature Education, Vol.* 3, No. 9, pp. 16-23

Lorenzen, K.; Olia, A.S.; Uetrecht, C.; Cingolani, G. & Heck, A.J. (2008). Determination of stoichiometry and conformational changes in the first step of the P22 tail assembly. *Journal of Molecular Biology*, Vol. 379, No.2, pp. 385–396.

Luo, E.A.; Frey, R.A.; Pfuetzner, A.L.; Creagh, D.G.; Knoechel , L.; Haynes, C.A., Finlay, B.B & Strynadka, N.C (2000). Crystal structure of enteropathogenic *Escherichia coli* intimin–receptor complex. *Nature*, Vol. 405, No.6790, pp. 1073–1077

Lurz, R.; Orlova, E.V.; Günther, D.; Dube, P.; Dröge, A.; Weise, F.; van Heel, M. & Tavares,
P. (2001), Structural organisation of the head-to-tail interface of a bacterial virus.
Journal of Molecular Biology, Vol. 310, No.5, pp. 1027-37.

Moody, M.F. (1973). Sheath of bacteriophage T4. 3. Contraction mechanism deduced from
partially contracted sheaths. *Journal of Molecular Biology*, Vol. 80, pp. 613–635.

Moody, M.F. & Makowski, L. (1981). X-ray diffraction study of tail-tubes from
bacteriophage T2L. *Journal of Molecular Biology*, Vol. 150, pp. 217–244.

Morais. M.C.; Choi, K.H.; Koti, J.S.; Chipman, P.R.; Anderson, D.L. & Rossmann, M,G.
(2005). Conservation of the capsid structure in tailed dsDNA bacteriophages, pp.
the pseudoatomic structure of phi29. *Molecular Cell*, Vol. 18, No.2, pp. 149-59.

Nakai, T. & Park, S.C. (2002). Bacteriophage therapy of infectious disease in aquaculture.
Research in Microbiology, Vol. 153, pp. 13-18.

Olia, A.S.; Al-Bassam, J.; Winn-Stapley, D.A.; Joss, L.; Casjens, S.R. & Cingolani, G. (2006).
Binding-induced stabilization and assembly of the phage P22 tail accessory factor
gp4. *Journal of Molecular Biology*, Vol. 363, pp. 558–576.

Olia, A.S.; Prevelige, P.E. Jr.; Johnson, J.E & Cingolani, G. (2011). Three-dimensional
structure of a viral genome-delivery portal vertex. *Nature Structural & Molecular
Biology*, Vol. 18, No.5, pp. 597-603.

Orlova, E.V.; Gowen, B.; Dröge, A.; Stiege, A.; Weise, F.; Lurz, R.; van Heel, M. & Tavares, P.
(2003), Structure of a viral DNA gatekeeper at 10 A resolution by cryo-electron
microscopy. *The EMBO Journal*, Vol. 22, No.6, pp.1255-62.

Parent. K.N.; Khayat, R.; Tu, L.H.; Suhanovsky, M.M.; Cortines, J.R.; Teschke, C.M.; Johnson,
J.E. & Baker, T.S. (2010). P22 coat protein structures reveal a novel mechanism for
capsid maturation, pp. stability without auxiliary proteins or chemical crosslinks.
Structure, Vol. 18, No.3, pp. 390-401.

Plisson, C.; White, H.E.; Auzat, I.; Zafarani, A.; São-José, C.; Lhuillier, S.; Tavares, P. &
Orlova, E.V. (2007). Structure of bacteriophage SPP1 tail reveals trigger for DNA
ejection. *The EMBO Journal*, Vol. 26, No.15, pp. 3720-3728.

Popa, M. P.; McKelvey, T. A.; Hempel, J. & Hendrix R. W. (1991). Bacteriophage HK97
structure: wholesale covalent cross-linking between the major head shell subunits.
The Journal of Virology, Vol. 65, pp. 3227-3237.

Prevelige, P.E. (2006). Bacteriophage P22. In *The Bacteriophages*, R. Calendar, ed. (Oxford, pp.
Oxford University Press), pp. 457–468.

Qin, L.; Fokine, A.; O'Donnell, E.; Rao, V.B. & Rossmann, M.G. (2010). Structure of the small
outer capsid protein, Soc, pp. a clamp for stabilizing capsids of T4-like phages.
Journal of Molecular Biology, Vol. 395, No.4, pp. 728-741.

Ray, K.; Ma, J.; Oram, M.; Lakowicz, J.R. & Black, L.W. (2010) .Single-molecule and FRET
fluorescence correlation spectroscopy analyses of phage DNA packaging, pp.
colocalization of packaged phage T4 DNA ends within the capsid. *Journal of
Molecular Biology*, Vol. 395, No.5, pp. 1102-1113.

Rossmann, M.G.; Mesyanzhinov, V.V.; Arisaka, F. & Leiman, P.G. (2004). The bacteriophage
T4 DNA injection machine. *Current Opinion in Structural Biology*, Vol. 14, No.2,
pp.171-180.

Rossmann, M.G.; Morais, M.C.; Leiman, P.G. & Zhang, W. (2005). Combining X-ray crystallography and electron microscopy. *Structure*, Vol. 13, No.3, pp. 355-362.

São-José, C.; Lhuillier, S.; Lurz, R.; Melki, R.; Lepault, J.; Santos, M.A. & Tavares, P. (2006), The ectodomain of the viral receptor YueB forms a fiber that triggers ejection of bacteriophage SPP1 DNA.*The Journal of Biological Chemistry*, Vol. 281, No.17, pp. 11464-11470.

Shors, T. (2008). *Understanding Viruses*. Jones and Bartlett Publishers. ISBN 0-7637-2932-9, Sudbury, USA

Simpson, A.A.; Tao, Y.; Leiman, P.G.; Badasso, M.O.; He, Y.; Jardine, P.J.; Olson, N.H.; Morais, M.C.; Grimes, S.; Anderson, D.L.; Baker, T.S. & Rossmann, M.G. (2000). Structure of the bacteriophage phi29 DNA packaging motor. *Nature*, Vol. 408, pp. 745–750.

Smith, D.E.; Tans, S.J.; Smith, S.B.; Grimes, S.; Anderson, D.L. & Bustamante, C. (2001). The bacteriophage straight phi29 portal motor can package DNA against a large internal force. *Nature*, Vol. 413, No.6857, pp .748-752.

Tang, J.; Olson, N.; Jardine, P.J.; Grimes, S.; Anderson, D.L. & Baker, T.S. (2008). DNA poised for release in bacteriophage phi29. *Structure*, Vol. 16, No.6, pp. 935-43.

Tang, J., Lander, G. C., Olia, A., Li, R., Casjens, S. R., Prevelige, P., Cingolani, G., Baker, T. S., & Johnson, J. E. (2011) Peering down the barrel of a bacteriophage portal: the genome packaging and release valve in P22. *Structure* Vol.19, pp. 496-502.

Tang, L.; Marion, W.R.; Cingolani, G.; Prevelige, P.E. & Johnson, J.E. (2005). Three-dimensional structure of the bacteriophage P22 tail machine. *The EMBO Journal*, Vol. 24, pp. 2087–2095.

Tao, Y.; Strelkov, S.V.; Mesyanzhinov, V.V. & Rossmann, M.G. (1997). Structure of bacteriophage T4 fibritin, pp. a segmented coiled coil and the role of the Cterminal domain. *Structure*, Vol. 5, pp. 789-798.

Tavares, P.; Lurz, R.; Stiege, A.; Rückert, B. & Trautner, T.A. (1996). Sequential headful packaging and fate of the cleaved DNA ends in bacteriophage SPP1. *Journal of Molecular Biology*, Vol. 264, No.5, pp. 954-67.

The Bacteriophages, 2nd edition (2006). Richard Calendar, Oxford University Press http, pp.//www.thebacteriophages.org/chapters/0020.htm

Thuman-Commike, P.A.; Greene, B.; Malinski, J.A.; Burbea, M.; McGough, A.; Chiu, W. & Prevelige, P.E. Jr. (1999). Mechanism of scaffolding-directed virus assembly suggested by comparison of scaffolding-containing and scaffolding-lacking P22 procapsids. *Biophysical Journal*, Vol. 76, No.6, pp. 3267-77.

Wikoff, W.R.; Liljas, L.; Duda, R.L.; Tsuruta, H.; Hendrix, R.W. & Johnson, J.E. (2000). Topologically linked protein rings in the bacteriophage HK97 capsid. *Science*, Vol. 289, No.5487, pp. 2129-33.

Wommack, K.E. & Colwell, R.R. (2000). Virioplankton, pp. viruses in aquatic ecosystems. *Microbiology and Molecular Biology Reviews*, Vol. 64, pp. 69–114.

Xiang, Y.; Morais, M.C.; Battisti, A.J.; Grimes, S.; Jardine, P.J.; Anderson, D.L. & Rossmann, M.G. (2006). Structural changes of bacteriophage phi29 upon DNA packaging and release. *The EMBO Journal*, Vol. 25, No.21, pp. 5229-39.

Zhang, R.; Hryc, C.F.; Cong, Y.; Liu, X.; Jakana, J.; Gorchakov, R.; Baker, M.L.; Weaver, S.C. & Chiu. W. (2011). 4.4 Å cryo-EM structure of an enveloped alphavirus Venezuelan equine encephalitis virus. *The EMBO Journal*, Vol. 30, No.18, pp. 3854-63.

Zhou Z.H. (2011) Atomic resolution cryo electron microscopy of macromolecular complexes. *Advances in protein chemistry and structural biology*, Vol. 82, pp. 1-35.

2

Gels for the Propagation of Bacteriophages and the Characterization of Bacteriophage Assembly Intermediates

Philip Serwer

The University of Texas Health Science Center, San Antonio,
USA

1. Introduction

Advances in biochemistry-based analysis of biotic systems depend on improved procedures for the sorting of the non-covalently joined macromolecular assemblies that are obtained by the expelling of cellular contents. The development of preparative ultracentrifugation, for example, made possible the discovery of both ribosomes and various intracellular organelles (reviewed in Alberts et al., 2002). Some biotic systems depend on assembly more than one might conclude at first (reviewed in Kurakin, 2007). Improving this biochemistry-based analysis is an ongoing process, given that (1) multi-molecular, assembly-derived mechanisms are not yet understood, including (especially) the mechanisms of biological motors (reviewed in Howard, 2009; Myong & Ha, 2010) and (2) one reasonably projects that components of current biotic systems have abiotic ancestors that were also non-covalently joined assemblies, the understanding of which is potentially essential for understanding the origins of life (Koonin, 2009; Serwer, 2011).

Separately, advances in analysis of environmental microbial systems depend on improved procedures for the propagation and sorting of individual microorganisms following their extraction from the environment. Improving microbial propagation/isolation/sorting remains an ongoing process (for example, Ferrari et al., 2008; Sait et al., 2002; Serwer et al., 2009), given that only a small fraction (< 0.01) of environmental microbes have been propagated (Ferrari et al., 2008 and included references). I review here some advances in the use of agarose and agar gels for both (1) biochemistry-based sorting of macromolecular assemblies and (2) detection, propagation and sorting of unusual environmental viruses, with focus on bacterial viruses (bacteriophages), abbreviated phages.

1.1 Basics

Fractionation-based sorting and characterization of macromolecular assemblies is a strategy complementary to biochemical assay-based determining of the activities of single, unassembled macromolecules. The unassembled macromolecules include many that function when assembled with molecules of other types. The use of these two strategies is illustrated by the analysis of mechanisms of DNA replication. The proteins involved invariably include a single protein with DNA polymerase activity. In contrast to the original thought, the process of DNA replication is so complex that numerous proteins of other types are also involved via a multi-molecular complex, usually called a replisome (reviews: Hamdan & Richardson, 2008; Langston et al., 2009). However, replisomes are difficult to isolate from cells and tend to dissociate during isolation. Thus, investigators usually replace sorting-based biochemistry with *in vitro* replisome assembly, followed by analysis of the transitions that occur *in vitro*. Nonetheless, if appropriate procedures can be developed, in theory, one might (1) isolate replisomes from a cell, (2) fractionate the replisomes by the extent of DNA replication and then (3) perform a biochemical/biophysical analysis of the replisome at each of the various stages of replication. Once this is done, a fluorescence-based signature can be developed for each state observed in the sorted replisomes. The signatures would provide a way to observe the progression of replisome-associated DNA replication *in vivo*.

To develop the tools needed for a sorting-based analysis of any biochemical system, we have focused on a multi-molecular, biotic system for which sorting has been relatively productive, because the multi-molecular complexes involved (1) are relatively stable, (2) are relatively uniform in surface characteristics and (3) can be fractionated by the extent of the biochemical process being analyzed. This system packages the double-stranded DNA of a phage after the phage DNA has been replicated in a phage-infected cell. For all studied double-stranded phages, DNA packaging is initiated by a protein capsid (procapsid) pre-assembled without interaction with DNA. This procapsid is called capsid I in the case of the related phages, T3 and T7. Capsid I converts to a larger, more phage-like capsid during DNA packaging (capsid II for T3/T7; Figure 1,a-b) and can be made to package DNA *in vitro*, after isolation of capsid I from a lysate of infected cells (reviews: Aksyuk & Rossmann, 2011; Catalano, 2000; Fujisawa, & Morita, 1997; Serwer, 2010). Subsequent steps in DNA packaging begin with capsid II and some are described in Section 1.2.

Returning to the topic of propagation-/isolation-based analysis of environmental phages, we have focused on developing and using procedures to propagate and isolate phages that are not isolated by conventional procedures. Our new procedures isolate phages that are unconventional in that they have one or more of the following characteristics: (1) propagation-associated aggregation, (2) unusually large size, (3) inactivation by dilution and (4) absence of sufficient propagation in liquid culture to produce cellular lysis in visible amount. The core procedure is based on initial isolation by incubation of soil samples in a medium-containing, dilute (0.1-0.2%) agarose overlay (Figure 2a; reviewed in Serwer et al., 2009). This procedure continues by platinum needle transfer only, as illustrated in Figures 2,b-d. As described in Section 5, the pores of the agar gels typically used for a phage plaque-supporting gel are too small for propagating some (maybe most) large or aggregating phages.

After presentation of some details of both DNA packaging and phage isolation in the remainder of Section 1, I will describe the studies of the gels and gel electrophoresis used for these studies. Understanding of these gels was (and presumably will be) essential to improving the use of them.

1.2 Some details

Past work on the sorting of infected cell-derived macromolecular assemblies has produced a hypothesis for the sequence of T3/T7 DNA packaging events. The solid arrows of Figure 1 indicate this proposed sequence, as derived from the fractionation-based sorting of particles

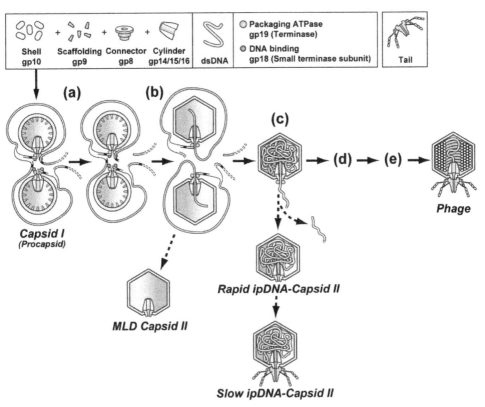

Fig. 1. Bacteriophage T3/T7 DNA packaging. The solid arrows connect intermediates that have been deduced from observed intermediates. Dashed arrows indicate the observed intermediates.

produced during packaging and found in lysates of infected cells. These latter particles will be called intermediates whether or not altered during fractionation. The isolated, fractionated intermediates include incompletely packaged DNA (ipDNA)-containing capsids (ipDNA-capsids). The ipDNA-capsids were originally sorted by ipDNA length via buoyant density-based ultracentrifugal fractionation (Fang et al., 2008). The ipDNA-capsids were then identified and the capsid further characterized by agarose gel electrophoresis (AGE) of intact particles (Fang et al., 2008), as further described in Section 3. A dashed arrow in Figure 1 connects a fractionated and characterized ipDNA-capsid to an intermediate proposed to exist during DNA packaging *in vivo*. The detection of ipDNA-capsid II by AGE was simplified by the fact that the AGE-migration of ipDNA-capsid II is independent of the length of ipDNA. In general, the migration of any particle during AGE depends only on the characteristics of the particle's surface, not on what is packaged inside. This point is discussed in more detail in Section 3.

The use of advanced procedures of AGE revealed that the protein shell of at least some ipDNA-capsid II does eventually undergo changes as ipDNA becomes longer. These changes produce intermediates at (d) and (e) in Figure 1. In brief, the changes at (d) and (e) suggest that the T3/T7 DNA packaging motor has two cycles, the second of which changes the capsid's shell and acts as a back-up cycle when the first stalls. The details (Serwer et al., 2010; Serwer & Wright, 2011) are not reviewed here because they are complex enough to be distracting to the main objectives.

2. Gel-forming polymers

2.1 Basics

Gels have spaces through which molecules migrate either by thermal motion or by response to an external potential gradient. The potential gradient is typically, but not necessarily, electrical. Agar gels, although initially (and still: Rasmussen & Morrissey, 2007) used as a supporting matrix for food, were subsequently found to be similarly useful as a supporting matrix for bacterial colonies. The bacterial colonies typically grew on the surface, but use of dilute (0.4%) agar in the presence of a complex medium was found to permit *Salmonella typhimurium* to swim through the gel. This swimming was used to assay transfer of genes needed for motility (Stocker et al., 1953). Thus, the pore size of the medium-containing 0.4% agar gel, while not precisely defined by these studies, could be estimated to be at least as large as the width of the bacteria, assumed to be ~ 500 nm. Smaller pores would not have allowed the bacteria to migrate to the interior of the gel.

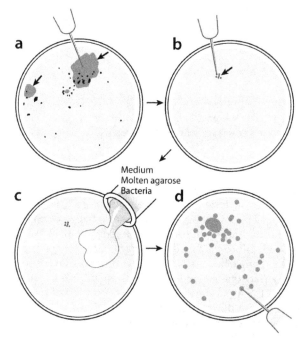

Fig. 2. Phage isolation. (a) Post-incubation initial plate with soil (irregular black objects) embedded in a dilute agarose overlay and phage-induced zones of clearing (arrows) in a host lawn. (b) Needle transfer via stabs (arrow) to the bottom agar of a new Petri plate. (c) Pouring of a new overlay. (d) Post-incubation secondary plate with single phage plaques one of which is being used for cloning. Further details are reviewed in Serwer et al. (2009)

The cause of the relatively large pores of some polysaccharide gels is the lateral aggregation of the polysaccharide polymer to form multi-chain "pillars" that provide gel strength while forming a mesh with relatively large pores. The existence of the pillars is qualitatively confirmed by simply observing the light scattering of a gel. Agarose gels are typically turbid because of the pillars (Rees, 1972). In contrast, polyacrylamide gels of the same total concentration and conventional cross-linker concentrations are not turbid (Chen and Chrambach, 1979). The turbidity of an agarose gel decreases as (1) the temperature of gelation and buffer ionic strength decrease, and (2) the agarose molecular weight increases (Griess et al, 1993; Griess et al., 1998; Serwer & Griess, 1999), with an associated decrease in the radius of the effective pore (P_E). If one adds consideration of the agarose source-, purification- and derivatization-dependence of P_E (Griess et al., 1989; Griess et al., 1998), one can only conclude that P_E-dependent results from different studies cannot be compared quantitatively unless one is willing to tolerate the likelihood of P_E errors of at least 100%. In general, quantitative comparisons should be performed with internal standards.

2.2 Some details

The gels to be discussed here are cast by cooling solutions of either agar or agarose that had been dissolved by boiling. Agar is a β-linked alternating co-polymer of two sugars; negatively charged groups are attached in variable amount to the sugars. Agar is obtained from red seaweed. Agarose is a sub-fraction of agar that has a relatively low density of charged groups (reviewed in Rees, 1972). The extent of residual charge is often used to name agarose preparations via the field-induced flow of buffer that gel-attached charged groups cause (electro-osmosis, abbreviated EEO; Griess et al., 1989). The minimum agarose concentration for gel formation varies somewhat with agarose EEO, and agarose chain length but can be as low as 0.03% for a high-strength agarose, when the gel is supported at its sides by embedding in a more concentrated gel (Serwer et al., 1988).

If one extrapolates previous determinations of P_E to 0.03% agarose, one finds that micron-sized particles can enter agarose gels. If entry into the gel is to be driven by a potential gradient, the entry will become limited by trapping of a micron-sized particle in the pores that are relatively small, if the potential gradient is too high in magnitude. The trapping occurs because of the relatively low thermal motion of particles this large (Serwer et al., 1988; Serwer & Griess, 1998). Nonetheless, by use of an electrical potential gradient relatively low in magnitude (0.5 V/cm; 2.0 V/cm is too high), intact (alive) cells of the bacterium, *Escherichia coli*, have been subjected to agarose gel electrophoresis and fractionated by length (Serwer et al., 1988).

On the other hand, one lowers P_E by raising agarose gel concentration, a process that is assisted, if necessary, by lowering the average agarose chain length and, therefore, reducing viscosity (Griess et al., 1993). The studies reported below have not been limited by difficulties in attaining any P_E needed.

3. Gel electrophoresis

3.1 Electrophoretic principles and some of their applications

Fractionation by gel electrophoresis depends on two characteristics of a roughly spherical particle being fractionated. The first characteristic is the average, per area, of the particle's surface electrical charge that is not counter ion-neutralized (σ). The force produced by application of an electrical potential is proportional to σ (Shaw, 1969; Stellwagen et al., 2003). Therefore, the terminal velocity (v) induced by an electrical potential is also proportional to σ. The magnitudes of σ and v typically decrease as the concentration of counter ions increases, because of the increase in surface charge neutralization, as described by the Debye-Hückel theory (Bull, 1971). The ionic strength of electrophoresis is kept relatively low both to increase the magnitude of v and to lower the heat produced during electrophoresis. A result is that adding any salt to an electrophoresis buffer reduces the force on the particle and, therefore, v.

This effect causes band spreading when a sample is in a relatively high ionic strength solution and electrophoresis is to be conducted at lower ionic strength. The relatively high ionic strength of the sample, coupled with diffusion of sample ions into the electrophoresis buffer, will cause band spreading because the leading edge of the sample will initially have a v higher in magnitude than v of the rest of the sample. Because v is usually (not always) proportional to the electrical potential gradient (E), I will sometimes refer to the v/E ratio, rather than v. The v/E ratio is also called the electrophoretic mobility (μ).

This band spreading becomes important when, to avoid loss of particles during dialysis, one wants perform AGE of particles that are in concentrated (2-4 M) solutions of cesium chloride. This situation arises after preparative fractionation by ultracentrifugation in a cesium chloride density gradient. Band spreading is avoidable, however, if AGE is performed with the gel submerged beneath the electrophoresis buffer (submerged gel electrophoresis). Submerged gel electrophoresis is a standard procedure with which I assume that the reader is familiar. After loading samples for submerged gel AGE, one avoids sample salt-induced band spreading by waiting for 1.0-1.5 hours before starting electrophoresis. In this time, salt ions dialyze into the electrophoresis buffer. This procedure was based on the previous observation that dialysis of 2-4 M cesium chloride from 0.5 inch dialysis tubing is complete by 30 minutes, as judged by measuring the refractive index of the cesium chloride solution after removing it from the dialysis tubing (unpublished data).

Although submerged gel AGE of nucleic acids is almost always done without attempting to control pH gradients, this absence of control is not a good idea when proteins are the samples. Without a counter-measure, a pH gradient is unavoidable because hydrogen gas is released at the cathode, thereby raising the pH, and oxygen gas is released at the anode, thereby lowering the pH. Proteins titrate much more than nucleic acids in the pH range of the pH gradient generated during submerged gel AGE. The result of this protein titration is likely to be disastrous. Informally, I have been told of failures of AGE and, in some cases, pH gradients would probably have caused failure even if other aspects were in order.

The most efficient way to prevent a pH gradient with a submerged agarose gel is to circulate the electrophoresis buffer from one buffer tank to the other. The buffer flows back to the source tank across the surface of the submerged gel. One can also reduce the pH gradient by reducing the height of the buffer, but not so much that the cross-sectional area of the buffer starts to fluctuate. To avoid buffer circulation-induced washing of the sample out of sample wells, the circulation is started after the electrophoresis. The details of timing and circulation speed are empirically determined for each system. We circulate at ~ 100 ml/minute, beginning at 30 minutes after the start of electrophoresis at 1 V/cm, with a phosphate buffer, pH 7.4 and a buffer height of about 0.8 cm.

3.2 Other factors that determine procedure

Any particle can be fractionated by AGE if the particle (1) is electrically charged in the buffer used, (2) is small enough to fit into the pores of the gel and is not electrophoretically trapped, (3) does not adhere to the gel and (4) is not damaged or dissociated by the process of electrophoresis. Particle-gel adherence and particle dissociation are the most likely causes of failure if the aspects from the previous section are in order. Particle-gel adherence is generally the case when particles either are found either to be broadly distributed near the origin or to form a sharp band at the origin edge of the agarose gel.

Responses to particle-gel adherence include changing the composition of the gel, overloading the binding sites by increasing particle concentration (Serwer & Hayes, 1982) and proteolytic cleavage of the gel-binding region of the particle (Serwer et al., 1982). Responses to dissociation include cross-linking, which is necessary in the case of microtubules, for example. Cross-linked microtubules do migrate during AGE, but dissociate if not cross-linked (Serwer et al., 1989).

3.3 Sieving during AGE

As solid, spherical particles migrate through a gel, they experience both hydrodynamic and steric effects of the presence of fibers that form the gel. If a sphere is almost as large as the effective pore of the gel, motion will be restricted to the point that the particle hardly moves. As the particle becomes smaller, the "sieving" effect of the fibers decreases and the particle undergoes more rapid motion. Eventually, while never zero, the sieving effect becomes so small that it changes almost imperceptibly with a percentage change in particle radius that caused a large change in sieving for larger spheres. So, to increase the sieving-based resolution by radius of a spherical particle, one decreases P_E (increases gel concentration), but stops before the pores are so small that the particles do not migrate. One pays for the increase in sieving-based resolution with an increase in the time of fractionation. A quantitative analysis of these effects for spheres is in Griess et al. (1989).

The effect of particle shape on gel sieving has been investigated for rod-shaped viruses. Without discussing the quantitative details, the lessons learned are the following. (1) Sieving effects do not discriminate a rod from a sphere when the 0.5xrod length/P_E ratio is below ~1 (Griess et al., 1990). (2) For a rod-shaped particle with length in this range, the effective radius that best describes the gel electrophoretic sieving is determined by assuming that the rod has a surface area (in contrast to either a length or a volume) equal to that of a sphere that exhibits the same sieving (Griess et al., 1990). (3) At smaller P_E values, a rod (unlike a sphere) has a gel electrophoretic μ that increases in magnitude as the magnitude of E increases. This effect can be exploited to help identify rod-shaped particles after AGE (Serwer et al., 1995).

3.4 One-dimensional gel electrophoresis

A frequent application of gel electrophoresis is sodium dodecylsulfate (SDS) polyacrylamide gel electrophoresis (SDSPAGE) of proteins (Studier, 2000). This procedure starts with boiling of the proteins in the presence of SDS, a negatively charged ionic detergent. The SDS binds to the proteins and produces a surface that has a σ that is assumed be the same for all proteins, based on empirical measurements of SDS binding (Reynolds & Tanford, 1970). Thus, even though both σ and sieving determine μ, the assumed uniformity of σ makes possible the interpretation of SDSPAGE patterns via sieving only. That is why SDSPAGE, even though one-dimensional, is useful for estimating the molecular weight of a protein.

Uniformity of σ is usually also assumed during the AGE of DNA and RNA. This assumption is based on the uniformity of the phosphate backbone and remains accurate until end-effects occur as double-stranded DNA fragments are shortened (Stellwagen et al., 2003). Thus, DNA fractionations are usually interpreted via sieving, without considering possible changes in σ. The sieving effects are complicated by flexibility and, in some cases, by either branched or circular conformation (Åkerman, & Cole, 2002; Brewer, & Fangman, 1991). Conformation-dependent effects on ion binding and, therefore, σ, also occur for unusually bent DNA molecules (Stellwagen et al., 2005). Both circular DNA and DNA bound to solid objects undergo elevated E-induced trapping effects. These trapping effects are the basis for pulsed field-based separations (Åkerman, & Cole, 2002; Gauthier & Slater, 2003, for example) that are outside of the area of this review.

One electrophoretic direction (dimension) is usually used for both SDSPAGE and nucleic acid gel electrophoresis (1d-AGE in the case of agarose gels). However, the electrophoretic profile does not have a unique interpretation, if both σ and particle dimension vary among particles subjected to 1d-AGE. To achieve a unique interpretation, based on both σ and particle dimensions (effective radius for a sphere), a second dimension of electrophoresis is added.

3.5 Separate analysis of σ and effective radius: a second dimension

To separately measure both the effective radius (R_E) and the σ of roughly spherical particles fractionated by AGE, one must add a second dimension (2d-AGE). Figure 3 illustrates a 2d-AGE procedure whereby one performs the first dimensional electrophoresis in a relatively

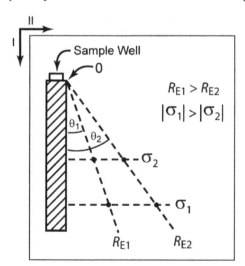

Fig. 3. Illustration of 2d-AGE. A sample is layered in the sample well and subjected to electrophoresis (arrow I indicates direction) through a dilute first dimension gel (diagonal bars) and, then, at a right angle (arrow II indicates direction) through a more concentrated second dimension gel.

dilute agarose gel so that μ is determined primarily by σ. The dilute, first dimensional gel is stabilized by embedding it in a more concentrated gel used for the second dimension. To perform the second dimensional electrophoresis, the field/gel angle is rotated by 90° and electrophoresis is repeated with a second dimensional gel that is much more concentrated than the first dimensional gel.

The key to the 2d-AGE-based analysis of R_E is that the percentage change in μ is independent of σ, when one compares μ in the second dimension with μ in the first, as first empirically confirmed in Serwer et al. (1986). Geometrically, this relationship implies that all particles of any given R_E are on one line (called a size line) that extends from the effective origin of electrophoresis (O in Figure 3) through the center of the band formed by a particle. As the angle (θ) between this line and the direction of the first electrophoresis decreases, R_E increases, as illustrated in Figure 3. The value of σ is proportional to the distance migrated in the first dimension, as illustrated in Figure 3. With the use of standards of size known by small-angle x-ray scattering (Serwer et al., 1986; Serwer et al., 1989), differences in R_E as small as 0.5% have been resolved by using P_E values close to the R_E's of the particles analyzed (Casjens et al., 1992).

An advantage of 2d-AGE is that patterns can be interpreted for particles heterogeneous in either σ or R_E (or both). This aspect was originally demonstrated for vaccine conjugates heterogeneous in R_E (Serwer & Hayes, 1986) and has been developed in quantitative detail, given that these conjugates are of high utility (Tietz, 2007, 2009). More recently, the use of 2d-AGE with particles heterogeneous in both σ and R_E has been used to detect ipDNA-capsids of new type. These new ipDNA-capsids are at positions (d) and (e) in the pathway of Figure 1. Details are in Serwer et al. (2010) and Serwer and Wright (2011).

4. Use of 1d-AGE to determine the kinetics of assembly *in vivo*

Major advantages of AGE are (1) the efficiency of fractionation of multiple samples and (2) the efficiency and accuracy of the quantification, via either autoradiography or fluorography, of the amount of radioisotope in each of several fractionated and detected particles. Thus, one can observe the kinetics of the passage of radiolabel through various intermediates, for the purpose of both determining the order of intermediate appearance and analyzing the mechanism of assembly and associated function. The efficiency makes this analysis possible not only for the wild type process, but also for the same process as it occurs for a mutant, with the 1d-AGE typically performed in a single agarose slab gel (see Serwer & Watson, 1982).

4.1 An example of information previously obtained

This strategy was previously used for determining the effects on the assembly of phage T7 capsid I of removing the protein (called the connector or portal protein; Figure 1) that connects the gp10-containing shell with the tail of the mature phage. T7 proteins are labeled by gp, followed by gene number, as reviewed in Pajunen et al. (2002); comparable genes in T3 and T7 are given the same number. The T3 and T7 connectors are 12-mers of gp8. Although gp8 was in a position that suggested a role in nucleating shell assembly (Figure 1), genetic removal of gp8 had no detectable effect on the initial kinetics of capsid I assembly. That is to say, gp8 is not part of the nucleus for shell assembly. However, capsid I assembly terminated prematurely in the absence of gp8 (Serwer & Watson, 1982). In these experiments, we analyzed completely unfractionated lysates of T7-infected *E. coli* by 1d-AGE.

Connector-independent shell assembly nucleation was subsequently also observed for phage P22, by use of rate zonal centrifugation in a sucrose gradient, rather than 1d-AGE, to assay for procapsids (Bazinet & King, 1988). Phages P22 and T7 are basically unrelated, but both have icosahedral shells with a triangulation number of 7 (P22: Chang et al., 2006; T7: Fang et al., 2008). Thus, the connector apparently has evolved after the shell. I note that the nucleus for shell assembly, whatever it is (a proposal is in Serwer, 1987), is likely to have at least 6 independent components, which implies a 6th order nucleation reaction, at least. The formation of a nucleus will, therefore, have a very high dependence on effective capsid protein concentration and, therefore, on excluded volume. Therefore, studies of *in vitro* shell assembly must be performed under conditions that mimic *in vivo* assembly, if any interpretation of what happens *in vivo* is intended. This *in vivo*-first priority cannot logically be reversed.

Finally, I note that herpes simplex virus also has a connector and that the herpes simplex virus connector (portal) is also not the nucleus for shell assembly, based on experiments similar in concept to those performed for phages (Newcomb et al., 2005). The work on phages preceded the work on herpes simplex virus by about 20 years largely because of the relative simplicity and speed of propagating and performing genetics with phages. Combining the simplicity and speed of work on phages with the simplicity and speed of AGE is a powerful addition to genetics.

4.2 Utilizing chromatographical effects during electrophoresis

Although the basics of 1d-AGE and 2d-AGE are well defined and standardized, the transformations of macromolecular assemblies have complex determinants and are generally unpredictable. Thus, 1d-AGE and 2d-AGE analysis of macromolecular assemblies should be interpreted with as little bias as possible. The data are primary and have sometimes been surprising, as illustrated in the previous section. The following, additional surprise occurred while we were determining the kinetics of T7 capsid I assembly by 1d-AGE. We found that some of the assembled, radiolabeled capsid protein appeared only near the origin of the agarose gels. This observation appeared, at first, to be a liability in that one could not initially characterize this material. However, the apparent liability rapidly became an asset when we discovered that these particles were capsid-like, as found by digestion with protease. Protease digestion converted these "agarose gel adherent" particles to particles that migrated as capsid II. One can only anticipate that non-standardized responses, such as this one, will be needed for most, if not all, comparable analyses of the biochemistry of multi-molecular complexes.

Knowing that the T7 agarose-adherent particles were capsid-like, we determined the kinetics of their formation in the presence and absence of the gp8 connector. The results were the following (Serwer & Watson, 1982). (1) In the presence of the gp8 connector, the agarose-adherent particles first increased in amount and then decreased, i.e., the agarose-adherent particles behaved as though either they or, more likely, related capsid I-like *in vivo* particles (that decayed to agarose adherent particles), were intermediates in the assembly of capsid I. (2) In the absence of the gp8 connector, the agarose-adherent particles increased in amount progressively, before and after the assembly of capsid I was terminated, as though the agarose-adherent particles had now become an end product of abortive assembly. These observations supported the previous conclusion that the gp8 connector did not nucleate shell assembly. In addition, the apparent conversion from intermediate to abortive end product in the absence of the gp8 connector was interpreted by the assumption that the connector was necessary for correction of errors of shell assembly. A role of the connector in assembly error correction explains why mis-assembled shells continued to accumulate in the absence, but not in the presence, of the connector (Serwer et al., 1982; Serwer & Watson, 1982). Apparently, analysis of *in vivo* procapsid assembly has not subsequently advanced past this point (recent review: Aksyuk & Rossmann, 2011).

To give some idea of the uniqueness of the data that can be obtained by AGE, I mention that (unpublished) efforts to purify the agarose adherent particles in large amount failed because these particles are lost as aggregates when the scale of the lysates was increased. We made this observation by mixing radiolabeled particles (from a small lysate) with relatively large lysates; the radiolabeled particles had been partially purified by rate zonal sucrose gradient centrifugation. That is to say, 1d-AGE (with and without protease digestion) is, thus far, the only way to identify these particles.

5. Propagation of large phages: a use of P_E values

Recent surprises in microbiology include the discovery of "giant" eukaryotic, double-stranded DNA viruses with genomes larger than 1 million base pairs (reviews: Claverie et al., 2009; Colson & Raoult, 2010). These giant viruses, originally thought to be cells, really are viruses based on the packaging of the genome in an icosahedral capsid shell assembled from subunits that have sequence similarity with the major shell protein of other viruses, including *Paramecium Bursaria Chlorella Virus* 1 (Azza et al., 2009). The first of these viruses, *Acanthamoebae polyphaga* Mimivirus, has shell-associated spikes that extend to an outer radius of 373 nm (Klose et al., 2010). This large radius is presumably the reason that no plaque assay was initially reported and, to my knowledge, has still not been reported. Based on previous determinations of P_E, a plaque assay should be possible for these viruses, as described in detail in Section 5.1, below.

The work on gel concentration-dependence of the formation of plaques by relatively large viruses began with the largest known phage, called phage G (for giant). Phage G makes only very small plaques in the traditional plaque supporting, 0.4-0.7% agar gels. However, the dimensions of phage G were roughly the same as the P_E of the plaque-supporting agar gels. When relatively dilute agarose gels are used to propagate phage G, this phage was found to make large plaques that, when confluent, produce an overlay that is an excellent preparative source of phage G (Serwer et al., 2009).

5.1 P_E values

The considerations of Sections 2.1 and 2.2 already suggest that P_E values of plaque-supporting gels can be made high enough to form plaques of Mimivirus and related giant viruses. The culmination of a series of sieving-based measurements of P_E produced the following equation that describes the relationship between P_E (in nm) and the percentage, $\underline{A}$, of the LE (low EEO) agarose usually used; gels were formed at ~25 °C in buffer that contained 0.025 M sodium phosphate, pH 7.4, 0.001 M $MgCl_2$ (Griess et al., 1989). $P_E = 148\underline{A}^{-0.87}$. This relationship predicts that LE agarose gels with $\underline{A}$ below about 0.34% are dilute enough for plaque formation by mimivirus; a concentration ~ 2x lower is appropriate for an initial test. This relationship also predicts that a 0.25%LE agarose gel is the most concentrated LE agarose gel with a P_E value large enough so that mobile *Salmonella* cells, assumed to have a diameter of about 500 nm, migrate through the gel. This $\underline{A}$ value is lower than the 0.4% used by Stocker et al. (1953) for bacterial migration, as discussed in Section 2.1. Higher EEO agarose preparations, which should better mimic the agar used in Stocker et al. (1953), form gels with an even lower maximal $\underline{A}$ for migration of bacteria (Griess et al., 1989). Thus, the medium present in the 0.4% agar gels of Stocker et al. (1953) appears to have caused an increase in P_E.

As discussed in Section 2.2, the value of P_E increases as the temperature of gelation increases, as judged by both sieving during gel electrophoresis and electron microscopy of thin sections. We have applied this principle to the propagation of a phage that is both large (shell radius ~ 50 nm; tail length ~ 486 nm) and aggregating and found that, indeed, plaque size increases as the temperature of gelation increases for the plaque-supporting gel (Serwer et al., 2009). That is to say, (1) the electrophoretic sieving and the apparent sieving during plaque formation move, as expected, in the same direction with change in P_E and (2) a plaque assay should be a possibility for Mimivirus, unless a trapping effect is encountered.

In the case of Mimivirus, however, the possibility of gravitational field-induced arrest of motion exists. If 1g sedimentation causes trapping of Mimivirus in gels, then buoying Mimivirus should make plaque formation possible in appropriately dilute agarose, plaque-supporting gels.

5.2 Gel-supported propagation of new phages: large and aggregating phages

Evidence exists that the viruses thus far isolated and propagated are not any more than 1% and probably much less of the total in the environment. This evidence includes the sequences of environmental viral RNA and DNA obtained without propagating the viruses involved (metagenomics: reviewed in Casas & Rohwer, 2007). We have used dilute agarose gel propagation to isolate several phages that cannot be propagated in any other way, including propagation in traditional agar gels and liquid enrichment culture. Several of these phages undergo extensive aggregation during plaque formation (Serwer & Wang, 2005; Serwer et al., 2009), which suggests that these phages would not be detected by metagenomics, because of loss during procedures (filtration, low speed centrifugation, for example) that are used to remove bacteria.

Virus aggregation is a well-known phenomenon, potentially important to new frontiers in virology. Historically, virus aggregation was important because of its potential (occasionally realized) to inhibit antibody neutralization of several eukaryotic viruses (Wallis, C. & Melnick, 1967). The need for revised procedures in the isolation of some, including aggregating, phages suggests that revised procedures will also be needed for the isolation of some not-yet-isolated eukaryotic viruses. As is usually the case, the advances needed are most rapidly explored with phages.

6. Acknowledgments

Recent work in the author's laboratory was supported by the Welch Foundation (AQ-764).

7. In Memoriam

Dr. Gary A. Griess made large contributions to our current knowledge gel electrophoresis, as apparent from the attached manuscript. These contributions were in several areas, perhaps most notably in the areas of the structure and sieving of gels. Gary died from complications of cancer on April 28, 2008. He had received an undergraduate degree in physics from MIT in 1962 and a PhD in biophysical chemistry from the University of Massachusetts Amherst (Advisor, John F. Brandts) in 1970. By 1985, Gary and I had established a collaboration that began with work on the biophysical characterization of phages. We soon developed a focus on the structure and sieving of gels. Gary provided essential computational, biophysical and experimental aspects of this work, much of which depended on his creativity and ingenuity. He was also very generous with his assistance to others in all laboratories of our department. We all miss him. This manuscript is dedicated to Gary.

8. References

Aksyuk, A.A. & Rossmann, M.G. (2011) Bacteriophage assembly. *Viruses* 3, 172–203.

Åkerman, B. & Cole, K.D. (2002) Electrophoretic capture of circular DNA in gels. *Electrophoresis* 23, 2549–2561.

Alberts, B., Johnson, A., Lewis, J., Raff, M., Keith Roberts, K. & Peter Walter, P. (2002) *Molecular Biology of the Cell, 4th edition*, Garland Science, New York.

Azza, S., Cambillau, C., Didier, R., & Suzan-Monti, M. (2009) Revised Mimivirus major capsid protein sequence reveals intron-containing gene structure and extra domain. *BMC Molecular Biology* 10, 39.

Bazinet, C., and King, J. (1988). Initiation of p22 procapsid assembly *in vivo*. *Journal of Molecular Biology* 202, 77-86.

Brewer, B.J. & Fangman, W.L. (1991) Mapping replication origins in yeast chromosomes. *Bioessays* 13, 317-322.

Bull, H.B. (1971) *An Introduction to Physical Biochemistry*. F.A. Davis, Philadelphia, Chapter, 3.

Casas, V, & Rohwer, F. (2007) Phage metagenomics. *Methods in Enzymology* 421, 259-268.

Casjens, S., Wyckoff, E., Hayden, M., Sampson, L., Eppler, K., Randall, S., Moreno, E.T. & Serwer, P. (1992) Bacteriophage P22 portal protein is part of the gauge that regulates packing density of intravirion DNA. *Journal of Molecular Biology* 1992, 224, 1055-1074.

Catalano, C.E. (2000) The terminase enzyme from bacteriophage lambda: a DNA-packaging machine. *Cellular and Molecular Life Sciences* 57, 128-148.

Chang, J., Weigele, P., King, J., Chiu, W. & Jiang, W. (2006) Cryo-EM asymmetric reconstruction of bacteriophage P22 reveals organization of its DNA packaging and infecting machinery. *Structure* 14, 1073-1082.

Chen, B. & Chrambach, A. (1979) Estimation of polymerization efficiency in the formation of polyacrylamide gel, using continuous optical scanning during polymerization. *Journal of Biochemical and Biophysical Methods* 1, 105-116.

Claverie, J.M., Abergel, C. & Ogata, H. (2009) *Current Topics in Microbiology and Immunology* 328, 89-121.

Colson, P. & Raoult, D. (2010) Gene repertoire of amoeba-associated giant viruses. *Intervirology* 53, 330-343.

Fang, P.A., Wright, E.T., Weintraub, S.T., Hakala, K., Wu, W., Serwer, P. & Jiang, W. (2008) Visualization of bacteriophage T3 capsids with DNA incompletely packaged *in vivo. Journal of Molecular Biology* 384, 1384-1399.

Ferrari, B.C., Winsley, T., Gillings, M. & Binnerup, S. (2008). Cultivating previously uncultured soil bacteria using a soil substrate membrane system. *Nature Protocols* 3, 1261-1269.

Fujisawa, H. & Morita, M. (1997) Phage DNA packaging. *Genes Cells* 2, 537-545.

Gauthier, M.G. & Slater, G.W. (2003) An exactly solvable Ogston model of gel electrophoresis: X. Application to high-field separation techniques. *Electrophoresis* 24, 441-451.

Griess, G.A., Edwards, D.M., Dumais, M., Harris, R.A., Renn, D.W. & Serwer, P. (1993) Heterogeneity of the pores of polysaccharide gels: Dependence on the molecular weight and derivatization of the polysaccharide. *Journal of Structural Biology* 111, 39-47.

Griess, G.A., Guiseley, K.B., Miller, M.M., Harris, R.A. & Serwer, P. (1998) The formation of small-pore gels by an electrically charged agarose derivative. *Journal of Structural Biology* 123, 134-142.

Griess, G.A., Moreno, E.T., Easom, R.A. & Serwer, P. (1989) The sieving of spheres during agarose gel electrophoresis: quantitation and modeling. *Biopolymers* 28, 1475-1484.

Griess, G.A., Moreno, E.T., Herrmann, R. & Serwer, P. (1990) The sieving of rod-shaped viruses during agarose gel electrophoresis. I. Comparison with the sieving of spheres. *Biopolymers* 29, 1277-1287.

Hamdan, S.M. & Richardson, C.C. (2008). Motors, switches, and contacts in the replisome. *Annual Review of Biochemistry* 78, 205-243.

Howard, J. (2009) Motor proteins as nanomachines: The roles of thermal fluctuations in generating force and motion. *Séminaire Poincaré* 12, 33-44.

Klose, T., Kuznetsov, Y.G., Xiao, C., Sun S., McPherson, A. & Rossmann, M.G. (2010) The three-dimensional structure of Mimivirus. *Intervirology* 53, 268-273.

Koonin, E.V. (2009) On the origin of cells and viruses: Primordial virus world scenario. *Annals of the New York Academy of Sciences* 1178, 47–64.

Kurakin, A. (2007) Self-organization *versus* watchmaker: Ambiguity of molecular recognition and design charts of cellular circuitry. *Journal of Molecular Recognition* 20, 205-214.

Langston, L.D., Indiani, C. & O'Donnell, M. (2009) Whither the replisome: emerging perspectives on the dynamic nature of the DNA replication machinery. *Cell Cycle* 8, 2686-2691.

Myong, S. & Ha, T. (2010) Stepwise translocation of nucleic acid motors. *Current Opinion in Structural Biology* 20, 121-127.

Newcomb, W.W., Homa, F.L. & Brown, J.C. (2005) Involvement of the portal at an early step in herpes simplex virus capsid assembly. *Journal of Virology* 79, 10540-10546.

Pajunen, M.I., Elizondo, M.R., Skurnik, M., Kieleczawa, J. & Molineux, I.J. (2002) Complete nucleotide sequence and likely recombinatorial origin of bacteriophage T3. *Journal of Molecular Biology* 319, 1115-1132.

Rasmussen, R.S. and Morrissey, M.T. (2007) Marine biotechnology for production of food ingredients. *Advances in Food and Nutrition Research* 52, 237-292.

Rees, D.A. (1972) Shapely polysaccharides. The eigth Colworth medal lecture. *Biochemical Journal* 126, 257-273.

Reynolds, J.A. & Tanford, C. (1970) Binding of dodecyl sulfate to porteins at high binding ratios. Possible implications for the state of proteins in biological membranes. *Proceedings of the National Academy of Sciences of the United States of America*, 66, 1002-1007.

Sait, M., Hugenholtz, P. & Janssen, P.H. (2002) Cultivation of globally distributed soil bacteria from phylogenetic lineages previously only detected in cultivation-independent surveys. *Environmental Microbiology* 4, 654–666.

Serwer, P. (1987) The mechanism for producing two symmetries at the head-tail junction of bacteriophages: a hypothesis. *Journal of Theoretical Biology* 127, 155-161.

Serwer, P. (2010) A hypothesis for bacteriophage DNA packaging motors. *Viruses* 2, 1821-1843.

Serwer, P. (2011) Proposed ancestors of phage nucleic acid packaging motors (and cells). *Viruses* 3, 1249-1280.

Serwer, P. & Griess, G.A. (1998) Adaptation of pulsed-field gel electrophoresis for the improved fractionation of spheres. *Analytica Chimica Acta* 372, 299-306.

Serwer, P. & Griess, G.A. (1999) Advances in the separation of bacteriophages and related particles. *Journal of Chromatography B* 722, 179-190.

Serwer, P. & Hayes, S.J. (1982) Agarose gel electrophoresis of bacteriophages and related particles. I. Avoidance of binding to the gel and recognizing of particles with packaged DNA. *Electrophoresis* 3, 76-80.

Serwer, P. & Hayes, S.J. (1986) Two-dimensional agarose gel electrophoresis. In: (Dunn, M.J., Ed.) *Electrophoresis '86*, VCH, Weinheim, pp. 243-251.

Serwer, P. & Wang, H. (2005) Single-particle light microscopy of bacteriophages. *Journal of Nanoscience and Nanotechnology* 5, 2014-2028.

Serwer, P. & Watson, R.H. (1982) Function of an internal bacteriophage T7 core during assembly of a T7 procapsid. *Journal of Virology* 42, 595-601.

Serwer, P. & Wright, E.T. (2011) Agarose gel electrophoresis reveals structural fluidity of a phage T3 DNA Packaging intermediate. *Electrophoresis*, 33, 352-365.

Serwer, P., Easom, R.A., Hayes, S.J. & Olson, M.S. (1989) Rapid detection and characterization of multimolecular cellular constituents by two-dimensional agarose gel electrophoresis. *Trends in Biochemical Sciences* 14, 4-7.

Serwer, P., Hayes, S.J. & Griess, G.A. (1986) Determination of a particle's radius by two-dimensional agarose gel electrophoresis. *Analytical Biochemistry* 152, 339-345.

Serwer, P., Hayes, S.J., Thomas, J.A., Demeler, B. & Hardies, S.C. (2009) Isolation of large and aggregating bacteriophages. *Methods in Molecular Biology* 1, 55-66.

Serwer, P., Khan, S.A. & Griess, G.A. (1995) Non-denaturing gel electrophoresis of biological nanoparticles: viruses *Journal of Chromatography A* 698, 251-261.

Serwer, P., Moreno, E.T. & Griess, G.A. (1988) Agarose gel electrophoresis of particles larger than 100 nm: Fractionation of intact *Escherichia coli*. In: (Schafer-Nielson, C., Ed.) *Electrophoresis '88*, VCH, Weinheim, pp. 216-227.

Serwer, P., Watson, R.H. & Hayes, S.J. (1982) Detection and characterization of agarose-binding, capsid-like particles produced during assembly of a bacteriophage T7 procapsid. *Journal of Virology* 42, 583-594.

Serwer, P., Wright, E.T., Hakala, K., Weintraub, S.T., Su, M. & Jiang, W. (2010) DNA packaging-associated hyper-capsid expansion of bacteriophage T3. *Journal of Molecular Biology* 397, 361-374.

Shaw, D.J. (1969) *Electrophoresis*, Academic Press, London, p. 4-26.

Stellwagen, E., Lu, Y. & Stellwagen, N.C. (2003) Unified description of electrophoresis and diffusion for DNA and other polyions. *Biochemistry* 42, 11745-11750.

Stellwagen, E., Lu, Y. and Stellwagen, N.C. (2005) Curved DNA molecules migrate anomalously slowly in free solution. *Nucleic Acids Research* 33, 4425-4432.

Stocker, B.A.D., Zinder, N.D., & Lederberg, J. (1953) Transduction of flagellar characters in *Salmonella*. *Journal of General Microbiology* 9, 410-433.

Studier, F.W. (2000) Slab-gel electrophoresis. *Trends in Biochemical Sciences* 25, 588-590.

Tietz, D. (2007) Computer-assisted 2-D agarose electrophoresis of *Haemophilus influenzae* type B meningitis vaccines and analysis of polydisperse particle populations in the size range of viruses: a review. *Electrophoresis* 28, 512-524.

Tietz, D. (2009) An innovative method for quality control of conjugated *Haemophilus influenzae* vaccines: A short review of two-dimensional nanoparticle electrophoresis. *Journal of Chromatography A* 1216, 9082-9033.

Wallis, C. & Melnick, J.L. (1967) Virus aggregation as the cause of the non-neutralizable persistent fraction. *Journal of Virology* 1, 478-488.

3

Some Reflections on the Origin of Lambdoid Bacteriophages

Luis Kameyama[1], Eva Martínez-Peñafiel[1], Omar Sepúlveda-Robles[1],
Zaira Y. Flores-López[1], Leonor Fernández[2],
Francisco Martínez-Pérez[3] and Rosa Ma. Bermúdez[1]
*[1]Departamento de Genética y Biología Molecular,
Centro de Investigación y de Estudios Avanzados del IPN,
[2]Facultad de Química, Universidad Nacional Autónoma de México D. F.
[3]Laboratorio de Microbiología y Mutagénesis Ambiental, Escuela de Biología,
Universidad Industrial de Santander, Bucaramanga,
[1,2]México
[3]Colombia*

1. Introduction

Monophyletic theory of the origin of life postulates that all cellular organisms have evolved from a common ancestor. This is based on nucleotide sequence analyses of rRNA genes, which all cellular organisms possess (Woese et al., 1975). On this basis and supported by other observations, the first common ancestor dates back to ~ 3.7 billion years. However, the scenario is not equivalent for the viruses, since they lack these genetic elements. In fact, there is such substantial diversity in viral genome structures (dsDNA, dsRNA, ssDNA, ssRNA) that it has proven extremely difficult to answer several key evolutionary questions. Do they co-evolve with their hosts? How do viruses first infect a new species? When was the first virus? Thus, a polyphyletic origin of viruses has been proposed (Bamford, 2003). It has been proposed that this could have been even before the appearance of the first cell. Although Boyer et al. (2010) have suggested that the eukaryote viruses may have appeared just after or simultaneously with the emergence of modern eukaryote lineages. However, there are other proposals which state that as new species appeared (of any of these three domains: Eukarya, Archaea and Bacteria), and after a certain period of time, their new infective viruses emerged. Nevertheless, information about common ancestor(s) related to viruses is still an enigma. Similarly, if we ask the same question for bacteriophages, or simply, when did the first lambdoid phage emerge? The answer is also unknown. Even with the support of bioinformatics and phage genomic knowledge, and quite possibly due to the lack of specific strategies and/or design methodologies, and phage genomic complexity the problem has not yet been resolved.

In this chapter, in an attempt to address this issue in *grosso modo*, we propose the analysis of two genomic regions of lambdoid phages, regions that are dissimilar regarding their nucleotide variability and stability: (1) The variable and not essential region that confers immunity to the lysogenic bacteria against phage superinfection counterparts, and is related to repressor; (2) The conserved and essential region for development that is related to the gene encoding the "Receptor-Binding Protein" (RBP), and which is involved in the process of infection onset. To begin to understand even a fraction of what was the common ancestor of lambdoid phages and the changes that had to occur to generate the diversity of lambdoid phages could be informative both of lambda biology and virus evolutionary processes in general.

2. Diversity of immunity regions in lambdoid phages

We reported the isolation and characterization of a collection of 47 lambdoid phages from human fecal samples (Kameyama et al., 1999). To determine the immunity group to which each phage belonged to; their lysogenic strains (lysogens) were constructed and then challenged with each lambdoid phage. The physiological study of growth indicates that for two phages belonging to the same immunity group, each lysogen should prevent growth of the other phage. For example, as we know each lysogens A and B are resistant to their respective phages, and if the lysogen A is resistant to phage B infection, and lysogen B to phage A, then both phages (A and B) belong to the same immunity group (Fig. 1D).

A	phage A	phage B
	-2 -4 -6 -8	-2 -4 -6 -8
lysogen (A)	- - - -	+ + + +
lysogen (B)	+ + + +	- - - -

B	phage A	phage B
	-2 -4 -6 -8	-2 -4 -6 -8
lysogen (A)	- - - -	- - - -
lysogen (B)	+ + + +	- - - -

C	phage A	phage B
	-2 -4 -6 -8	-2 -4 -6 -8
lysogen (A)	- - - -	+ + + +
lysogen (B)	- - - -	- - - -

D	phage A	phage B
	-2 -4 -6 -8	-2 -4 -6 -8
lysogen (A)	- - - -	- - - -
lysogen (B)	- - - -	- - - -

A) When lysogens are sensitive to the other phages but are resistant to their homologous phages they are referred as heteroimmune. B) The lysogen A is resistant to phage B, this indicates that lysogen A must have another exclusion system than the repressor. C) Similar example as B, but it is referred to the lysogen B. D) If both lysogens (A and B) are resistant to phages A and B infections, this result strongly suggests that they belong to the same immunity group. The numbers -2, -4, -6 and -8 represent different dilutions of phage lysate that can be tested on the strain. Presence (+) or absence (-) of phage growth can be determined by infecting with a series of phage lysate dilutions.

Fig. 1. Possible combinations between phages (A and B) and their lysogens.

It is simple to understand when two lysogens A and B are resistant to their respective homologous phages, but are sensitive to the phages B an A, respectively, this indicates that both phages (A and B) belong to different immunity group (Fig. 1A). A different scenario would be if lysogen A is resistant to phage B, this would indicate that lysogen A must have another exclusion system different to the repressor (Fig. 1B). The same case is applied for lysogen B (Fig.1C). From this study (Kameyama et al., 1999), it was possible to classify 19 different immunity groups (Fig. 2), of which 9 out of 19 (~ 50%) had to represent a unique individual.

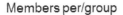

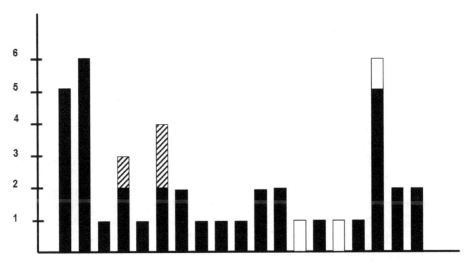

Phages were classified into nineteen immunity groups following the phage-lysogen cross test as previously reported by Kameyama et al. (1999). Phage groups were classified according to phage number per group. Black bars represent FhuA-receptor dependent phages; striped and white bars are phages that were unable to grow in strain MCR106 (ΔlamB) and MH760 (ompC⁻) cells, respectively (Taken from Hernandez-Sanchez, et al., 2008).

Fig. 2. Frequency distribution of nineteen lambdoid phage-infection immunity groups.

It is noteworthy that the immunity group XVIII (lambda phage belongs to this group) is comprised by 6 individuals. To determine whether the lambda specific repressor CI was present in all of them, we proceeded to evaluate its physiological function. For this, the strain LK1683, derived from the *E. coli* W3110 with a cryptic lambda prophage and the genotypic main feature: N::Kan, cI_{857} (Kameyama et al., 1999) was used. As expected, lambda phage was not developed at 32 °C, since at this temperature the CI_{857} repressor is active, but lambda phage developed normally at 42 °C, as CI_{857} from lysogenic strain is heat-inactivated. If the 5 individuals of the group XVIII all had the same lambda CI repressor, one would expect that their behavior were similar to that shown by lambda. Indeed, 4/5 phages of this group were unable to develop at 32 °C, while at 42 °C they did. One phage did not develop because it is temperature-sensitive, since it did not develop even in the wild type strain at 42 °C. In addition, Degnan et al. (2007) sequenced more than 5,000-bp of the immunity region of several lambdoid phages and among them the mEp234 and mEp332 (belonging to the immunity group XVIII, from our collection). They found that the mEp234 and mEp332 sequences coded for repressors almost identical to that of lambda CI with equivalent function, although respect to the Rex Region sequences they were different. The CI functional assay findings, are supported by genetic and sequencing data, therefore the classification of immunity lambdoid phage groups is reliable.

3. The gene coding for the "Receptor-Binding Protein" (RBP), which recognizes FhuA is borne by most lambdoid phages

The structural morphology of lambda and most lambdoid phages is characterized by a non-contractile and flexible tail, which is necessary for infection. At the onset and in order to carry out infection, gpJ lambda protein [equivalent to the Receptor-Binding Protein (RBP) for other phages], and located in the distal part of the tail, recognizes an outer membrane protein (OM), the trimeric maltoporin LamB (Gurnev et al., 2006). In other lambdoid phages, such as ϕ80, HK022, mEp167, they all recognize the ferrichrome-Fe^{3+} receptor (FhuA) (Guihard et al., 1992; Uc-Mass et al., 2004). In an approach to identify the most common receptors FhuA, LamB and OmpC used by lambdoid phages from our collection, we used three different deficient E. coli mutants fhuA-, lamB- and ompC- for the physiological assay. It was found that 37 out of 43 phages (~85%) used FhuA, since they are not able to infect strain C600 (fhuA-), but they can whenever this strain is complemented with a plasmid expressing FhuA (Hernández-Sánchez et al., 2008). These results clearly indicate that most of the lambdoid phages require FhuA to penetrate into bacteria.

4. The cor gene is present in half of the lambdoid phages population

cor gene product is involved in phage exclusion, in those that require FhuA receptor for infection. Thus, cor excludes lambdoid phages ϕ80, HK022, mEp167, etc., and non lambdoid phages T1, T5, etc. (Kozyrev & Rybchin, 1987; Malinin et al., 1993; Matsumoto et al., 1985; Uc-Mass et al., 2004), being all of them FhuA dependent. cor gene and the gene encoding RBP (gene p21 for phage N15 and gene p23 to HK022), are separated by two ORFs (Wietzorrek et al., 2006), and these are located in the cluster of genes that encode tail proteins. Because of this tight physical and functional association, we asked how many of the lambdoid phages contain cor? The answer was obtained by amplifying a 155 bp intragenic region of cor by PCR. We found that 25 out of 43 (~58%) phages bore it. To verify that the products corresponded to cor region, 4 PCR fragments were taken randomly and sequenced. Alignment analysis confirmed that the amplified region corresponded to cor gene (Hernández-Sánchez et al., 2008).

5. Identification of Nus-dependent non-lambdoid phages group

During the characterization of the first isolated phages from our collection, a new group of phages emerged. As part of the selection strategy, the potential lambdoid phages were challenged with 4 isogenic nus mutant strains nusA1, nusB5, nusD026 and nusE71. Pre-selection of potential lambdoid phages was carried out considering those phages that failed to grow at least in a couple of these mutants (Kameyama et al., 1999). Of these, 97 phages were selected. However, in the course of the characterization, a group of them (48 phages) did not recombine, nor hybridize with the lambda DNA, nor were recognized by antibodies directed against lambda structural proteins, nor their prophages were induced with light UV, and most failed to develop at 32 °C (Kameyama et al., 2001). However, this group of non-lambdoids shares an essential feature with lambdoid phages and that is the requirement for Nus factors to grow, suggesting that these phages may have an anti-termination mechanism homologous to that reported for lambdoid phages. Regarding growth cross-test assay, unlike the great diversity found in lambdoid phages, all of them had a single immunity! It is amazing how any of these lysogens has the ability to exclude any of the 48 phages of this group.

6. Discussion

We can infer from phage-lysogen cross test that lambdoid phage immunity groups are diverse and rich. If we consider that 9 out of the 19 groups had a unique representative, this could indicate that the number of different groups of immunity should be much larger. However, taking into account that the sample of the population of phage analyzed is small, it is not possible to infer probabilistically the number of possible different immunities in the region. A completely different scenario was obtained when testing the requirement of different bacterial receptors. It was found that 37 out of 43 require the E. coli FhuA receptor for infection. As mentioned above, the gene encoding RBP is essential, therefore nucleotide changes in this gene may be deleterious, and then it can be considered highly conserved. On the other hand, the bacterium E. coli use the FhuA receptor for iron assimilation through the ferrichrome-Fe^{3+} transport system. Interestingly, although the bacterium E. coli contains the genes fhuA, fhuB, fhuC and fhuD in an operon, for the ferrichrome-Fe^{3+} transport, it lacks of the genes for the biosynthesis of ferrichrome. Hence, in nature, the ferrichrome is produced by other species such as Ustilago maydis, and is taken in by E. coli for growth. This argument suggests that E. coli had to acquire the fhu operon at some stage of its evolution. Provided that most of lambdoid phages are FhuA dependent and considering that the gene encoding RBP would be highly conserved, as its product requires a perfect match with its receptor, it is likely that the first lambdoid phage had to require the FhuA receptor to infect its host E. coli. In addition, if we consider the argument that the fhu operon was acquired at some stage during E. coli evolution, then, the origin of the first lambdoid phage must not be older than that of its host E. coli! This idea though highly speculative, if true, it would support the proposal that viruses appeared after new cellular species emerged.

On the other hand, even having a great variety of phage immunity groups, it is still not possible to propose a putative origin of the first repressor, because the major constraint is present in the sample population. However, the dynamics of changes can be appreciated by the wide range of immunities provided.

Also it is interesting the analysis of cor gene implication. It has been proposed that cor is a moron that at some point of phage evolution was obtained (Juhala et al., 2000). Morons are autonomous genetic modules that are expressed from the repressed prophage probably acquired by horizontal transfer (Juhala et al., 2000). However, given the proximity to the gene encoding RBP and the high percentage that is present in the population (more than 50%), it would be more likely that *cor* associated with the RBP gene were acquired together in the formation of new phage. Unlike the RBP gene, *cor* is not essential, making easier to explain why *cor* is missing in a sector of the population.

It is also interesting to take into account the other group of 48 phages that emerged as a new group with a unique immunity, and knowing that this region must be very variable (as well as has been indicated for lambdoid phages), this may suggest that this group was recently created. However, other explanations are possible. For example the acquisition of the unique immunity region as a possible recombination with a different phage, since its repressor must be different to that of lambdoid phages, in which this is not inducible by UV light. Based on the phage numbers, one can infer that they are successful as lambdoid phages in nature. It is also a notable observation that only a single group or family of phages in *Brucella abortus* has been observed (personal communication of Flores, V). This idea complements the proposal that viruses appeared after emerging of the species.

New data will be needed to generate more precise and convincing answers. It should be noted that host participation can be critical in certain tasks, and finally given the great diversity of the viruses, these studies should be carried out according to each one of the family or group of viruses concerned.

It is clear that this chapter would be subject to polemic, as would any different or relatively new idea proposed to explain viral evolution. Indeed, it will serve to enhance, refine, or change approaches to shed more precise answers in this topic.

7. Acknowledgments

This work was supported in part by Instituto de Ciencia y Tecnología del Distrito Federal (ICyT-DF) PICSA11-107 and CONACyT # 82622, México grants.

8. References

Bamford, D. (2003). Do viruses form lineages across different domains of life? *Research in Microbiology,* Vol.154, No.4, (May), pp. 231-236, ISSN 0923-2508.

Boyer, M.; Madoui, M.; Gimenez, G.; La Scola, B. & Raoult, D. (2010). Phylogenetic and phyletic studies of informational genes in genomes highlight existence of a 4th domain of life including giant viruses. *PLoS ONE,* Vol.5, pp. e15530, ISSN 1932-6203.

Degnan, P.; Michalowski, C.; Babic, A.; Cordes, M. & Little, J. (2007). Conservation and diversity in the immunity regions of wild phages with the immunity specificity of phage lambda. *Molecular Microbiology,* Vol.64, No.1, (April), pp. 232-244, ISSN 1365-298.

Guihard, G.; Boulanger, P. & Letellier, L. (1992). Involvement of phage T5 tail proteins and contact sites between the outer and inner membrane of *Escherichia coli* in phage T5

DNA injection. *Journal of Biological Chemistry*, Vol.267, No.5, (February), pp. 3173-3178, ISSN 1083-351X.

Gurnev, P; Oppenheim, A.; Winterhalter, M. & Bezrukov, M. (2006). Docking of a single phage lambda to its membrane receptor maltoporin as a time-resolved event. *Journal of Molecular Biology*, Vol.359, No.5, (June), pp. 1447-1455, ISSN 0022-2836.

Hernandez-Sanchez, J.; Bautista-Santos, A.; Fernandez, L.; Bermudez-Cruz, R.; Uc-Mass, A.; Martínez-Peñafiel, E.; Martínez, M.; Garcia-Mena, J.; Guarneros, G. & Kameyama, L. (2008). Analysis of some phenotypic traits of feces-borne temperate lambdoid bacteriophages from different immunity groups: a high incidence of cor+, FhuA-dependent phages. *Archives of Virology*, Vol.153, No.7, (May), pp. 1271-1280, ISSN 0304-8608.

Juhala, R.J.; Ford, M.E.; Duda, R.L.; Youlton, A.; Hatfull, G.F. & Hendrix, R.W. (2000). Genomic sequences of bacteriophages HK97 and HK022: pervasive genetic mosaicism in the lambdoid bacteriophages. *Journal of Molecular Biology*, Vol. 299, No.1, (May), pp. 27-51, ISSN 0022-2836.

Kameyama, L.; Fernández, L.; Calderon, J.; Ortíz-Rojas, A. & Patterson, T. (1999). Characterization of wild lambdoid bacteriophages: Detection of a wide distribution of phage immunity groups and identification of a Nus-dependent, nonlambdoid phage group. *Virology* Vol.263, No.1, (October 10) pp. 100-111, ISSN 0042-6822.

Kameyama, L.; Fernández, L.; Bermúdez, R.; García-Mena, J.; Ishida, C. & Guarneros, G. (2001). Properties of a new coliphage group from human intestinal flora. *Recent Research Developments in Virology* Vol.3, Part I. Trivandrum, India: Transworld Research Network, pp. 297-303, ISBN 81-86846-99-9.

Kozyrev, D. & Rybchin, V. (1987). Lysogenic conversion caused by phage phi 80. III. The mapping of the conversion gene and additional characterization of the phenomenon. *Genetika*, Vol.23, No.5, (May), pp. 793-801, ISSN 1608-3369.

Malinin, A.; Vostrov, A.; Vasiliev, A. & Rybchin, V. (1993). Characterization of the N15 lysogenic conversion gene and identification of its product. *Genetika*, Vol.29, No.7, (July), pp. 257-265, ISSN 1608-3369.

Matsumoto, M.; Ichikawa, N.; Tanaka, S.; Morita, T. & Matsushiro, A. (1985). Molecular cloning of the phi 80 adsorption-inhibiting *cor* gene. *The Japanese Journal of Genetics*, Vol.60, No.5, pp. 475-483, ISSN 1880-5787.

Uc-Mass, A.; Jacinto-Loeza, E.; de la Garza, M.; Guarneros, G.; Hernández-Sánchez, J. & Kameyama, L. (2004). An orthologue of the *cor* gene is involved in the exclusion of temperate lamboid phages. Evidence that Cor inactivates FhuA receptor functions. *Virology*, Vol.329, No.2, pp.425-433, ISSN 0042-6822.

Wietzorrek, A.; Schwarz, H.; Herrmann, C. & Braun, V. (2006). The genome of the novel phage Rtp, with a rosette-like tail tip, is homologous to the genome of phage T1. *Journal of Bacteriology*, Vol.188, No.4, (February), pp. 1419-1436, ISSN 0021-9193.

Woese, C.; Fox, G.; Zablen, L.; Uchida, T.; Bonen, L.; Pechman, K.; Lewis, B. & Stahl, D.
(1975). Conservation of primary structure in 16s ribosomal RNA. *Nature*, Vol.254,
(Mar 6), pp. 83-86, ISSN 0028-0836.

Part 2

Bacteriophages as Contaminants and Indicator

4

Bacteriophages in Dairy Industry: PCR Methods as Valuable Tools

Beatriz del Río, María Cruz Martín, Víctor Ladero, Noelia Martínez,
Daniel M. Linares, María Fernández and Miguel A. Alvarez
Instituto de Productos Lácteos de Asturias, IPLA-CSIC,
Spain

1. Introduction

Microorganisms have been empirically used since ancestral times to produce fermented dairy products from milk. In the actual dairy industry, milk is subjected to large scale fermentation processes that involve microorganisms mostly belonging to the Lactic Acid Bacteria (LAB) group. Bacteriophages that infect LAB have been claimed as one of the principal sources of fermentation failure (spoilage or delay) on the manufacture of many dairy products (Brüssow et al., 1998; Josephsen & Neve, 1998; Garneau & Moineau, 2011). Some estimates assume that virulent phages are the primary direct responsible of the largest-economic loss of dairy factories, since they affect negatively up to the 10% of all milk fermentations (Moineau & Levesque, 2005).

Starter cultures consisting in selected bacterial strains are added to the fermentation vats to enhance the fermentative process and also to improve or influence the flavor and texture of the cultured products. The starter culture population grows through the fermentation process and reaches high levels inside the industrial vat. This is the perfect environment where bacteriophages can infect sensitive bacteria. The lysis of the infected host-bacteria can decrease several folds the total number of starter cells with consequences ranging from the delay of the acidification with quality changes of the final product, to the total failure of the fermentation.

Even with frequent cleaning, disinfestations and sterilization of all the facilities, the total absence of phages in the dairy plants is a utopia. The number and types of phages that are introduced within the system, presumably as a consequence of the constant supply of wild phages, is very variable and different subpopulations can prevalence as soon as a susceptible strain is introduced in the fermentation scheme (Neve et al., 1995; Bruttin et al., 1997; Chibani-Chennoufi et al., 2004; Kleppen et al., 2011). In fact, bacteriophages have been detected in variable titer in the milk, appliances of the factory, additives and starter cultures.

Since dairy bacteriophages are one of the mayor dairy threats, great research efforts have been made to reduce its load on dairy plants and to design new monitoring methods for their early detection (Magadán et al., 2009; Garneau & Moineau, 2011). Classical microbiological assays are routinely used on dairy plants to test the presence of bacteriophages. The spot/plaque assay and the turbidity/growth test are the two methods more frequently employed (Svensson & Christiansson, 1991; Neve et al., 1995; Capra et al., 2006, 2009; Atamer et al., 2009). These microbiological assays are based on the inhibitory effect of the bacteriophages in the host strain growth. The culture activity is measured in such a way that the presence of a bacteriophage in the tested sample inhibits the growth of the host strain and also decreases the production of lactic acid.

Although the microbiological detection methods are economically accessible, they have undesirable features such us the long processing time, since they take at least 24 hours to be completed. In addition, bacteriophages are extremely host-specific, so one phage can only infect one or few strains of bacteria, forcing to maintain a big collection of LAB strains that can be tested. Furthermore, a negative result on the spot/plaque assay or the turbidity/growth test does not guarantee the sample is phage-free, but it could also indicate that the host strain chosen is not sensitive to the bacteriophage on the sample.

Due to all these disadvantages, classical microbiological methods have being ousted by more sensitive, faster, and accurate genetic tools as the polymerase chain reaction (PCR). This is a specific and sensitive technology used to detect and identify (traditional PCR), and also to quantify (quantitative PCR) minimal amounts of bacteriophage DNA from samples of different dairy sources. Therefore, the PCR is a fast and reliable approach to screen a great number of samples for phage DNA. In most cases, low quantities of the sample (milk, whey, yogurt, etc) can be directly checked for bacteriophage presence by PCR.

Since the traditional PCR generates DNA fragments of specific size, different protocols have been designed not only to detect but to identify, by fragment size discrimination, bacteriophages that infect the main dairy LAB species used in the dairy industry such as *Lactobacillus casei/paracasei*, *Lactobacillus delbrueckii*, *Lactobacillus helveticus*, *Streptococcus thermophilus*, and *Lactococcus lactis* (Labrie & Moineau, 2000; Binetti et al., 2005; Quiberoni et al., 2006, 2010; Zago et al., 2006, 2008; del Río et al., 2007). One of the advantages of the PCR-based methods is that they can be applied in the raw milk received in the fermentation plant and within hours, each milk batch can be classified for its phage type content and it could be rapidly distributed. For example, if *Lb. delbrueckii* phages are detected in the raw milk, the batch cannot be used for the elaboration of yoghurt but the milk could be redirected to elaborate cheese, buttermilk, bottled to drink milk or dried to skim milk, thus allowing for an efficient distribution of the raw material. PCR-based approaches have been also applied for routine checking of lysogenic strains in starter cultures, which contain hidden temperate bacteriophages that have the potential to become lytic and compromise the fermentation process (Martín et al., 2006).

A step forward is the use of the quantitative real-time PCR (qPCR), which allows the quantification of phage particles in the sample, is faster and more sensitive than the traditional PCR. Additionally, the use of different fluorochromes enables the identification of multiple bacteriophage species in the same reaction (del Río et al., 2008; Martín et al., 2008; Verreault et al., 2011).

The detection by PCR-based methods might be easily incorporated into dairy industry routines to monitor the presence of phages and be included as part of the prevention strategy for controlling phage contamination. This methodology allows the detection of bacteriophages directly in a small input volume of industrial samples. It also reduces the screening time, since the sample can be directly checked for bacteriophages, avoiding previous enrichment steps with the sensitive host, needed for most of the microbiological methods. The correct and rapid identification and quantification of bacteriophages potentially able to attack starter cultures allows for rapid decisions with regard to the destination of contaminated milk that can be used for elaboration processes in which the phages detected do not constitute a threat. The possibility to reduce the milk storage time plays an important strategic role with economic implications for the dairy industry.

2. Traditional PCR

The PCR (polymerase chain reaction) is a broadly used molecular biology technique developed in 1983 by Kary Mullis and collaborators (Saiki et al., 1985; Mullis & Faloona, 1987), in which a single or a few copies of a specific DNA fragment present in a sample are amplified by a DNA polymerase in a thermal cycler to produce up to 2^{35} copies in few hours. At the end of the amplification reaction, the product is analyzed on an agarose gel and the DNA fragments are separated by size. To ensure the specificity of the reaction, both the target DNA sequence and the flanking primers are chosen in such a way that only samples containing the specific DNA are PCR-positives and yield DNA fragments of known size. Fig. 1 shows a graphic representation of the process to detect and identify bacteriophages that infect different LAB species from dairy samples by PCR amplification and subsequent size discrimination on agarose gel.

During the last fifteen years, the PCR-based methodology has been applied in several stages of dairy product manufacturing to rapid detect, identify and characterize different bacteriophages that infect LAB strains (e.g. *S. thermophilus, Lactobacillus* sp. and *Lb. casei/paracasei*). The technique has been adapted to detect phages in milk samples (Binetti et al., 2005; Dupont et al., 2005; del Río et al., 2007), in cheese whey (Brüssow et al., 1994; Labrie and Moineau, 2000), in the equipment within the factory (Verreault et al., 2011, Kleppen et al., 2011) or even from air samples (Verrault et al., 2011) with a detection limit of 10^3-10^4 phage particles per milliliter.

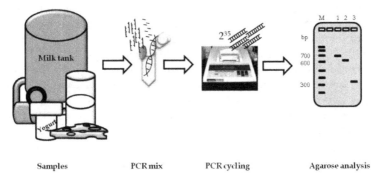

Fig. 1. Detection of bacteriophages in dairy samples by PCR. A variety of dairy samples can be checked for bacteriophage presence. The sample is loaded in a tube with the PCR mix and subjected to amplification in a thermocycler. At the end of the process, the amplified DNA is analyzed in an agarose gel and the bacteriophage is identified by the fragment size. Lane 1: *S. thermophilus* phage (750 bp), Lane 2: *Lb. delbrueckii* phage (650 bp), Lane 3: *Lc. lactis* phage (300 bp).

S. thermophilus bacteriophages have been identified in industry milk fermentation samples as yogurt and cheese whey, after phage DNA extraction and subsequent PCR (Brüssow et al., 1994) or by direct PCR of milk (Binetti et al., 2005; Dupont et al., 2005, del Río et al., 2007). A PCR protocol to detect *Lb. casei/paracasei* bacteriophages from milk and other commercial samples as fermented milk and cheese whey was also designed (Binetti et al., 2008). In addition, *Lb. delbrueckii* bacteriophages have been detected in milk and yogurt samples (del Río et al., 2007). Moreover, several protocols have been published to detect and identify the three species of *Lc. lactis* bacteriophages most frequently found in milk plants (c2, 936 and 335), in cheese whey samples (Labrie & Moineau, 2000; Dupont et al., 2005; Deveau et al., 2006; Szczepańska et al., 2007; Suárez et al., 2008; Kleppen et al., 2011), milk (del Río et al., 2007) and even in the factory equipment (Verreault et al., 2011; Kleppen et al., 2011) or air samples (Verrault et al., 2011).

2.1 Multiplex PCR

The manufacture of a great number of cultured milk products implies the addition of starter cultures. The composition of modern dairy starter cultures is increasing on complexity and they are actually made of a variety of strains. This variety of strains raises the total number of different bacteriophage species that might affect the starter culture and therefore the quality of the final product. Usually, one PCR protocol is designed to specifically amplify and detect only one species of bacteriophage in a sample, so more than one PCR would be needed to detect all the phage types that could be present in a sample of a product manufactured with a complex starter culture.

An elegant and efficient solution for such a problem has been proposed by different authors who have developed multiplex PCR assays to check for more than one phage species in a single reaction. There are different multiplex assays towards multi-type bacteriophage detection in the literature (Labrie & Moineau, 2000; Quiberoni et al., 2006; del Río et al., 2007). In 2000, Labrie and Moineau designed a multiplex PCR method to detect in a single reaction the presence of the three *Lc. lactis* phages species considered as problematic in dairy companies, namely 936, c2, and P335. The assay was optimized for detection and identification of the three phage types from whey samples. Later on, del Río et al. (2007) extended the PCR for the detection of two additional phage groups: bacteriophages infecting *S. thermophilus* and *Lb. delbrueckii*, starters of common use for the elaboration of yoghurt. This simple and rapid multiplex PCR method detects the presence of bacteriophages infecting LAB species most commonly used as starters in dairy fermentations: the three genetically distinct groups of *Lc. lactis* phages species (P335, 936 and c2) plus phages infecting *S. thermophilus* and *Lb. delbrueckii*.

2.2 PCR-detection of lysogenic strains in starter cultures

Phages are ubiquitous organisms, in the particular case of dairy fermentations, the main source of phages is the raw milk, from where they spread and contaminate the dairy facilities. Other external sources of phage contamination that must be considered are the starter strains that carry prophages into their genome. The analysis of bacterial genomes revealed that prophages are more widespread than previously thought (Canchaya et al., 2003; Mercanti et al., 2011).

Numerous LAB belonging to different genera are lysogenic bacteria, meaning that carry one or several inducible prophages integrated into their genomes. Functional prophages has been identified in species usually used in dairy fermentations such as *Lc. lactis* subsp. *lactis*, *Lc. lactis* subsp. *cremoris*, *S. thermophilus*, *Lb. delbrueckii*, *Lb. casei*, *Lactobacillus rhamnosus* or *Bifidobacterium longum* (Chopin et al., 2001; Desiere et al., 2002; Proux et al., 2002; Ventura et al., 2006, 2007, 2009; Zago et al., 2007; Durmaz et al., 2008).

The lysogenic strains are introduced and maintained in fermentation vats for extended periods of time where they are subjected to the stressful fermentation conditions. This environment may induce prophages into lytic cycle and release the viral. The risk for prophage induction and its consequences on the final product quality must be carefully evaluated when developing industrial fermentation processes thorough the selection of suitable prophage-free strains as starters.

Ideally, bacterial strains should be tested in conditions that mimic industrial fermentation processes for the induction of putative prophages. In this context, PCR can be used as a tool for the rapid screening towards the identification of lysogenic strains in large culture collections (Martín et al., 2006). The potential lysogenic strains should be further tested for their capacity to release phage particles. In fact, PCR methods could, overcome frequent problems of prophage induction assays such as lack of detection of the bacteriophages released due to a low viral production, the high sensitivity of the screened strain to the induction agent or the lack of a suitable host strain for the phage.

The cells can be tested by a direct phage-specific PCR assay. If no amplification is obtained, the problem strains can be tagged as phage-free. However, if the strain gives positive PCR result, it can be further assayed for prophage induction by using a known induction agent as mitomycin C (MitC). The presence of viral particles in the cell-free supernatant can be evaluated again with the same phage-specific PCR assay. Only those strains able to liberate functional phage particles would result in a positive PCR amplification and consequently should be discarded as starter strains (Fig. 2).

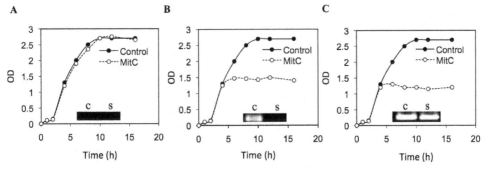

Fig. 2. PCR evaluation of putative lysogenic strain cultures induced with mitomycin C (MitC) (adapted from Martín et al., 2006). The optical density was measured in a control culture and in a culture induced with MitC after 4 hour. The Cells were PCR tested for the presence of prophages in the genome (c), and cleared supernatants of the induced cultures were also PCR tested (s). (A) Strain with no prophages. (B) Strain with no inducible prophages but inhibited by MitC. (C) Strain with inducible prophages.

Those strains carrying not inducible prophages could be used as starters, although their use it is not recommended since they could constitute a pool of genes that can be transferred to incoming lytic phages by homologous recombination, thus expanding their host range (Bouchard & Moineau, 2000; Durmaz & Klaenhammer, 2000).

2.3 Bacteriophages typification by PCR and amplicon sequencing

The rotation of starter cultures that share similar technological properties but with a different pattern of phage susceptibility is the usual and most successful strategy to reduce the impact of phage attacks in dairy industries (Edmon & Moineau, 2007; Kleppen et al., 2011). Therefore, the establishment of a system that allows not only to detect, but to classify the detected phages based on their host range would permit a rational modification of the strain rotation scheme.

In some detection methods, once a sample is confirmed for bacteriophage presence by PCR amplification, it is possible to get additional information from the amplicon nucleotide sequence. In this sense it has been described a PCR assay that correlates the host range of a bacteriophage and the nucleotide sequence of the amplified fragment (Binetti et al., 2005). This test is based on the amplification of the VR2 variable region of *orf18*, the antireceptor gene of *S. thermophilus* phages that was claimed as responsible for host specificity (Duplessis & Moineau, 2001). The sequence of the VR2 variable sequence can be used to classify the detected phages within a host range and consequently to establish a rational rotation of the available starter cultures.

When a phage is detected by PCR the obtained sequence will classify it in a VR2 type (Binetti et al., 2005; Guglielmotti et al., 2009). This VR2 type would indicate those strains that are sensitive and those that are resistant to that phage in particular. The susceptible strains cannot be introduced in the elaboration routine and should be substituted for other similar strains that are not included in the same VR2 type thus being resistant to that particular phage. In other words, it is possible to know which strains are susceptible to be infected by the detected phages. This is a very useful tool for the dairy industry, since it allows preventing phage attack by designing a rational starter rotation system based on the phage types detected.

3. Quantitative PCR

Even though the PCR-based technology is a fast and sensitive approach to detect bacteriophages, traditional PCR has an important disadvantage since it is a qualitative but not a quantitative method for bacteriophage screening. In this sense it is important to note that 10^5 phage particles per millilitre of milk has been estimated as the threat threshold (Neve & Teuber, 1991; Emond & Moineau, 2007; Magadán et al., 2009). Therefore, the mere presence of phage particles might be not enough to guaranty the useless of the material and quantitative approaches would avoid the disposal of milk that could still be useful.

The traditional PCR methodology is being changed towards a faster, more sensitive and useful technique, the quantitative PCR (qPCR). This is a new technology based on the traditional PCR that not only adds specificity, accuracy and speed to former PCR screening methods, but also allows the quantification of the number of copies of the target DNA, in this case bacteriophage particles in a dairy sample. Additionally, the qPCR collects the data in real time throughout the reaction and not just at the end point as the conventional PCR. This feature could be of great value for the dairy industry since it saves time and speed up the decision towards the use of the analyzed material, in case it is contaminated with bacteriophages.

The qPCR technology monitors the increase of fluorescence emitted during the synthesis of the newly synthesized DNA fragment in each cycle. The reactions are characterized by the cycle in which the amplification of a target DNA results in a fluorescent signal that reaches the detection level of the equipment; this cycle is namely the threshold cycle (Ct) and it is directly proportional to the initial copies of target DNA in the sample (Logan et al., 2001). Absolute numbers are obtained by comparing the sample Ct value against a standard curve that is prepared with templates carrying a known titer of bacteriophages (Fig. 3).

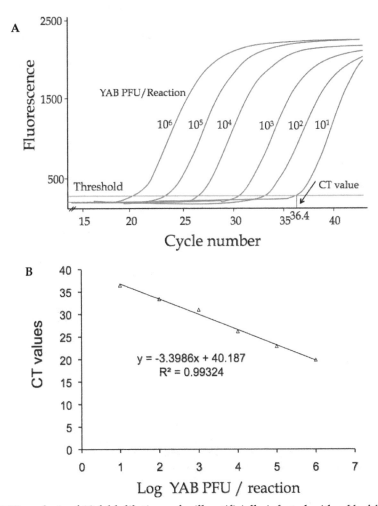

Fig. 3. qPCR analysis of 10-fold dilutions of milk artificially infected with a *Lb. delbrueckii* bacteriophage (YAB phage) (adapted from Martín et al., 2008). (A) Amplification plot of bacteriophage DNA in serial diluted samples starting at 10 PFU/sample. (B) A linear correlation is established between the C_T value and the logarithm of the number of phage particles on the sample.

To detect the amplified DNA, non-specific fluorescent dyes that bind to any double-stranded DNA (e.g. SYBR® Green) or specific DNA probes (e.g. TaqMan® probes) could be used. The non-specific fluorescent dyes are flexible, inexpensive and accurate, but the formation of primers dimmers could led to detect false positives and the use of an internal positive control is not possible. Compare to DNA binding dyes, the fluorescence probe technology provides an additional level of specificity to the reaction, since in addition to the PCR primers a third oligonucleotide labeled with a fluorochrome (the probe) is added. Other significant advantage of using probe chemistry is that different probes can be labeled with different reporter dyes and be combined in a single reaction (multiplex qPCR). As each fluorescent signal is individually detected, the qPCR technology could be used as a qualitative tool to identify different bacteriophage species present in a sample. In addition, it allows the addition of an internal positive control to the PCR mix which is especially useful in dairy sample analysis, since in some food matrixes as milk many PCR inhibitors can be present and some screened samples could be annotated as false negatives (for a review see Martínez et al., 2011).

Several qPCR protocols based on DNA fluorescent probes have been described to detect, quantify and identify bacteriophages present in dairy samples. Martín et al. (2008) developed a qPCR method to quantify *Lb. delbrueckii* bacteriophages present in milk samples. The assay combines two different TaqMan MGB probes, one which identify the phage and another for an internal positive control that is added to all the samples. The assay allows the quantification of 10^2 *Lb. delbrueckii* bacteriophage particles per reaction and it could be also applied to test other dairy niches such as starter cultures and fermented milks.

As was described above, a multiplex qPCR assay has been applied to identify in a single reaction more than one different bacteriophage type. In this context, del Río et al. (2008) designed a multiplex assay to detect two types of *S. thermophilus* bacteriophages (*cos* and *pac*) in milk samples combining three probes individually labeled with different fluorochromes (one for each bacteriophage and another one for the internal control). The assay shown to be highly specific, since no false-positive or false-negative results were obtained even when the analyzed milk samples were artificially contaminated with *Lb. delbrueckii* or *Lc. lactis* bacteriophages.

Concerning to *Lc. lactis* phages, Verreault et al. (2011) recently developed a qPCR method using the SYBR green fluorescence technology to quantify bacteriophages belonging to the 936 and c2 groups in aerosol and surface samples of a typical cheese manufacturing plant.

Even though the qPCR technology has shown to be a great platform to detect and quantify virulent phages in dairy plants, further advancements are needed in order to improve the automation of the process, the sensitivity of the detection in dairy samples susceptible of contain PCR inhibitors and to expand the possibility to target simultaneously a bigger number of different bacteriophages in a single reaction.

4. Discussion/conclusions

Bacteriophages that attack bacteria used as starters cause great economic losses to global dairy industry. Even though strict sanitation programs and rigorous culture handling are established on the dairy plants, the bacteriophage presence is ubiquitous in all the facilities in variable titer. PCR-based methods for early and fast detection and quantification of bacteriophage that might jeopardize the survival of the culture starters are shown to be effective and might be incorporated as routine control of bacteriophages in the dairy industry.

The machinery, surfaces and aerosols within the facilities of the dairy plant can be checked to ensure the absence or at least the low titer of bacteriophages that otherwise would be incorporated on the fermentative process. Surely, PCR can be applied to determine the potential uses of a batch of milk, but also could be applied to samples collected all through the fermentative process, starting with the raw milk and ending with the final product. Since PCR protocols are much faster than traditional microbiological methods, the raw milk can be checked for bacteriophages in an acceptable time, allowing decisions for the better use of contaminated material, which could be diverted toward procedures in which bacteriophages are deactivated or do not require a fermentative step with LAB starters. In consequence, whole tanks of raw milk are not kept for big periods of time and also the subsequent fermentative process is not on risk.

The presence of PCR inhibitors such calcium ions, plasmin, proteins, fat..., is a fact that should be taken into account in the design of PCR protocols to analyze dairy products. Special care must be taken when dairy samples are used directly as templates, because the presence of inhibitors may interfere with the PCR amplification and leads false negative results (Wilson, 1997; Ercolini et al., 2004). Magnetic capture hybridization (Dupont et al., 2005) or a previous DNA isolation from the sample, as is described for detecting pathogens in dairy samples (Cremonesi et al., 2007), are some of the additional steps that could be included in the PCR methods to remove PCR inhibitors and/or reduce the effect of the components of the dairy matrix.

Given that some qPCR-based protocols can confirm the bacteriophage presence in just 30 minutes, they are a great tool for fast detection of bacteriophage breakouts in a failed fermentative process and may help in the substitution of starter strain for another resistant to the detected phage. The qPCR methods save performing time compared to the traditional PCR, since the post-amplification processing is not generally needed. However, when the SYBR® Green is the qPCR technology of choice, a further processing of the amplicon (melting point or dissociation curve analysis) is essential in order to detect non-specific amplifications or primer-dimmers. The SYBR® Green chemistry could be the cheapest option for the screening of a large number of samples, but requires an extensive optimization of the protocols designed and also the complete analysis of the samples would take longer than the probe-based methods.

Both, traditional and real-time PCR techniques are extremely specific and their reliability depends on the design of the protocol to specifically target one bacteriophage species. However, a critical point and a potential limitation to design new specific PCR protocols, is the need of some prior information about the nucleotide sequence of the bacteriophage target gene. In this context, the new sequencing techniques are a valuable tool to characterize new bacteriophages isolated from the raw milk or within the facilities and hence, to increase the available phage genomic data. qPCR protocols should be updated with the new sequences as they become available.

The fact that a sample can be quickly assessed for bacteriophage presence by a qPCR assay do not necessary implies its classification as infective or non-infective for the starter cultures used in a particular dairy plant. The host range of bacteriophages is usually defined by traditional microbiological tests, using a collection of potential bacterial host. That process is cost- and time-inefficient for the dairy industry. Nevertheless, traditional PCR and amplicon sequencing have been successfully applied in dairy phages infecting *S. thermophilus* (Binetti et al., 2005) and *Lc. lactis* (Stuer-Lauridsen et al., 2003; Dupont et al., 2004), to correlate a specific nucleotide sequence with the host range of each phage. qPCR and PCR methods can be sequentially combined, applying first the qPCR method for a fast screening of a large batch of samples, followed by a traditional PCR in which the host range of the phage could be assessed and the potential host strains identified and discarded as starters for fermentative processes.

Continuous efforts are being made in order to improve the PCR-based protocols for the detection and quantification of bacteriophages from samples of any origin within the dairy plant. PCR-based methods could be included in the Hazard Analysis and Critical Control Points (HACCP) protocol to prevent phage accumulation niches and reduce their impact in dairy fermentations.

5. Acknowledgments

This work was performed with financial support from the Ministry of Science and Innovation, Spain (AGL2010- 18430). B. del Río and N. Martínez are beneficiary of a JAE DOC-CSIC contract (Spain). D. M. Linares is beneficiary of a FICYT contract (Asturias, Spain).

6. References

Atamer, Z.; Dietrichb, J.; Müller-Merbacha, M.; Neveb, H.; Hellerb, K.J. & Hinrichsa, J. (2009). Screening for and characterization of *Lactococcus lactis* bacteriophages with high thermal resistance. *International Dairy Journal*, Vol.19, No.4, (April 2009), pp. 228-235, ISSN 0958-6946.

Binetti, A.G.; del Rio, B.; Martin, M.C. & Alvarez, M.A. (2005). Detection and characterization of *Streptococcus thermophilus* phages based on the antireceptor gene sequence. *Applied and Environmental Microbiology*, Vol.71, No.10, (October 2005), pp. 6096-103, ISSN 0099-2240.

Binetti, A.G.; Capra, M.L.; Alvarez, M.A. & Reinheimer, J.A. (2008). PCR method for detection and identification of *Lactobacillus casei/paracasei* bacteriophages in dairy products. *International Journal of Food Microbiology*, Vol.124, No.2, (May 2008), pp. 147-153, ISSN 0168-1605.

Bouchard, J.D. & Moineau, S. (2000). Homologous recombination between a lactococcal bacteriophages and the chromosome of its host strain. *Virology*, Vol.270, No.1, (April 2000), pp. 65–75, ISSN 0042-6822.

Brüssow, H.; Freimont, M.; Bruttin, A.; Sidoti, J.; Constable, A. & Fryder, V. (1994). Detection and classification of *Streptococcus thermophilus* bacteriophages isolated from industrial milk fermentation. *Applied and Environmental Microbiology*, Vol.60, No.12, (December 1994), pp. 4537-4543, ISSN 0099-2240.

Brüssow, H.; Bruttin, A.; Desiere, F.; Lucchini, S. & Foley S. (1998). Molecular ecology and evolution of *Streptococcus thermophilus* bacteriophages--a review. *Virus Genes*. Vol.16, No.1, (January 1998), pp. 95-109, ISSN 0920-8569.

Bruttin, A.; Desiere, F.; d'Amico, N.; Guérin, J.P.; Sidoti, J.; Huni, B.; Lucchini, S. & Brüssow, H. (1997). Molecular ecology of *Streptococcus thermophilus* bacteriophage infections in a cheese factory. *Applied and Environmental Microbiology*, Vol.63, No.8, (August 1997), pp. 3144-3150, ISSN 0099-2240.

Canchaya, C.; Proux, C.; Fournous, G.; Bruttin, A. & Brüssow, H. (2003). Prophage genomics. *Microbiology and Molecular Biology Reviews*, Vol.67, No.2, (June 2003), pp. 238–276, ISSN 1092-2172.

Capra, M.L.; Del, L.; Quiberoni, A.; Ackermann, H.W.; Moineau, S. & Reinheimer, J.A. (2006). Characterization of a new virulent phage (MLC-A) of *Lactobacillus paracasei*. *Journal of Dairy Science*, Vol.89, No.7, (July 2006), pp. 2414-2423, ISSN 0022-0302.

Capra, M.L.; Binetti, A.G.; Mercanti, D.J.; Quiberoni, A. & Reinheimer, J.A. (2009). Diversity among *Lactobacillus paracasei* phages isolated from a probiotic dairy product plant. *Journal of Applied Microbiology*, Vol.107, No.4, (October 2009), pp. 1350-1357, ISSN 1364-5072.

Chibani-Chennoufi, S.; Bruttin, A.; Dillmann, M.L. & Brüssow, H. (2004). Phage-host interaction: an ecological perspective. *Journal of Bacteriology*. Vol.186, No.12, (June 2004), pp. 3677-3686, ISSN 0021-9193.

Chopin, A.; Bolotin, A.; Sorokin, A.; Ehrlich, S.D. & Chopin, M. (2001). Analysis of six prophages in *Lactococcus lactis* IL1403: different genetic structure of temperate and virulent phage populations. *Nucleic Acids Research*, Vol.29, No.3, (February 2001), pp. 644-651, ISSN 0305-1048.

Cremonesi, P., Perez, G.; Pisoni, G.; Moroni, P.; (2007). Detection of enterotoxigenic *Staphylococcus aureus* isolates in raw milk cheese. *Letters in Applied Microbiology*, Vol.45, No.6, (December 2007), pp: 586-591, ISSN 0266-8254.

del Río, B.; Binetti, A.G.; Martín, M.C.; Fernández, M.; Magadán, A.H. & Alvarez M.A. (2007). Multiplex PCR for the detection and identification of dairy bacteriophages in milk. *Food Microbiology*, Vol.24, No.1, (February 2007), pp. 75-81, ISSN 0740-0020.

del Río, B.; Martín, M. C.; Martínez, N.; Magadán, A.H. & Alvarez, M.A. (2008). Multiplex fast real-time polymerase chain reaction for quantitative detection and identification of cos and pac *Streptococcus thermophilus* bacteriophages. *Applied and Environmental Microbiology*, Vol.74, No.15, (August 2008), pp. 4779-4781, ISSN 0099-2240.

Desiere, F.; Lucchini, S.; Canchaya, C.; Ventura, M. & Brüssow, H. (2002). Comparative genomics of phages and prophages in lactic acid bacteria. *Antonie Van Leeuwenhoek*, Vol.82, No.1-4, (August 2002), pp. 73-91, ISSN 0003-6072.

Deveau, H.; Labrie, S.J.; Chopin, M.C. & Moineau, S. (2006). Biodiversity and classification of lactococcal phages. *Applied and Environmental Microbiology*,Vol.72, No.6, (January 2006), pp. 4338-4346, ISNN 0099-2240.

Duplessis, M. & Moineau, S. (2001). Identification of a genetic determinant responsible for host specificity in *Streptococcus thermophilus* bacteriophages. *Molecular Microbiology*, Vol.41, No.2, (July 2001), pp. 325-336, ISSN 0950-382X.

Dupont, K.; Vogensen, F.K.; Neve, H.; Bresciani, J. & Josephsen, J. (2004). Identification of the receptor-binding protein in 936-species lactococcal bacteriophages. *Applied and Environmental Microbiology*, Vo.70, No.10, (October 2004), pp. 5818–5824, ISSN 0099-2240.

Dupont, K.; Vogensen, F.K. & Josephsen, J. (2005). Detection of lactococcal 936-species bacteriophages in whey by magnetic capture hybridization PCR targeting a variable region of receptor-binding protein genes. *Journal of Applied Bacteriology*, Vol.98, No.4, (April 2005), pp. 1001-1009, ISSN 0021-8847.

Durmaz, E. & Klaenhammer, T.R. (2000). Genetic analysis of chromosomal regions of *Lactococcus lactis* acquired by recombinant lytic phages. *Applied and Environmental Microbiology*, Vol.66, No.3, (March 2000), pp. 895–903, ISSN 0099-2240.

Durmaz, E.; Miller, M.J.; Azcarate-Peril, M.A.; Toon, S.P. & Klaenhammer, T.R. (2008). Genome sequence and characteristics of Lrm1, a prophage from industrial *Lactobacillus rhamnosus* strain M1. *Applied and Environmental Microbiology*, Vol.74, No.15, (August 2008), pp. 4601-4609, ISSN 0099-2240.

Emond, E. & Moineau, S. (2007). Bacteriophages and food fermentations. In: *Bacteriophage: Genetics and Molecular Biology*, Mc Grath, S. & van Sinderen, D. (Eds.), pp. 93-124, Caister Academic Press, ISBN 978-1-904455-14-1, Portland, USA.

Ercolini, D.; Blaiotta, G.; Fusco, V. & Coppola, S. (2004). PCR-based detection of enterotoxigenic *Staphylococcus aureus* in the early stages of raw milk cheese making. *Journal of Applied Microbiology*, Vol.96, No.5 , (May 2004), pp. 1090–1098, ISSN 1364-5072.

Garneau, J.E. & Moineau, S. (2011). Bacteriophages of lactic acid bacteria and their impact on milk fermentations. *Microbial Cell Factories*, 10 (Suppl 1):S20, (August 2011), Retrieved from <http://www.microbialcellfactories.com/content/10/S1/S20>, ISSN 1475-2859.

Guglielmotti, D.M.; Binetti A.G.; Reinheimer, J.A. & Quiberoni, A. (2009). *Streptococcus thermophilus* phage monitoring in a cheese factory: Phage characteristics and starter sensitivity. *International Dairy Journal*, Vol.19, No.8, (August 2009), pp. 476-480, ISNN 0958-6946.

Josephsen, J. & Neve, H. (1998). Bacteriophages and lactic acid bacteria. In: *Lactic Acid Bacteria. Microbiology and Functional Aspects*. Eds. Salminen, S., and S. von Wright, Marcel Dekker Inc, (February 1998), pp. 385-436, ISBN 082470133X, New York.

Kleppen, H.P.; Bang, T.; Nes, I.F. & Holo, H. (2011). Bacteriophages in milk fermentations: Diversity fluctuations of normal and failed fermentations. *International Dairy Journal*, Vol.21, No.9, (September 2011), pp. 592-600, ISSN 0958-6946.

Labrie, S. & Moineau, S. (2000). Multiplex PCR for detection and identification of lactococcal bacteriophages. *Applied and Environmental Microbiology*, Vol.66, No.3, (March 2000), pp. 987-994, ISSN 0099-2240.

Logan, J.M.; Edwards, K.J.; Saunders, N.A. & Stanley, J. (2001). Rapid identification of *Campylobacter* spp. by melting peak analysis of biprobes in real-time PCR. *Journal of Clinical Microbiology*, Vol.39, No.6, (June 2001), pp. 2227-2232, ISSN 0095-1137.

Magadán, A.H.; Ladero, V.; Martínez, N.; del Río, B.; Martín M.C. & Alvarez, M.A. (2009). Detection of bacteriophages in milk. In *Handbook of Dairy Foods Analysis*, L.M.L. Nollet & F. Toldrá, (Eds.), pp. 469-482, CRC Press, Taylor & Francis Group, ISBN 978-1-4200-4631-1, Boca Raton, USA.

Martín, M.C.; Ladero, V. & Alvarez, M.A. (2006). PCR Identification of lysogenic *Lactococcus lactis* strains. *Journal of Consumer Protection and Food Safety*, Vol.1, No.2, (May 2006), pp. 121-124, ISSN 1661-5751.

Martín, M.C.; del Río, B.; Martínez, N.; Magadán, A.H. & Alvarez, M.A. (2008). Fast real-time polymerase chain reaction for quantitative detection of *Lactobacillus delbrueckii* bacteriophages in milk. *Food Microbiology*, Vol. 25, No.8, (December 2008); pp. 978-982, ISSN 0740-0020.

Martínez, N.; Martín, M.C.; Herrero, A.; Fernández, M.; Alvarez, M.A. & Ladero, V. (2011). QPCR as a powerful tool for microbial food spoilage quantification: Significance for food quality. *Trends in Food Science and Technology*, Vol.22, No.7, (July 2011), pp.367-376, ISSN 0924-2244

Mercanti, D. J.; Carminati, D.; Reinheimer, J.A. & Quiberoni, A. (2011). Widely distributed lysogeny in probiotic lactobacilli represents a potentially high risk for the fermentative dairy industry. *International Journal of Food Microbiology*, Vol.144, No.3, (January 2011), pp. 503-510, ISSN 0168-1605.

Moineau, S. & Levesque, C. (2005). Control of bacteriophages in industrial fermentations. In: *Bacteriophages: Biology and Applications*, E. Kutter & A. Sulakvelidze, (Eds.), pp. 285-296, CRC Press, ISBN 0-8493-1336-8, Boca Raton, USA.

Mullis, K.B. & Faloona, F.A. (1987). Specific synthesis of DNA in vitro via a polymerase-catalyzed chain reaction. *Methods in Enzymology*, Vol.155, (December 1987), pp. 335-350, ISNN 0076-6879.

Neve, H. & Teuber, M. (1991). Basic microbiology and molecular biology of bacteriophage of lactic acid bacteria in dairies. *Bulletin of the IDF*, Vol.263, (January 1991), pp. 3-15, ISSN 0250-5118.

Neve, H.; Berger, A. & Heller, K.J. (1995). A method for detecting and enumerating airborne virulent bacteriophage of dairy starter cultures. *Kieler Milchwirtschaftliche Forschungsberichte*, Vol.47, No.3, (March, 1995), pp. 193-207, ISSN 0023-1347.

Proux, C.; van Sinderen, D.; Suárez, J.; García, P.; Ladero, V.; Fitzgerald, G.F.; Desiere, F. & Brüssow, H. (2002). The dilemma of phage taxonomy illustrated by comparative genomics of Sfi21-like *Siphoviridae* in lactic acid bacteria. *Journal of Bacteriology*, Vol.184, No.21, (November 2002), pp. 6026-6036, ISSN 0021-9193.

Quiberoni, A.; Tremblay, D.; Ackermann, H.W.; Moineau, S. & Reinheimer, J.A. (2006). Diversity of *Streptococcus thermophilus* phages in a large-production cheese factory in Argentina. *Journal of Dairy Science*, Vol.89, No.10, (October 2006), pp. 3791-3799, ISSN 0022-0302.

Quiberoni, A.; Moineau, S.; Rousseau, G.M.; Reinheimer, J. & Ackermann, H.W. (2010). *Streptococcus thermophilus* bacteriophages. *International Dairy Journal*, Vol.20, No.10, (October 2010), pp. 657-664, ISSN 0958-6946.

Saiki, R.; Scharf, S.; Faloona, F.; Mullis, K.; Horn, G. & Erlich, H. (1985). Enzymatic amplification of beta-globin genomic sequences and restriction site analysis for diagnosis of sickle cell anemia. *Science*, Vol.230 , No.4732, (December 1985), pp. 1350-1354, ISSN 0036-8075.

Stuer-Lauridsen, B.; Janzen, T.; Schnabl, J. & Johansen, E. (2003). Identification of the host determinant of two prolate-headed phages infecting *Lactococcus lactis*. *Virology*, Vol.309, No.1, (April 2003), pp.10-17, ISSN 0042-6822.

Suárez, V.; Moineau, S.; Reinheimer, J.A. & Quiberoni, A. (2008). Argentinean *Lactococcus lactis* bacteriophages: genetic characterization and adsorption studies. *Journal of Applied Microbiology*, Vol.104, No.2, (February 2008), pp. 371-379, ISSN 1364-5072.

Svensson, V. & Christiansson, A. (1991). Methods for phage monitoring. *Bulletin FIL-IDF*, Brussels, Belgium, Vol.263, pp. 29–39, ISSN 0250-5118.

Szczepańska, A.K.; Hejnowicz, M.S.; Kołakowski, P. & Bardowski, J. (2007). Biodiversity of *Lactococcus lactis* bacteriophages in Polish dairy environment. *Acta Biochimica Polonica*, Vol. 54, No.1, (March 2007), pp. 151-158, ISSN 0001-527X.

Ventura, M.; Canchaya, C.; Bernini, V.; Altermann, E.; Barrangou, R.; McGrath, S.; Claesson, M. J.; Li, Y.; Leahy, S.; Walker, C.D.; Zink, R.; Neviani, E.; Steele, J.; Broadbent, J.; Klaenhammer, T.R.; Fitzgerald, G.F.; O'toole, P.W. & van Sinderen, D. (2006). Comparative genomics and transcriptional analysis of prophages identified in the genomes of *Lactobacillus gasseri*, *Lactobacillus salivarius*, and *Lactobacillus casei*. *Applied and Environmental Microbiology*, Vol.72, No.5, (May 2006), pp. 3130-3146, ISSN 0099-2240.

Ventura, M.; Zomer, A.; Canchaya, C.; O'Connell-Motherway, M.; Kuipers, O.; Turroni, F.; Ribbera, A.; Foroni, E.; Buist, G.; Wegmann, U.; Shearman, C.; Gasson, M.J.; Fitzgerald, G.F.; Kok, J. & van Sinderen, D. (2007). Comparative analyses of prophage-like elements present in two *Lactococcus lactis* strains. *Applied and Environmental Microbiology*, Vol.73, No.23, (December 2007), pp. 7771-7780, ISSN 0099-2240.

Ventura, M.; Turroni, F.; Lima-Mendez, G.; Foroni, E.; Zomer, A.; Duranti, S.; Giubellini, V.; Bottacini, F.; Horvath, P.; Barrangou, R.; Sela, D.A.; Mills, D.A. & van Sinderen, D. (2009). Comparative analyses of prophage-like elements present in bifidobacterial genomes. *Applied and Environmental Microbiology*, Vol.75, No.21, (November 2009), pp. 6929-6936, ISSN 0099-2240.

Verreault, D.; Gendron, L.; Rousseau, G.M.; Veillette, M.; Massé, D.; Lindsley, W.G.; Moineau, S. & Duchaine, C. (2011). Detection of airborne lactococcal bacteriophages in cheese manufacturing plants. *Applied and Environmental Microbiology*, Vol.77, No.2, (January 2011), pp. 491-497, ISSN 0099-2240.

Wilson, I.G. (1997). Inhibition and facilitation of nucleic acid amplification. *Applied and Environmental Microbiology*, Vol.63, No.10, (October 1997), pp. 3741–3751, ISSN 0099-2240.

Zago, M.; De Lorentiis, A.; Carminati, D.; Comaschi, L. & Giraffa, G. (2006). Detection and identification of *Lactobacillus delbrueckii* subsp. *lactis* bacteriophages by PCR. *Journal of Dairy Research*, Vol. 73, No.2, (May 2006), pp. 146-153, ISSN 0022-0299.

Zago, M.; Suárez, V.; Reinheimer, J.A.; Carminati, D. & Giraffa, G. (2007). Spread and variability of the integrase gene in *Lactobacillus delbrueckii* subsp. *lactis* strains and phages isolated from whey starter cultures. *Journal of Applied Microbiology*, Vol.102, No.2, (February 2007), pp. 344-351, ISSN 1364-5072.

Zago, M.; Rossetti, L.; Reinheimer, J.; Carminati, D. & Giraffa, G. (2008). Detection and identification of *Lactobacillus helveticus* bacteriophages by PCR. *Journal of Dairy Research*, Vol.75, No.2, (May 2008), pp. 196-201, ISSN 0022-0299.

5

Bacteriophages of *Bacillus subtilis* (*natto*) and Their Contamination in Natto Factories

Toshirou Nagai
National Institute of Agrobiological Sciences,
Japan

1. Introduction

Natto is a fermented soybean food, which is produced and consumed mainly in Japan (Nagai & Tamang, 2010). The Japanese usually eat natto with cooked rice, after mixing it with seasonings attached in a package or with soy sauce (Fig. 1). Natto has a characteristic odour of short-chain fatty acids and ammonia (Ikeda *et al.*, 1984; Kanno & Takamatsu, 1987), and a highly viscous polymer, poly-γ-glutamate (PGA, see Section 3).

Bacillus natto, named after "natto" when isolated from it for the first time, is the sole microorganism used for natto fermentation (Sawamura, 1906). However, the species was regarded as a "probable synonym" of *B. subtilis* in *Bergey's Manual of Determinative Bacteriology*, 8th Edition (Gibson and Gordon, 1974). This classification was supported by the fact that the chromosomal DNA of *B. natto* has a high level of homology with that of *B. subtilis* (Seki *et al.*, 1975). Phylogenetic analysis of *B. natto* (meaning *Bacillus* isolates from natto) and typical *B. subtilis* strains by sequencing of the 16S rRNA gene also showed that *B. subtilis* (*natto*) and *B. subtilis* are the same species (Tamang *et al.*, 2002). Although the scientific name "*B. natto*" was abandoned, the informal name "*B. subtilis* (*natto*)" is often used in the food industry, and even in the scientific field, to emphasize that *B. subtilis* (*natto*) isolates have the ability to produce natto unlike the type strain of *B. subtilis*.

Commercial *B. subtilis* (*natto*) starters for natto fermentation are sold by three companies (Miura, Naruse and Takahashi) in Japan. The key strains for natto fermentation were isolated from the starters and characterized (Kiuchi *et al.*, 1987; Sulistyo *et al.*, 1988), and the characteristics, including PGA production and flavor, were found to be very similar among the strains.

Until the early 20th century, natto had been produced by packing boiled soybeans in a bag made of rice straw, which *B. subtilis* (*natto*) inhabit as soil bacteria (Fig. 1 C). Since the discovery of *B. subtilis* (*natto*) by Sawamura (1906) and the development of sanitary containers (Fig. 1 A & B) as substitutes for bags made of rice straw, the process of natto fermentation has been modernized and automated.

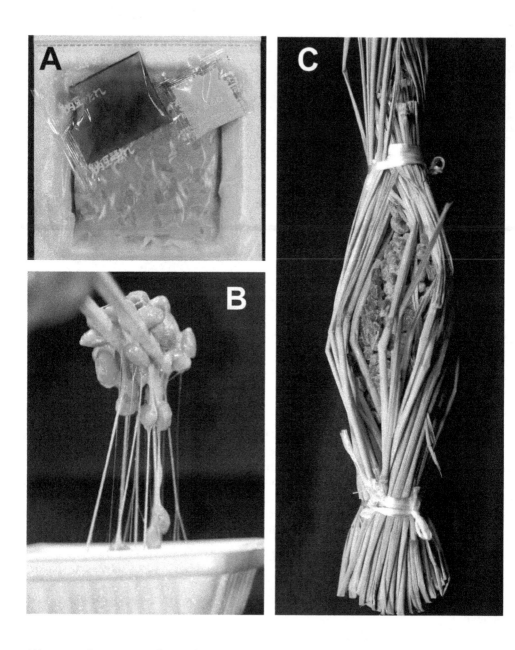

A) brown package, sauce; yellow package, mustard. B) Natto is very stringy because of PGA production (see Section 3). C) This natto is for a souvenir, so the straw is sterilized for hygien unlike the real classic type of natto. After sterilization, a spore suspension of *B. subtilis* (*natto*) is inoculated as shown in Fig. 2.

Fig. 1. Modern natto (A and B) and a classic type of natto (C)

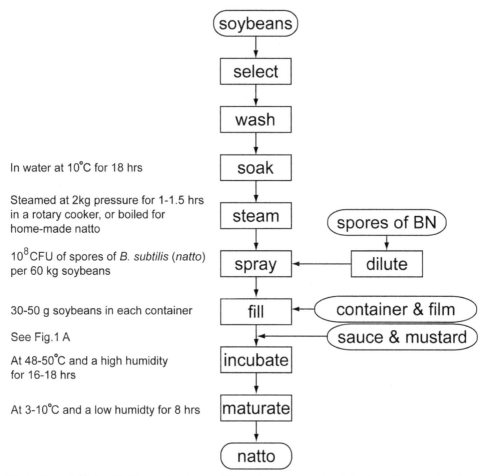

BN, *Bacillus subtilis (natto)*. The production conditions shown on the left of the flow chart are cited from reviews on natto fermentation (Ueda, 1989; Kiuchi & Watanabe, 2004).

Fig. 2. Process of natto fermentation

The process of natto fermentation is so simple that very small factories, even homes, can produce natto in 2 or 3 days (Ueda, 1989; Kiuchi & Watanabe, 2004) (Fig. 2). After being selected and washed, soybeans are soaked in water at 10°C for 18 hr. Soybeans are cooked (steamed or boiled) and a suspension of spores of *B. subtilis* (*natto*) is sprayed on the boiled soybeans while they are hot to prevent the soybeans from being contaminated with other kinds of bacteria or phages. The soybeans are packed in containers made of polystyrene paper together with packed seasonings (typically, sauce and mustard; Fig. 1 A) and incubated at 50°C for 16–18 hr and then kept at 3–10°C for 8 hr to mature the natto further. Natto products are delivered to markets or stores through a cold chain system, keeping the products at a low temperature until consumers buy them. This simplicity of the process of natto fermentation, including a limited number of starters of natto, can result in devastating damage to a factory that becomes contaminated with bacteriophages.

2. Classification of *Bacillus subtilis* (*natto*) phages

The viscous polymer PGA on natto is an important characteristic. However, natto products without the polymer were often found in the market, despite normal synthesis of PGA just after fermentation in a factory. In some cases, the viscosity of natto decreased rapidly while mixing with seasonings. Fujii *et al.* (1967) found a bacteriophage from such an abnormal natto and named it PN-1. Fermentation of soybeans with *B. subtilis* (*natto*) and PN-1 resulted in the production of natto with no viscous polymer, indicating that PN-1 is attributed to a loss of polymer on natto. This was the first report on a *B. subtilis* (*natto*) phage.

Yoshimoto *et al.* (1970) surveyed contamination by *B. subtilis* (*natto*) phages in natto factories all over Japan. Among 60 factories, 28 were contaminated with phages at densities ranging from 5 PFU/cm³ sample to 2000. Forty-two phages were isolated from the samples and classified into 4 groups based on host ranges, and finally into two serological groups, NP-4 and NP-38 groups, using four anti-phage serums.

Group	MAFF no.	Strain	Source & month and year of isolation
I	270104	JNCHUP	natto, Oct. 1980
	270105	JNDMP	natto, Feb. 1981
	270106	JNHMP	natto, Feb. 1981
II	270101	P-1	abnormal natto, Jan. 1980
	270102	DMP	natto, May 1980
	270103	MIP	natto, May 1980
	270107	MOP	abnormal natto, Jul. 1981
	270108	THP	abnormal natto, Jul. 1981
	270109	THAP	abnormal natto, Dec. 1981
	270110	SUP	abnormal natto, Dec. 1981
	270111	KKP	abnormal natto, Mar. 1982
	270112	KKP-GE	industrial sewage, Mar. 1982
	270113	SS1P	abnormal natto, Dec. 1983
	270114	SS2P	abnormal natto, Dec. 1983
	270115	ONPA	abnormal natto, Aug. 1984
	270116	ONPB	abnormal natto, Aug. 1984
	270117	ONPC	abnormal natto, Sep. 1985
	270118	FUKUSHOGUNP	abnormal natto, Oct. 1985
	270119	ONPD	abnormal natto, Aug. 1986
	270120	SUP-SS1P	not recorded

Note: These phage isolates are distributed by the NIAS Genebank (acronym, MAFF; web site, http://www.gene.affrc.go.jp/index_en.php).

Table 1. *Bacillus subtilis* (*natto*) phages used in classification

Twenty *B. subtilis* (*natto*) phages were isolated from abnormally fermented natto, effluent from natto factories and soil of paddy fields in Kyushu Island, southern Japan (Fujii *et al.*, 1975). The phages, including PN-1, were classified into three groups based on host ranges, immunological reaction, and morphologies. All tested *B. subtilis* (*natto*) strains were infected by 21 *B. subtilis* (*natto*) phages, and some strains of *B. subtilis* also were infected by 17 phages. Three representative phages, PN-3, PN-6 and PN-19, from three serological groups were studied using an electron microscope. They had a head (diameter, 80–90 nm) and a contractile tail (length, 165–175 nm). Although PN-3 and PN-6 did not resemble each other in shape, judging from their morphologies on somewhat obscure photographs, they were found to belong to the same group (Nagai & Yamasaki, 2009).

Nagai and Yamasaki (2009) classified 20 phages [deposited in the NIAS Genebank (Tsukuba, Japan) with accession numbers from MAFF 270101 to MAFF 270120 (Table 1)], mainly isolated from abnormally fermented natto, into two groups based on DNA-DNA hybridization (Fig. 3). No cross hybridized band is visible in the photograph, indicating that group I phages and group II phages are genetically independent of each other. Representative phages from the two groups were further characterized. Phage JNDMP (Group I) has a head (diameter, 60 nm) and a flexible tail (7 x 200 nm) (Fig. 4 A) and requires magnesium ions for amplification. Phage ONPA (Group II) has a head (diameter, 89 nm) and a contractile tail (9 x 200 nm) with a sheath (width, 23 nm) and does not require additional magnesium ions (Fig. 4 B). Plaques of ONPA were clearer than those of JNDMP. Other characteristics of JNDMP and ONPA are summarized in Table 2.

Group type	I	II
	JNDMP(MAFF 270105)	ONPA (MAFF 270115)
Head (nm)	60	89
Tail (nm)	7 x 200	9 x 200
Genome DNA (kb)	42	91
Latent time (min)	35	50
Burst size	46	72
Heat stability (°C)[1]	53	63
Mg ion requirement	+	-
Host range[2]		
Miura (MAFF 118100)	-	+
Naruse (MAFF 118103)	+	+
Takahashi (MAFF 118105)	+	+
Marburg	-	-

1)Temperature at which about 1% of phage particles in suspension can survive after 10-min heating in a water bath.
2) Miura, Naruse and Takahashi are commercial starters for natto fermentation.

Table 2. Other characteristics of JNDMP and ONPA

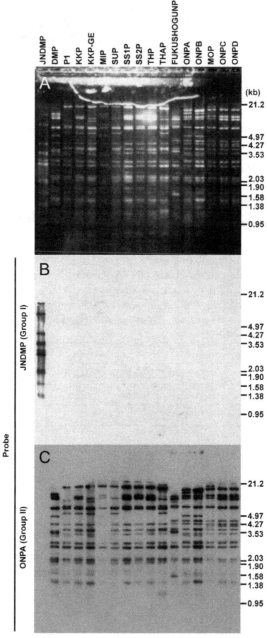

A) An agarose gel electrophoresis of fragments of phage DNA after digestion with a restriction enzyme, *Hind*III. B and C) Southern hybridization of the gel using genomic DNA of JNDMP (B) or ONPA (C) as a probe. (from Nagai & Yamasaki, 2009, with permission)

Fig. 3. Analysis of Southern hybridization of phage genome DNA

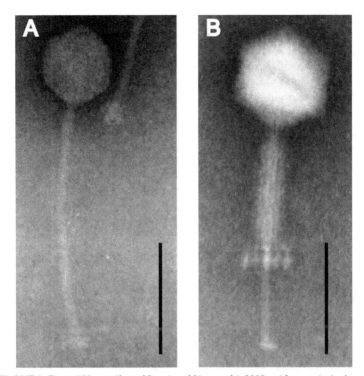

A) JNDMP, B) ONPA. Bar = 100 nm. (from Nagai and Yamasaki, 2009, with permission)

Fig. 4. Electron microscope photographs of *B. subtilis* (*natto*) phages

The phages of Yoshimoto's group (1970) had been discarded, and so genetic relationships among their two types of phages and ONPA or JNDMP could not be investigated. However, NP-4 and NP-38 had the same morphologies as JNDMP (Group I) and ONPA (Group II), respectively. Using DNA-DNA hybridization, type strains of Fujii et al. (1975), PN-3, PN-6 and PN-19, were found to belong to Group I, I and II, respectively (Nagai & Yamasaki, 2009).

3. Polyglutamate degrading enzyme

3.1 Polyglutamate

B. subtilis (*natto*) produces a very viscous polymer of DL-glutamic acid, which has two carboxyl groups on α- and γ-carbons (Fig. 5A), with extremely high degrees of polymerization. Unlike proteins, in which amino acid residues bind via α-carboxyl groups and amino groups (Fig. 5B), the glutamic acids in this polymer, poly-γ-glutamate (PGA, Fig. 5C), are synthesized by binding a γ-carboxyl group and an amino group of an adjacent glutamic acid via a hyperphosphorylated intermediate (Fig. 5D) (Ashiuchi et al., 2001). Genes related to the production of PGA were cloned and expressed well in *Escherichia coli* cells (Ashiuchi et al., 1999). The gene *pgsBCA* is homologous with genes for capsular PGA production of *B. anthracis* and codes for a membranous enzyme complex (Ashiuchi et al.,

2001). On the other hand, a regulatory gene for PGA production was cloned and found to be *comP*, which codes for a sensor protein kinase of the ComP-ComA two-component signal transduction system (Nagai *et al.*, 2000). Recently, PGA cross-linked by γ-radiation was found to hold a large quantity of water and to be useful for greening of desert areas by scattering the PGA resin in which seeds (e.g. soybeans) are embedded (Hara, 2006). Thus, PGA is becoming an important industrial material.

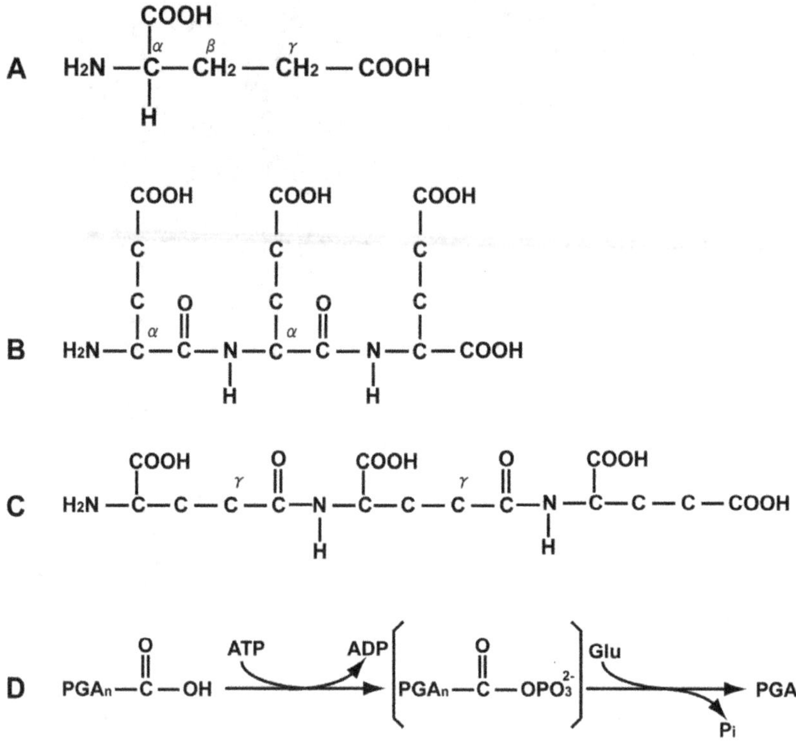

Fig. 5. Glutamic acid (A), tri-α-glutamic acid (B), tri-γ-glutamic acid (C), and the mechanism of synthesis of poly-γ-glutamic acid (D)

3.2 Polyglutamate depolymerase

It was found that a *B. subtilis* (*natto*) lysogenic strain could not accumulate PGA in the culture, whereas a nonlysogenic strain accumulated it under the same conditions (Hongo & Yoshimoto, 1968). In the culture of the lysogenic strain, bacteriophages were induced at a high density of 10^9 PFU/ml from the early stage of the experiment. Around the peak of production of phages, PGA depolymerase appeared to be released from the lysogenic strains to the culture. The enzyme was also produced extracellularly, when *B. subtilis* (*natto*) was infected by phages. After infection with phage NP-1 cl (Yoshimoto & Hongo, 1970), depolymerase was synthesized in parallel with the production of phage particles in host cells (Hongo & Yoshimoto, 1970a). The depolymerase digested PGA by endopeptidase-type action, resulting in the rapid loss of viscosity of PGA. Chromatographic studies showed that the products of the reaction were di-γ-glutamate and tri-γ-glutamate (Hongo & Yoshimoto, 1970b). Optimal pH and temperature of the depolymerase were pH 6–8 and 40–50°C, respectively. The depolymerase was stable at pH values ranging from 4 to 9 and below 70°C. The depolymerase was not linked to phage particles.

3.3 Poly-γ-glutamate hydrolase, PghP

Culture supernatant of *B. subtilis* (*natto*) infected with a phage ΦNIT1, which had been isolated from natto containing a small amount of PGA, caused the viscosity of PGA to decrease rapidly, and degraded PGA with a molecular weight of 10^6 Daltons to oligo-γ-glutamyl peptides (Kimura & Itoh, 2003). From the culture supernatant, a 25-kDa monomeric enzyme was purified through five column chromatographic steps and was named PghP for "γ-PGA hydrolase of phage". Analysis of the products of enzymatic reaction on PGA showed that they were tri-γ-glutamate, tetra-γ-glutamate and penta-γ-glutamate. PghP was inhibited with monoiodoacetate and EDTA and the activity inhibited by EDTA was restored by adding Zn^{2+} or Mn^{2+}, indicating that a cisteine residue(s) of PghP and these ions participated in the hydrolase reaction.

The gene for PghP was cloned based on a nucleotide sequence predicted from the N-terminal amino acid sequence of purified PghP, and sequenced (Kimura & Itoh, 2003). The predicted PghP was a 22.9-kDa protein with 203 amino acid residues, in which the first methionine was eliminated posttranslationally. PghP was a unique protein: similar proteins were not detected in the database by a BLAST search program. PghP is distributed in a variety of *B. subtilis* (*natto*) phages including some phages isolated from *B. subtilis* strains which produce no PGA. PghP of ΦNIT1 has a different substrate specificity from the PGA depolymerase in Section 3.2 (Hongo & Yoshimoto, 1970b).

ΦNIT1 could amplify in both encapsulated and non-encapsulated *B. subtilis* (*natto*). On the other hand, *B. subtilis* phage BS5 (Ackermann et al., 1995), which produced no PghP, could amplify only in non-encapsulated *B. subtilis* (*natto*). BS5, however, could amplify in encapsulated *B. subtilis* (*natto*) in the presence of additive PghP. These results indicate that *B. subtilis* (*natto*) produces PGA for physical protection from attacks by phages (Kimura & Itoh, 2003).

Apparently opposite results were also reported: that PGA production made the cells susceptible to *B. subtilis* (*natto*) phages (Hara et al., 1984). In their study, *B. subtilis* (*natto*) phages could infect *B. subtilis* (*natto*) producing PGA. After curing of plasmid pUH1, which harboured genes controlling PGA production, of *B. subtilis* (*natto*) strains, the cured strains could no longer be infected by the phages. *B. subtilis* Marburg, which was not infected by the *B. subtilis* (*natto*) phages and did not produce PGA by nature, and the cured *B. subtilis* (*natto*) strains became susceptible to the phages after transformation with DNA from *B. subtilis* (*natto*). Hara et al. thought that "PGA might be associated with phage absorption" from the results. However, a cured *B. subtilis* (*natto*) strain was susceptible to phage ΦBN100, indicating basically that the plasmid did not control phage absorption (Nagai & Itoh, 1997). The experiments conducted by Hara et al. examined three factors: the existence of pUH1-type plasmids, PGA productivity and γ-glutamyl transpeptidase (γ-GTP) (at the time, γ-GTP and pUH1 were thought to be a PGA synthesizing enzyme and a plasmid coding γ-GTP genes, respectively [Aumayr et al., 1981; Hara et al., 1981], but pUH1 was found to harbour no genes for PGA production [Nagai et al., 1997]), so the situation might make the results difficult to interpret accurately. This discrepancy remains to be elucidated.

Natto without PGA has no commercial value as a food on the market, but might have a value as an ingredient of natto fried rice or natto snack foods because of its ease of manufacturing in food factories (Kimura, 2008). Thus, PghP could be useful in the food industry.

4. Phage contamination in natto factories

The first report on contamination of natto by phages was reported by Fujii et al. (1967). At that time, the authors did not search the factory for phages where the contaminated natto had been made. In 1975, the authors investigated contamination by phages in the same factory. Phages were not detected in the factory, but in effluent from it because of modernization of the factory (Fujii et al., 1975).

Yoshimoto et al. (1970) searched natto factories throughout Japan for contamination by phages. Phages were detected in 28 factories (47%) among 60 factories at a density ranging from 5 to over 1000 PFU/cm^3 of sample. Phages were very often detected in old factories, the walls of which were made of clay (common in old Japanese buildings). The surfaces of clay walls have too many asperities to clean off soybean debris perfectly, resulting in amplification of phages in their host cells on soybean debris. The walls of modern factories are made of clean stainless steel, so phages were rarely detected. In factories contaminated by phages, the phages were detected most frequently in the fermentation rooms. Fujii et al. also reported that 25% of factories (2/8 factories in Kyushu Island, west Japan) were polluted with phages (Fujii et al., 1975).

Nakajima (1995) investigated contamination by B. subtilis (natto) phages in four natto factories in Ibaraki prefecture, central Japan (Fig. 6). Before inoculation of natto starter to soybeans, phages were detected only on the floor of the washing room in a factory and its detection rate was very low. The phages might have been brought in with raw soybeans or dust in the air. After inoculation, the detection rate rose to 100%. These results indicate that perfect cleaning to ensure that no soybean debris remains in machines or on floors is essential, especially after soybeans have been sprayed with a spore suspension of B. subtilis (natto). The author did not mention contamination by phages of natto products made by the four factories. Another report showed that phage contamination was not detected in any natto made in factories in Iwate prefecture, northern Japan (Yamamoto, 1986). In total, contamination of natto products by phages has decreased drastically, but B. subtilis (natto) phages still exist in natto factories, fields, and waste water.

Improvement of a factory highly polluted with phages and bacteria was reported (Takiguchi et al., 1999). When phages and bacteria were detected in natto made by the factory, a manual was compiled to ensure strict separation of the entrance and exit, hand-washing, removal of abnormal natto from the factory as soon as possible, dilution of starter with sterilized water, installation of UV lighting, replacement of wooden parts with those made of stainless steel, including periodical cleaning and hygiene education. After these efforts, no contamination was detected in the natto.

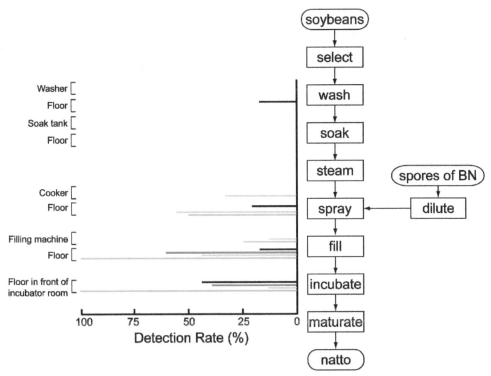

Four independent factories (indicated by different color bars) were surveyed for *B. subtilis* (*natto*) phages in several facilities in the process of natto fermentation (the right flow chart, also see Fig. 2) and the floors near the facilities.

BN, *Bacillus subtilis* (*natto*)

Detection rate (%) = no. of detections / no. of samples x 100

Fig. 6. Contamination by *B. subtilis* (*natto*) phages in natto factories (adapted from Nakajima, 1995)

The following disinfectants were effective against *B. subtilis* (*natto*) phages: benzalkonium chloride, chloramine-T, sodium hypochlorite, TEGO-51 and Vantocil IB (Fujii *et al.*, 1983). The most important measure is cleaning the machines and floors of natto factories to remove soybean debris on which *B. subtilis* (*natto*) and phages can propagate.

5. Other topics on *B. subtilis* (*natto*) phages

5.1 Generalized transducing phage for *B. subtilis* (*natto*)

For genetic transfer of DNA between *B. subtilis* (*natto*) strains by transduction, a phage ΦBN100 was screened in laboratory stock strains (Nagai & Itoh, 1997). The phage could transduce prototroph genes (for adenine, uracil or leucine requirement) to auxotrophs at rates ranging from 3.8×10^{-8} to 1.6×10^{-6} (number of transductants per phage particle). The phage was also used for analysis of transposon insertional mutagenesis on a gene responsible for the regulation of PGA production (Nagai & Itoh, 1997) and construction of mutants on production of branched short-chain fatty acids for preparation of odorless natto (Takemura *et al.*, 2000). ΦBN100 is a synonym for JNDMP (see Section 2).

5.2 *B. subtilis* (*natto*) phage PM1 and a phage detection system by PCR

A phage was newly isolated from natto producing no PGA and characterized (Umene *et al.*, 2009). The morphology of the phage, PM1, was very similar to that of JNDMP, and the size of its genomic DNA was found to be 50 kb using field inversion gel electrophoresis, 10 kb smaller than that of JNDMP. The genome of PM1 was a linear double-stranded DNA, and might be circularly permuted and have no definite termini, like T4 phage.

Based on a sequence of a 1.1-kb *Eco*RI fragment of genomic DNA, which did not have significant homology with any sequences deposited at the DNA database so far, the following pair of primers for PCR to amplify a 0.53-kb region in the 1.1-kb *Eco*RI fragment was designed:

5′-CGCACTGGAAGCAATCAAGTCGG-3′ (corresponding to nt 33–55)

5′-CAACCCTCTGACCGACTTTTCCC-3′ (corresponding to nt 538–560)

Among ten *B. subtilis* (*natto*) phage isolates in the authors' laboratory, eight were target sequences of amplified with the primer set, suggesting that PM1 phages are distributed over a large area of Japan.

5.3 *Bacillus* phage isolated from chungkookjang

Chungkookjang is a Korean soybean food fermented by *Bacillus subtilis*. From the fermented soybeans, a virulent *Bacillus* phage was isolated and named Bp-K2 (Kim *et al.*, 2011). Bp-K2 resembled ONPA in morphology, but had a smaller head (width, 80 nm) and genomic DNA (21 kb). Bp-K2 had a contractile tail with a sheath (85–90 nm x 28 nm), a tail fiber (80–85 nm x 10 nm) and a basal plate (29 nm x 47 nm). Bp-K2 could develop plaques on not only *B. subtilis* strains isolated from chungkookjang but also *B. subtilis* (*natto*).

5.4 Defective phage of *B. subtilis* (*natto*)

As *B. subtilis* Marburg strain produces a defective phage PBSX (Seaman *et al.*, 1964; Anderson & Bott, 1985; Zahler, 1993 for a review), which cannot amplify in host cells, *B. subtilis* (*natto*) IAM 1207 produces defective phage PBND8 after induction with bleomycin (Tsutsumi *et al.*, 1990). Although PBND8 resembled PBSX in morphology, the size of DNA contained in heads of PBND8 was 8 kb, 5 kb smaller than that of PBSX (13 kb). SDS-polyacrylamide gel electrophoresis of component proteins of the phage particles showed that PBND8 was clearly distinct from PBSX and PBSY, a defective phage from *B. subtilis* W23.

Seaman *et al.* (1964) also showed the production of PBSX-like particles from *B. natto*. The particles neutralized antiserum against PBSX particles, indicating that the PBSX-like particles from *B. natto* were very closely related to PBSX. At least two kinds of defective phages might be produced by strains belonging to a *B. subtilis* (*natto*) group (i.e., PBND8 and PBSX-like defective phage).

6. Conclusion

B. subtilis (*natto*) phages that have been isolated in Japan are classified into two groups (Groups I and II), which are genetically independent of each other judging from DNA-DNA hybridization analysis. Phage JNDMP (Group I) has a head (diameter, 60 nm) and a flexible tail (7 x 200 nm) and requires magnesium ions for amplification. Phage ONPA (Group II) has a head (diameter, 89 nm) and a contractile tail (9 x 200 nm) with a sheath (width, 23 nm) and does not require additional magnesium ions. JNDMP was found to be a generalized transducing phage for *B. subtilis* (*natto*). Natto contaminated with phages is not covered with PGA, which is an important factor of the quality of natto. The loss of PGA is attributed to PGA hydrolase, PghP, or its relevant enzyme, which is expressed from a gene on phage genomic DNA in infected host cells. The enzymes digest PGA by endopeptidase-type action, resulting in a rapid loss of viscosity of PGA. Contamination of natto products by phages can be prevented by cleaning the facilities and floors of natto factories. Until 1980, contamination by phages had caused devastating damage to natto factories, but such trouble is now rare thanks to the modernization of natto factories and hygiene education for workers.

7. References

Ackermann, H., Azizbekyan, R., Bernier, R., de Barje, H. , Saindouk, S. , Valéro, J. & Yu, M. (1995). Phage typing of *Bacillus subtilis* and *B. thuringensis*, *Research in Microbiology*, Vol. 146, Issue 8, (October 1995), pp. 643–657, ISSN 0923-2508

Anderson, L.& Bott, K. (1985). DNA Packaging by the *Bacillus subtilis* Defective Bacteriophage PBSX, *Journal of Virology*, Vol. 54, No.3, (June 1985), pp. 773-780, ISSN 0006-2960

Ashiuchi, M., Soda, K. & Misono, H. (1999). A poly-γ-glutamate synthetic system of *Bacillus subtilis* IFO 3336: Gene cloning and biochemical analysis of poly-γ-glutamate produced by *Escherichia coli* clone cells, *Biochemical and Biophysical Research Communications*, Vol.263, Issue 1, (September 1999), pp. 6-12, ISSN 0006-291X

Ashiuchi, M., Nawa, C., Kamei, T., Song, J., Hong, S., Sung, M., Soda, K. , Yagi, T. & Misono, H. (2001). Physiological and Biochemical Characteristics of Poly γ-glutamate synthetase complex of *Bacillus subtilis*, *European Journal of Biochemistry*, Vol. 268, No. 20, (October 2001), pp. 5321-5328, ISSN 0014-2956

Aumayr, A. , Hara, T. & Ueda, S. (1981). Transformation of *Bacillus subtilis* in polyglutamate production by deoxyribonucleic acid from *B. natto*. *Journal of General and Applied Microbiology*, Vol. 27, No. 2, (1981), pp. 115-123, ISSN 0022-1260

Fujii, H. , Oki, M. , Makihara, M. , Keshino, J. & Takeya, R. (1967). On the Formation of Mucilage by *Bacillus natto* ; Part VII. Isolation and Characterization of a Bacteriophage Active against "Natto"-Producing Bacteria, *Journal of the Agricultural Chemical Society of Japan*, Vol.41, No.1, (January 1967), pp. 39-43, ISSN 0002-1407 (in Japanese)

Fujii, H. , Shiraishi, A. , Kaba, K , Shibagaki, M. , Takahashi, S. & Honda, A. (1975). Abnormal Fermentation in Natto Production and *Bacillus natto* Phages, *Journal of Fermentation Technology*, Vol.53, No.7, (July 1975), pp. 424-428, ISSN 0367-5963 (in Japanese)

Fujii, H. , Shiraishi, A. , Kiryu, K. & Fujimoto, Y. (1983). Isolation and some characteristics of *Bacillus natto* phage PM, *Bulletin of the Faculty of Home Life Science, Fukuoka Women's University*, Vol. 14, (January 1983), pp. 1-5, ISSN 0288-3953 (in Japanese)

Gibson, T. & Gordon, R. (1974). Genus I. *Bacillus* Cohn 1872, In : *Bergey's Manual of Determinative Bacteriology, 8th ed.*, Buchanan, R. & Gibbons, N., (Ed.), pp. 529-550, The Williams & Wilkins Company, ISBN 0-683-01117-0, Baltimore, Md

Hara, T. , Aumayr, A. & Ueda, S. (1981). Characterization of plasmid deoxyribonucleic acid in *Bacillus natto*: Evidence for plasmid-linked PGA production. *Journal of General and Applied Microbiology*, Vol. 27, No. 4, (1981), pp. 299-305, ISSN 0022-1260

Hara, T. , Shiraishi, A. , Fujii, H. & Ueda, S. (1984). Specific Host Range of Bacillus subtilis (natto) Phages Associated with Polyglutamate Production, *Agricultural and Biological Chemistry*, Vol.48, No.9, (Augast 1984), pp. 2373-2374, ISSN 0002-1369

Hara, T. (2005) Desert Greening with Natto Resin – From Dream to Reality, *Kino Zairyo* , Vol. 26, No. 7, (July 2005), pp. 14-18, ISSN 0286-4835 (in Japanese)

Hongo, M. & Yoshimoto, A. (1968). Formation of Phage-Induced γ-Polyglutamic Acid Depolymerase in Lysogenic Strain of *Bacillus natto*, *Agricultural and Biological Chemistry*, Vol.32, No.4, (April 1968), pp. 525-527, ISSN 0002-1369

Hongo, M. & Yoshimoto, A. (1970a). Bacteriophages of *Bacillus natto*. Part II. Induction of γ-Polyglutamic Acid Depolymerase Following Phage Infection, *Agricultural and Biological Chemistry*, Vol.34, No.7, (July 1970), pp. 1047-1054, ISSN 0002-1369

Hongo, M. & Yoshimoto, A. (1970b). Bacteriophages of *Bacillus natto*. Part III. Action of Phage-induced γ-Polyglutamic Acid Depolymerase on γ-Polyglutamic Acid and the Enzymatic Hydrolyzates, *Agricultural and Biological Chemistry*, Vol.34, No.7, (July 1970), pp. 1055-1063, ISSN 0002-1369

Ikeda, H. & Tsuno, S. (1984) The componential changes during the manufacturing process of Natto (Part 1): On the amino nitrogen, ammonia nitrogen, carbohydrates and vitamin B2, *Journal of Food Science, Kyoto Women's University*, Vol. 39, (December 1984), pp. 19-24, ISSN 0289-3827 (in Japanese)

Kanno, A. &Takamatsu, H. (1987) Changes in the volatile components of "Natto" during manufacturing and storage (Studies on "Natto", Part IV), *Nippon Shokuhin Kogyo Gakkaishi*, Vol. 34, No. 5, (May 1987), pp. 330-335, ISSN 0029-0394

Kim, E. , Hong, J. , Yun, N. & Lee, Y. (2011). Characterization of *Bacillus* Phage-K2 Isolated from Chungkookjang, A Fermented Soybean Foodstuff, *Journal of Industrial Microbiology & Biotechnology*, Vol. 38, No. 1, (January 2011), pp. 39-42, ISSN 1367-5435

Kimura, K. & Itoh, Y. (2003). Characterization of Poly-γ-Glutamate Hydrolase Encoded by a Bacteriophage Genome: Possible Role in Phage Infection of *Bacillus subtilis* Encapsulated with Poly-γ-Glutamate, *Applied and Environmental Microbiology*, Vol.69, No. 5, (May 2003), pp. 2491-2497, ISSN 0099-2240

Kimura, K. (2008). γ-PGA Hydrolase of Phage, In : *Advanced Science on Natto : Japanese Soybean Fermented Foods*, K. Kiuchi, T. Nagai & K. Kimura, (Ed.), pp. 268-270, Kenpakusha, ISBN 978-4-7679-6123-1, Tokyo, Japan (in Japanese)

Kiuchi, K. , Taya, N. , Sulistyo, J. & Funane, K. (1987). Isolation and Identification of Natto Bacteria from Market-Sold Natto Starters, *Report of National Food Research Institute*, Vol.50, (March 1987), pp. 18-21, ISSN 0301-9780 (in Japanese)

Kiuchi, K. & Watanabe, S. (2004). Industrialization of Japanese Natto, In : *Industrialization of Indigenous Fermented Foods*, K. Steinkraus, (Ed.), pp. 193-246, Marcel Dekker, ISBN 0-8247-4784-4, New York, NY

Nagai, T. & Itoh, Y. (1997). Characterization of a Generalized Transducing Phage of Poly-γ-Glutamic Acid-Producing *Bacillus subtilis* and Its Application for Analysis of Tn*917*-LTV1 Insertional Mutants Defective in Poly-γ-Glutamic Acid Production, *Applied and Environmental Microbiology*, Vol.63, No.10, (October 1997), pp. 4087-4089, ISSN 0099-2240

Nagai, T. , Koguchi, K. & Itoh, Y. (1997). Chemical analysis of poly-γ-glutamic acid produced by plasmid-free *Bacillus subtilis* (*natto*) : Evidence that plasmids are not involved in poly-γ-glutamic acid production. *Journal of General and Applied Microbiology*, Vol. 43, No. 3, (1997), pp. 139-143, ISSN 0022-1260

Nagai, T. , Tran, L. , Inatsu,Y. & Itoh, Y. (2000). A New IS4 Family Insertion Sequence, IS*4Bsu1*, Responsible for Genetic Instability of Poly-γ-Glutamic Acid Production in *Bacillus subtilis*, *Journal of Bacteriology*, Vol.182, No.9, (May 2000), pp. 2387-2392, ISSN 0021-9193

Nagai, T. & Yamasaki, F. (2009). *Bacillus subtilis* (*natto*) Bacteriophages Isolated in Japan, *Food Science and Technology Research*, Vol.15, No.3, (May 2009), pp. 293-298, ISSN 1344-6606

Nagai, T. & Tamang, J. (2010). Fermented Legumes : Soybean and Non-Soybean Products, In : *Fermented Foods and Beverages of the World*, J. Tamang & K. Kailasapathy, (Eds.), pp. 191-224, CRC Press, ISBN 978-1-4200-9495-4, Boca Raton, FL

Nakajima, J. (1995). Natto hishitsu Furyo no Kaizen : Phage Osen Tyosa (Inprovement of Quality of Natto: Research on Contamination of *B. subtilis* (*natto*) Phages in Natto Factories), *Reports of the Ibaraki Prefectural Industrial Technology Center*, Vol.23, (1995), 59, ISSN 0912-9936 (in Japanese)

Sawamura, S. (1906) On the Micro-organisms of Natto, *Bulletin of the College of Agriculture, Tokyo Imperial University*, Vol. 7, (1906), pp. 107-110

Seaman, E. , Tarmy, E. & Marmur, J. (1964). Inducible Phages of *Bacillus subtilis*, *Biochemistry*, Vol. 3, No. 5, (May 1964), 607-613, ISSN 0006-2960

Seki, T. , Oshima, T. & Oshima, Y. (1975). Taxonomic Study of *Bacillus* by Deoxyribonucleic Acid-Deoxyribonucleic Acid Hybridization and Interspecific Transformation. International Journal of Systematic Bacteriology, Vol. 25, No. 3, (July, 1975), pp.258-270, ISSN 1466-5026

Sulistyo, J. , Taya, N. , Funane, K. & Kiuchi, K. (1988) Production of Natto Satarter, *Nippon Shokuhin Kogyo Gakkaishi*, Vol. 35, No. 4, (April 1987), pp. 278-283, ISSN 0029-0394

Takemura, H. , Ando, N. & Tsukamoto, Y. (2000). Breeding of Branched Short-Chain Fatty Acids Non-Producing Natto Bacteria and its Application to Production of Natto with Light Smells, *Nippon Shokuhin Kagaku Kogaku Kaishi*, Vol.47, No.10, (October 2000), pp. 773-779, ISSN 1341-027X (in Japanese)

Takiguchi, T. , Yoshino, I. , Yuasa, H. , Kawano, I. & Aoki, Y. (1999). Pollution of Fermentation Process of Natto by Some Kind of Bacteriophage, *Reports of Gunma Prefectural Industrial Technology Research Laboratory*, Vol. 1999, (1999), pp. 35-39, ISSN 1341-0245 (in Japanese)

Tamang, J. , Thapa, S. , Dewan, S. , Jojima, Y. , Fudou, R. & Yamanaka, S. (2002). Phylogenetic Analysis of *Bacillus* Strains Isolated from Fermented Soybean Foods of

Asia : Kinema, Chungkokjang and Natto, *Journal of Hill Resarch*, Vol.15, No.2, (2002), pp. 56-62, ISSN 0970-7050

Tsutsumi, T. , Hirokawa, H. & Shishido, K. (1990). A New Defective Phage Containing a Randomly Selected 8 Kilobase-pairs Fragment of Host Chromosomal DNA Inducible in a Strain of *Bacillus natto, FEMS Microbiology Letters*, Vol.72, Issue 1-2, (October 1990), pp. 41-46, ISSN 0378-1097

Ueda, S. (1989). Industrial Application of *B. subtilis*, In : *Bacillus subtilis : Molecular Biology and Industrial Application*, B. Maruo & H. Yoshikawa, (Ed.), pp. 143-161, Kodansha, ISBN 4-06-201778-4, Tokyo, Japan

Umene, K. , Oohasi, S. , Yamanaka, F. & Shiraishi, A. (2009). Molecular Characterization of the Genome of *Bacillus subtilis* (natto) Bacteriophage PM1, A Phage Associated with Disruption of Food Production, *World Journal of Microbiology and Biotechnology*, Vol. 25, No. 10, (October 2009), pp. 1877-1881, ISSN 0959-3993

Yoshimoto, A. & Hongo, M. (1970). Bacteriophages of *Bacillus natto*. 1. Some Characteristics of Phage NP-1, *Journal of the Faculty of Agriculture, Kyushu University* Vol. 16, No. 2, (July 1970), pp. 141-158, ISSN 0023-6152

Yoshimoto, A. , Nomura, S. & Hongo, M. (1970). Bacteriophages of *Bacillus natto* (IV) Natto Plant Pollution by Bacteriophages, *Journal of Fermentation Technology*, Vol.48, No.11, (November 1970), pp. 660-668, ISSN 0367-5963 (in Japanese)

Yamamoto, T. (1986). Shihan Natto no Phage Osen Tyosa Hokoku (Report on the Phage contamintion to Commercial Natto), *Iwate-ken Jozo Shokuhin Shikenjyo Hokoku*, Vol. 20, (1986), pp. 106-107, ISSN 0387-4966 (in Japanese)

Zahler, S. (1993). Temperate Bacteriophages, In : *Bacillus subtilis and Other Gram-Positive Bacteria : Biochemistry, Physiology, and Molecular Genetics*, Sonenshein, A., Hoch, J. & Losick, R., (Ed.), pp. 831-842, American Society for Microbiology, ISBN 1-55581-053-5, Washington DC

6

Bacteriophages as Surrogates for the Fate and Transport of Pathogens in Source Water and in Drinking Water Treatment Processes

Maria M.F. Mesquita and Monica B. Emelko
Department of Civil and Environmental Engineering, University of Waterloo, Canada

1. Introduction

Less than 1% of the world's fresh water accessible for direct human uses is found in lakes, rivers, reservoirs and those underground sources that are shallow enough to be tapped at an affordable cost. Only this amount is regularly renewed by rain and snowfall, and is therefore available on a sustainable basis (Berger, 2003).

More than a billion people have limited access to safe drinking water; over 2 million die each year from water-related diarrhea, which is one of the leading causes of mortality and morbidity in less economically developed countries (UNICEF and WHO, 2009). In more economically developed countries, increasing demands on water resources raise concerns about sustainable provision of safe drinking water. In 2008, supply and protection of water resources was identified as the top strategic priority of North American water professionals (Runge and Mann, 2008). This is not surprising given the rapidly expanding competition for existing water supplies from industrial, agricultural and municipal development, as well as the vital needs to protect human health and ecosystem functions. The challenge of sustaining supply is further exacerbated by changes in water quality and availability as a direct or indirect result of population growth, urban sprawl, climate change, water pollution, increasing occurrence of natural disasters, and terrestrial and aquatic ecosystem disturbance.

Most of the world population depends on groundwater for their supplies. Due to the proximity of groundwater to sources of microbial contamination, the increasing occurrence of extreme climate events and the lack of adequate disinfection, groundwater is responsible for a large percentage of the waterborne outbreaks of disease worldwide (WHO, 2004; 2011). For example, between 1999 and 2000, 72% of drinking water outbreaks of disease were associated with groundwater. Although the number of groundwater-associated disease outbreaks associated in the United States decreased during 2001–02, the proportion of outbreaks associated with groundwater increased to 92% from 87% (Tufenkji and Emelko, 2011). As a result of such outbreaks and the economic implications of waterborne illness, stricter water quality regulations to protect public health have been implemented in many countries. Significant examples of such regulations include the Surface Water Treatment Rules (SWTR -1989a; 2002) and the Ground Water Rule (2006) by the U.S. Environmental

Protection Agency (USEPA); the revised Bathing Water Directive (2006/7/EC) and the Water Framework Directive (200/60/EC) by the European Union. The pressure generated by such regulations has increased the need to quantitatively understand and describe microbial pathogen transport and survival in various natural and engineered environments, including treatment systems.

Monitoring the fate and transport of all of the various microorganisms that can cause outbreaks of waterborne disease is cost prohibitive; accordingly, representative organisms such as "indicators" of pathogenic contamination or "surrogates" for the transport and survival of pathogens in various environments are sought. While indicators often originate from the same source and act as signals of pathogen presence, surrogates may or may not be derived from the same source as pathogens and are often introduced into natural and engineered environments to pseudoquantitatively assess pathogen fate and transport. Commonly used surrogates for such investigations include several bacteria, aerobic and anaerobic bacterial endospores, numerous bacteriophages, microbe-sized microspheres, chemically inactivated protozoa, and nonpathogenic, fluorescently labeled bacteria and protozoa (Tufenkji and Emelko, 2011). Bacteriophages meet many of the requirements of "ideal" surrogates because they have many characteristics that are similar to those of mammalian viral pathogens (i.e., size, shape, morphology, surface chemistry, isoelectric points, and physiochemistry), are unlikely to replicate in environments such as the subsurface due to a lack of viable hosts and other limiting factors, pose little risk to the health of humans, plants, and animals, and are easier and less expensive to isolate and enumerate relative to enteric viruses (Tufenkji and Emelko, 2011). All of these factors contribute to the utility of bacteriophages as surrogates for microbial pathogen transport and fate in source waters and in drinking water treatment processes.

This chapter focuses on the utility of bacteriophages as surrogates for the fate and transport of microbial pathogens of health concern in source and drinking waters, with particular reference to: (1) indicating the presence of enteric viruses in natural waters, (2) contributing to microbial source tracking, (3) evaluating the effectiveness of water treatment processes such as disinfection and filtration, and (4) elucidating the mechanisms involved in the fate and transport of enteric viruses in natural or engineered filtration media. Present knowledge acquired through laboratory and field approaches is reviewed and further research needs are identified to respond to current and future challenges in this field.

1.1 Major waterborne microbial pathogens of concern

Although water-transmitted microbial pathogens include bacteria, protozoa, helminthes and viruses, the groups of major threat to human health in freshwater supplies are pathogenic protozoa and enteric viruses (Schijven and Hassanizadeh, 2000) (Table 1). The protozoans *Cryptosporidium* and *Giardia* are among the major causal agents of diarrhoeal disease in humans and animals worldwide, and can even potentially shorten the life span of immunocompromised hosts (WHO, 2004). Their resistant forms (cysts or oocysts) are shed in large numbers by infected animals or humans and are ubiquitous in surface water. They are resistant to harsh environmental conditions and to chemical disinfectants at concentrations commonly used in water treatment plants to reduce bacterial contamination (LeChevallier et al., 1991; Rose, 1997; Karanis et al. 2002; Aboytes et al., 2004). Their small size (*Giardia* cysts 8-13 μm and *Cryptosporidium* oocysts 4-6 μm) and infectious dose (as low as a single organism -

Health Canada, 2004), also contribute to waterborne disease transmission. Several studies have revealed little or no correlation between bacterial fecal indicator and protozoan (oo)cyst densities in source surface waters (reviewed by Health Canada, 2004). These observations highlight the need for: (1) routine monitoring of surface waters for protozoan (oo)cysts or for reliable indicators of their presence and infectivity, and (2) implementation of improved drinking water technologies to effectively protect public health.

Group	Pathogen	Disease
Enteric viruses	Poliovirus	Meningitis, paralysis, fever
		Echovirus Meningitis, diarrhea, rash, fever, respiratory disease
	Coxsackievirus A	Meningitis, herpangina, fever, respiratory disease
	Coxsackievirus B	Myocarditis, congenital heart anomalies, pleurodynia, respiratory disease, fever, rash, meningitis
	New enteroviruses (types 68-71)	Meningitis, encephalitis, acute hemorrhagic conjunctivitis, fever, respiratory disease
	Hepatitis A	Hepatitis
	Enterovirus 72	Infectious hepatitis
	Norovirus	Diarrhea, vomiting, fever
	Calcivirus	Gastroenteritis
	Astrovirus	Gastroenteritis
	Reovirus	Not clearly established
	Rotavirus	Diarrhea, vomiting
	Adenoviruses	Respiratory disease, eye infections, gastroenteritis
	Snow mountain agent	Gastroenteritis
	Epidemic non-A non B hepatitis	Hepatitis
Enteric Protozoa		
	Acanthamoeba spp	Amoebic encephalitis or keratitis
		Cryptosporidium parvum
	Entamoeba histolytica	amoebic dysentery
		Giardia lambia Giardiasis (gastrointestinal disease)
	Naegleria fowleri	Amoebic meningoencephalitis
	Toxoplasm gondii Toxoplasmosis	

Table 1. Water-transmitted microbial pathogens of major concern in drinking water (adapted from: Azadpour-Keeley et al., 2003; CDC, 2003).

The collective designation "enteric viruses" includes more than 140 serological types that multiply in the gastrointestinal tract of both humans and animals (AWWA, 2006). Enteric viruses associated with human waterborne illness include noroviruses, hepatitis A virus (HAV), hepatitis E virus (HEV), rotaviruses and enteroviruses (polioviruses, coxsackieviruses A and B, echoviruses and four ungrouped viruses numbered 68 to 71) (AWWA, 2006). Enteric viruses are widespread in sewage and some have been detected in wastewater, surface water and drinking water (Gerba and Rose 1990; Payment and Franco, 1993; AWWA, 2006). Although they cannot multiply in the environment, they can survive for several months in fresh water and for shorter periods in marine water (Health Canada, 2004).

Enteric viruses are the most likely human pathogens to contaminate groundwater because they are shed in enormous quantities in feces of infected individuals (10^9 to 10^{10}/g) (Melnick and Gerba, 1980) and their extremely small size (20 to 100 nm) allows them to infiltrate soils, eventually reaching aquifers (Borchardt et al., 2003) (Fig. 1). Depending on physicochemical and virus-specific factors (e.g. size and isoelectric point), viruses can move considerable distances in the subsurface environment (Vaughn et al., 1983; Bales et al. 1993) and persist for several months in soils and groundwater (Keswick et al., 1982; Gerba and Bitton, 1984; Yates et al., 1985; Sobsey et al., 1986; Gerba and Rose, 1990; John and Rose, 2005). Enteroviruses also have been shown to be more resistant to disinfection than indicator bacteria (Melnick and Gerba, 1980; Stetler, 1984; IAWPRC, 1991).

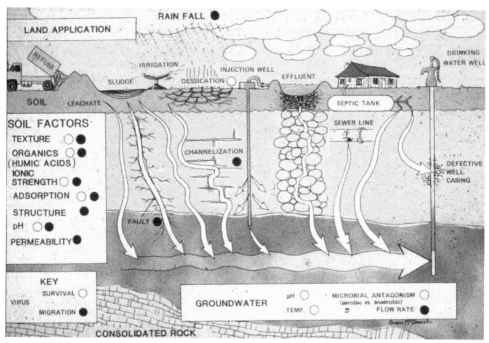

Fig. 1. Migration and survival of viruses and protozoa in the subsurface (adapted from Keswick and Gerba 1980 with permission).

1.2 Source water protection and treatment

In general the multiple-barrier approach to water treatment including watershed or wellhead protection, optimized treatment including disinfection, a well-maintained distribution system, monitoring the effectiveness of treatment, and safe water storage, is the best approach for reducing the risk of infection to acceptable or non-detectable levels (Health Canada, 2004). Surface and groundwater protection from microbial contamination largely depend on adequate land use policies related to: (1) waste and wastewater management practices, (2) the interaction of contaminated surface water with groundwater supplies (including artificial recharge with treated wastewater) and (3) the effective placement and protection of drinking water wells. Pathogenic protozoa and enteric viruses are considered priority microbial contaminants in drinking water legislation because of the significant role they play in waterborne disease outbreaks and the associated risks to public health, their extended survival in the environment, their considerable resistance to conventional water disinfection processes compared to bacteria, and the often poor or lacking correlation with traditional bacterial water quality indicator numbers.

Commonly used free chlorine concentrations and contact times applied in drinking water treatment are effective in inactivating enteric viruses (Thurston-Enriquez et al. 2003; Health Canada 2004). Ozone is generally considered more efficient against both protozoa and enteric viruses than chlorine or chlorine dioxide (Erickson and Ortega 2006). UV light disinfection, although highly effective for inactivation of protozoa, is not as efficient at inactivating viruses as more traditional chlorine-based disinfection processes (Health Canada, 2004). More recently, the combined performance of UV light and chlorine has been suggested as more effective for reclaimed water disinfection than the use of each process separately (Montemayor et al., 2008).

Effective "green" ways to remove existing and emerging pathogens and produce safe drinking water at lower cost have received much attention in recent years. These include the passage of surface water and/or groundwater through porous media in the subsurface during processes such as riverbank filtration, dune recharge, aquifer storage and recovery, and deep well injection. The need to develop regulations to protect public health coupled with the infeasibility of concentration-based criteria for all known waterborne pathogens has resulted in the evolution of regulatory approaches for water quality and treatment that rely on performance indicators and surrogates and assume specific levels of pathogen reduction through well-operated treatment systems (Tufenkji and Emelko, 2011).

1.3 Global quest for an effective pathogen indicator

Because routine monitoring for pathogens is usually costly and often unrealistic, the use of surrogate parameters (i.e. microbial indicators) to predict the presence of pathogens in water and model their behavior has long been pursued. For decades fecal bacterial indicators (e.g. fecal coliforms and *E.coli*) have been useful to identify fecal contamination to indicate the probable presence of microbial pathogens in water (Payment and Locas, 2011). However, their concentrations rarely correlate well with those of pathogens. Thus, bacterial indicators may signal the probable presence of pathogens in water, but they cannot predict precisely their level of occurrence (Payment and Locas, 2011). They are also not reliable pathogen surrogates because when compared with both virus and protozoa, bacterial indicators are less persistent in the aquatic environment and less resistant to disinfection and removal by other water treatment processes (IAWPRC 1991; Payment and Franco, 1993).

Some enteroviruses have been evaluated for monitoring environmental waters and tracking sources of water pollution (Metcalf, 1978; Goyal, 1983; Payment et al., 1985). However, the limitations associated with their use soon became apparent: (1) they are not constant inhabitants of the intestinal tract and are excreted only by infected individuals and small children, (2) laboratory methods for their detection and quantification are time-consuming, expensive, require high expertise and are restricted to some enteroviruses subgroups, and (3) virion size, surface characteristics and resistance to external agents such as disinfectants vary among subgroups. Some studies have suggested using adenoviruses as an index of human pollution because they have been shown to be more persistent and present in greater numbers than enteroviruses in sewage and fecal contaminated aquatic environments (Pina et al. 1998, Thurston-Enriquez et al. 2003).

When sewage is the source of enteric viruses and protozoa, spores of the anaerobic bacterium *Clostridium perfringens* have been suggested as suitable indicators of the presence and behavior of these pathogens in aquatic environments (Payment and Franco, 1993). Both *Bacillus* spp. aerobic endospores and *Clostridium perfringens* spores have been used as models for the removal of protozoa (oo)cysts and enteric viruses by drinking water treatment processes (Payment and Franco 1993, Rice et al. 1996).

Increasing awareness of the shortcomings of fecal bacteria as indicators of the presence of pathogenic viruses and protozoa in the environment has attracted attention to the potential value of bacteriophages that infect enteric bacteria as indicators and surrogates for evaluating the presence and behavior of human pathogenic viruses in aquatic environments and during water treatment (Noonan and McNabb, 1979; Stetler, 1984; Gerba, 1987; Havelaar, 1987; Havelaar et al., 1993). However, while phage meet many of the requirements as surrogates for enteric viruses and are useful in certain situations, they are not universal indicators, models or surrogates for enteric viruses in water environments because several disadvantages can be associated with their use (further discussed in section 3). For example, enteric viruses have been detected in treated drinking water supplies that yielded negative results for phages, even in presence–absence tests on 500 mL water samples (Ashbolt et al., 2001).

Many years of research gradually elucidated that variations in pathogen input, dilution, retention, and die-off in water environments result in conditions in which relationships/correlations between any pathogen and any indicator may be random, site-specific, and/or time-specific (Grabow, 1996; Payment and Locas, 2011). As a consequence, the present general scientific consensus is that there is no universal indicator of microbial water quality. Each specific situation, set of conditions, and objectives of study require a great deal of judgment to select the best group(s) of pathogen indicator(s) and/or surrogate(s) to be used most effectively (Table 2). Improved molecular detection techniques (e.g. PCR amplification or hybridization) based on host specificity of targeted viral and protozoan pathogens and surrogates in environmental samples may soon enable more reliable source tracking and improved public health surveillance (Scott et al. 2002; Fong and Lipp, 2005). Similarly, in-line microbial and chemical analytical systems installed at critical treatment points may replace microbial indicators and may provide continuous monitoring and reliable data, facilitating decision making. To further assist in process evaluation, efforts also have been made to eliminate ambiguities in the term "microbial indicator". Several subgroups based on function have been recognized and are now commonly used in the literature, such as: process indicators or surrogates (useful for demonstrating the efficiency of a process), fecal indicators (that indicate the presence of fecal contamination and imply that pathogens may be present), and index or virus models (indicative of pathogen presence and behavior respectively) (Ashbolt et al. 2001).

Group	Use (publisher use indents)
E. coli	- Indicator of recent fecal pollution and of <u>potential</u> presence of enteric pathogens in water
Enterococci	- Indicators of fecal pollution and indirectly of the <u>potential</u> presence of enteric viruses in groundwater
Somatic coliphages	- Index of sewage contamination - Process indicators - helpful as viral surrogates in evaluating efficiency of drinking water treatment - Some useful as pathogenic viruses models and tracers in transport studies in the subsurface and groundwater
F-RNA phages	- Index of sewage contamination - Index and models of human enteric viruses in contaminated freshwater and shellfish - Process indicators- helpful as viral surrogates in evaluating efficiency of drinking water treatment - Useful in microbial source tracking - Some useful as pathogenic viruses models and tracers in transport studies in the subsurface and groundwater
Phages of *B.fragilis*	- Indicators of human fecal pollution - Useful in microbial source tracking
C. perfringens spores	- Fecal indicators of both recent and past contamination in surface waters. - Process indicators - helpful as viral and protozoan (oo)cysts surrogates in evaluating drinking water treatment efficiency (e.g. disinfection)

Table 2. Most commonly used pathogen surrogates and their uses (Sources: Havelaar et al. 1993, Health Canada 2004, Payment and Locas 2011)

2. Multifunctionality of bacteriophages

Estimated to be the most widely distributed and diverse entities in the biosphere (McGrath and van Sinderen, 2007), bacterial virus, bacteriophages or phage can be found in all environments populated by bacterial hosts, such as soil, water and animal guts. Their unique characteristics bring several advantages to their use as pathogen surrogates (Table 3). Phages have been successfully used in a variety of environmental applications as follows:

- As **fecal indicators** - the environmental occurrence and persistence of some groups relate to health risks associated with fecal pollution and the potential occurrence of enteric pathogens in aquatic environments (Havelaar, 1987; IAWPCR, 1991; Leclerc et al., 2000; Morinigo et al., 1992; Lucena et al., 2006; Lucena and Jofre, 2010). As a result

Advantages

i.   Have no known impact on the environment
ii.   Are non-toxic and non-pathogenic for humans animals or plants
iii. Have a specific affinity to their bacterial host
iv.   Are reasonably similar to mammalian viral pathogens in size, shape, morphology, surface properties, mode of replication and persistence in natural environments
v.   Are colloidal in nature which makes them more adequate virus models then dissolved tracers
vi.   Are stable over periods of several months under laboratory conditions,
vii. Can be detected and enumerated by rapid and inexpensive methods with low detection limits (1 to 2 phage per mL)
viii. Can be prepared in large quantities at high concentrations
ix.   Specific phage groups are similar to specific pathogenic viral groups allowing the use of phage cocktails to simultaneously target several groups of concern.

Disadvantages

i.   Are excreted by a certain humans and animals all the time while pathogenic viruses are excreted by infected individuals for a short period of time (depending on the epidemiology of viruses, outbreaks of infection, and vaccination). Consequently there is no direct correlation between numbers of phages and viruses excreted by humans
ii.   A wide range of different phage can be detected by methods for somatic coliphages
iii. At least some somatic coliphages may replicate in water environments
iv.   Enteric viruses have been detected in water environments in the absence of coliphages
v.   Pathogenic human enteric viruses are excreted almost exclusively by humans, while bacteriophage used in water quality assessment are excreted by humans and animals.
vi.   The microbiota of the gut, diet and physiological state of animals seems to affect the numbers of coliphages in their feces
vii. The composition and numbers of phages excreted by humans is variable (e.g. patients under antibiotic treatment excrete lower numbers than healthy or non- medicated individuals)
viii. As water flows through porous media in the subsurface or engineered filtration processes phage can attach, detach, and re-attach by physico-chemical filtration mechanisms.

Table 3. Advantages and disadvantages of the use of bacteriophages as viral pathogen surrogates and tracers in aquatic environments (Sources: Havelaar et al., 1993; Ashbolt et al., 2001; Bateman et al., 2006).

phage infecting enteric bacteria are now accepted as useful indicators in water quality control and included in some regulations as required parameters. For example, coliphages are used in the US Water Ground Rule (USEPA, 2006), the drinking water quality regulation for the Canadian Province of Quebec (Anonymous, 2001) and a few USA states regulations regarding required quality for reclaimed water for certain uses (USEPA, 2003).

• In **microbial source tracking** (MST) or identification of fecal contamination sources by genotypic, phenotypic, and chemical methods, phage have proven useful based on their host specificity (Hsu et al. 1995; Hsu et al., 1996; Simpson et al., 2003; Jofre et al., 2011). By identifying problem sources (animal and human) and determining the effect of implemented remedial solutions MST is of special interest in waters used for recreation (primary and secondary contact), public water supplies, aquifer protection, and protection and propagation of fish, shellfish and wildlife (Simpson et al., 2003).

- As **process indicators** phage groups are often successfully employed as enterovirus surrogates in evaluating the effectiveness of water treatment processes and final product quality. This is the case with filtration and disinfection (Stetler et al., 1984; Payment et al., 1985; Havelaar et al., 1993; Durán et al., 2003; Davies-Colley et al., 2005; Persson et al., 2005; Abbaszadegan et al., 2008).

- As comprehensive pathogenic **virus indices,** phages are not very useful. This is because their numbers seldom seem to correlate to pathogenic viruses numbers in water samples when conventional statistics are applied (Lucena and Jofre, 2010). However, in the future the application of advanced mathematical models to new databases may reduce uncertainty and provide better information about relationships between phage and pathogenic virus numbers (Lucena and Jofre, 2010).

- As viral **models and tracers**, bacteriophages are often used at both field and laboratory scales as biocolloids to estimate the fate and transport of pathogenic viruses in surface and subsurface aquatic environments and through natural and manmade saturated and unsaturated porous media. This use of phage as surrogates for pathogen transport applies to protection of surface and groundwater supplies from microbial contamination, assessment of potential health risk from pathogens in groundwater and design of more efficient treatment systems in removing pathogens from drinking water supplies (Sen, 2011).

3. Main bacteriophage groups used in environmental studies

Three bacteriophage groups, somatic coliphages, male-specific F-RNA phages and *Bacteroides fragilis* phages, have been proposed and are frequently used as surrogates for pathogenic viruses in environmental studies (IAWPRC, 1991; WHO, 2004; Lucena and Jofre, 2010). However, because each group has its pros and cons as a representative of enteric virus presence and behavior in aquatic environments and water treatment processes, no agreement has been reached on which of the three groups best fulfills the index/indicator function.

3.1 Somatic coliphages

Somatic coliphages are the most numerous and most easily detectable phage group in the environment. It is a heterogeneous group whose members infect host cells (*E.coli* and other *Enterobactereacea*) by attaching to receptors located in the bacterial cell wall. Their numbers are low in human feces (often <10 g⁻¹), but abundant in untreated domestic sewage (10^4 to 10^5 particles g⁻¹) and in animal feces (Havelaar et al., 1986).

Somatic coliphages are not usually considered good fecal indicators because some of their hosts are unlikely to be of fecal origin (Hsu et al. 1996), and some of these phage are able to multiply in waters not subjected to fecal pollution (Gerba, 2006). However, some authors argue that the number of somatic phage that replicate in environmental waters is negligible (Jofre, 2009). Moreover, they are not predictive indicators of virus presence or absence in groundwater (Payment and Locas, 2011), though some somatic phage such as T-4, T-7, ΦX174, and PRD-1 have proven useful as viral surrogates of fate and transport in laboratory investigations, pilot trials, and validation testing (WHO, 2004; Lucena and Jofre, 2010).

Phage	Family name	Type	Lipid (%)	pH$_{zpc}$	Hosts	Phage Size/Shape
T2, T4, T6	*Myoviridae*	Somatic Linear ds-DNA	0	—	*E. coli* and other *Enterobateriaceae*	Cubic capsid (icosahedral or elongated), long contractile tail, 95 x 65 nm (EM)
T5, λ	*Siphoviridae*	Somatic Linear ds-DNA	0	—	*E. coli* and other *Enterobateriaceae*	Cubic capsid (icosahedral), long non-contractile tail (150 nm), 54-60 nm (EM)
T3, T7	*Podoviridae*	Somatic Linear ds-DNA	0	—	*E. coli* and other *Enterobateriaceae*	Cubic capsid (icosahedral), short non-contractile tail, 54-61 nm (EM)
PM2	*Corticoviridae*	Somatic Linear ds-DNA	13	7.3	*Pseudomonas sp., Pseudoalteromona s sp.*	Cubic capsid (icosahedral), with spikes in vertices, no tail, 60 nm (EM)
PRD-1 *	*Tectiviridae*	Somatic Circular ds-DNA	16	4.5	*S. typhimurium* and other *Enterobactereaceae*	Cubic capsid (icosahedral), no tail, 63 nm (EM) 82 ± 6 nm (DLS)
PR772**	*Tectiviridae*	F-specific Linear ds-DNA	—	3.8-4.2	*E. coli* and other *Enterobateriaceae*	Cubic capsid (icosahedral), no tail, 63 nm (EM)
MS2, Qβ	*Leviviridae*	F-specific Linear ss-RNA	0	3.9; 5.2	*E. coli* and *Salmonella sp.*	Cubic capsid (icosahedral), no tail, 20-30 nm (EM)
φX174	*Microviridae*	Somatic Circular ss-DNA	0	6.6	*Pseudomonas sp., Pseudoalteromona s sp.*	Cubic capsid (icosahedral), with spikes in vertices, no tail, 27 nm(EM)
SJ2, fd, M13	*Inoviridae*	F-specific Circular ss-RNA	0	—	*E. coli* and *Salmonella sp.*	Filamentous or rod-shaped, 810 x 6 nm (EM)
Bacteroides fragillis phages	*Siphoviridae*	Linear ds-DNA	0	—	*Bacteroides fragillis* HSP40	Icosahedral head (60 nm), flexible non-contractile tail, 150 x 8 nm (EM)

EM - electron microscopy analysis (measures physical diameter of dry particles)
DLS - dynamic light scattering analysis (measures hydrodynamic size of particles in a fluid)
pHzpc - zeta potential charge

Table 4. Characteristics of bacteriophages commonly used as pathogenic virus surrogates in environmental studies (adapted from Mesquita et al., 2010).

Bacteriophage PRD-1 (Table 4) in particular has emerged as an important viral model for studying microbial transport through a variety of subsurface environments. Its popularity is due to its similarity to human adenoviruses in size (~62nm) and morphology (icosahedric), its relative stability over a range of temperatures and low degree of attachment in aquifer sediments (Harvey and Ryan, 2004; Ferguson et al., 2007).

3.2 F-RNA bacteriophages

F or male specific RNA bacteriophages are a homogeneous group of phage that attach to fertility fimbriae (F-pili or sex-pili) produced by male bacterial cells (possessing an F-plasmid) in certain stages of their growth cycle. Since the F-plasmid is transferable to a wide range of Gram-negative bacteria, F-specific bacteriophages may have several hosts besides *E.coli* (Havelaar 1987). This group ranks second in abundance in water environments although its persistence in surface waters, mainly in warm climates is low (Chung and Sobsey, 1993; Mocé-Llivina et al., 2005).

F-RNA bacteriophages have been most extensively studied due to their similarity (in size, shape, morphology and physiochemistry) to many pathogenic human enteric viruses, namely enteroviruses, caliciviruses, astroviruses and Hepatitis A and E virus (Jofre et al., 2011) (Table 4). These phages are infrequently detected in human and animal feces (10^3 g^{-1}) or in aquatic environments despite their frequent detection in wastewater (10^3 to 10^4 mL^{-1}) (Havelaar et al., 1986; Gerba, 2006). Further research is needed to clarify if their consistently higher concentrations in sewage relative to feces are the result of direct environmental input or multiplication. If the latter is true, F-RNA bacteriophages may not be acceptable fecal pollution indicators (Havelaar et al., 1990). Jofre et al. (2011) suggested that the environmental multiplication of these phages is unlikely, however, because F-pili production only occurs at temperatures above 25ºC and replication does not occur in nutrient-poor environments and requires a minimum host density of 10^4 colony forming units (cfu) per mL.

The presence of F-RNA phage in high numbers in wastewater and their resistance to chlorination contribute to their usefulness as process indicators, indices of sewage pollution, and conservative models of human viruses in water and shellfish (Havelaar et al., 1993; Havelaar, 1993; Love and Sobsey, 2007). They are also promising in microbial source tracking since they can be subdivided in four antigenically distinct serogroups. Because those predominating in humans (groups II and III) differ from those predominating in animals (groups I and IV), it is possible to distinguish between human (higher public health risk) and animal wastes by serotyping or genotyping F-RNA coliphage isolates (Hsu et al., 1995; Hsu et al., 1996; Scott et al., 2002).

F-RNA bacteriophages MS2 and f2 (Table 4) are morphologically similar to enteroviruses and are frequently used to study viral resistance to environmental stressors, disinfection and other treatment processes (Havelaar, 1986, Havelaar et al., 1993; WHO, 2004). These phage have been shown to attach poorly to soil particles and survive relatively well in groundwater (Goyal and Gerba, 1979; Yates et al., 1985; Powelson et al., 1990). As a result, Havelaar (1993) described F-RNA phage as a "worst case" virus model for virus transport in soil. Bacteriophage transport in the subsurface is reviewed in section 5 of this chapter.

Together, somatic and F-specific bacteriophages counts in water samples are usually designated as "total coliphage count". Some bacterial strains can be used to enumerate both simultaneously (Guzmán et al., 2008). Their enumeration may be a good alternative for determination of viral contamination in poorly contaminated waters such as groundwater and drinking water or in double disinfection water treatments (Lucena and Jofre, 2010).

3.3 Bacteriophages of *Bacteroides fragilis*

Bacteriophages of *Bacteroides fragilis* and other *Bacteroides* species rank third in abundance in natural waters. They have been suggested as potential indicators of human viruses in the environment by Tartera and Jofre (1987). Their host *Bacteroides fragilis* is a strict anaerobic bacterium abundant in human feces. These bacteriophages attach to the host bacteria cell wall and have narrow host range. They occur only in human feces (10^8 g^{-1}) and in environmental samples contaminated with human fecal pollution (Havelaar et al., 1986). Consequently they are useful in microbial source tracking, helping to differentiate human from animal contamination (Ebdon et al. 2007; Lucena and Jofre, 2010). In contrast with other phage they are absent from natural habitats and unable to multiply in the environment (Tartera et al., 1989). They also decay in the environment at a rate similar to that of enteric viruses. The main drawbacks associated with their use as routine fecal indicators, are that: (1) their host is a strict anaerobe requiring complex and tedious cultivation methodology, (2) their numbers in water may be low requiring concentration from large volumes, and (3) different hosts are needed for different geographic areas. Within this group, the most commonly used bacteriophages in environmental and treatment resistance studies are B40-8 and B56-3 (Lucena and Jofre, 2010).

4. Available methodology for bacteriophage detection, enumeration and propagation

Relatively simple and reliable methods for detection, isolation, enumeration and characterization of bacteriophages from natural sources are available in the literature. These include classic culture-based techniques using liquid or solid bacteriological media, as well as more recent physico-chemical, immunological, immunofluorescence, electron microscopy, and molecular methods. However, a lack of methodology standardization and quality control has for decades limited the use of phage data for comparison studies. This situation has improved since the publication of standardized plaque assays and presence/absence methods in the USA and Europe. For somatic coliphages (APHA, EWWA, and WEF, 2005; EPA, 2001a; 2001b), F-specific RNA phages (ISO, 1995; ISO, 2000; EPA, 2001a; 2001b) and bacteriophages infecting *Bacteroides fragilis* (ISO, 2001).

Sobsey et al. (1990) developed a simple, inexpensive and practical procedure for the detection and recovery of F-RNA bacteriophages from low turbidity water using mixed cellulose and acetate filters with 47 mm diameter and 0.45 um pore size. A slightly modified version of this method has shown excellent performance for recovery of somatic and F-specific phages, and bacteriophages of *Bacteroids fragilis* in up to 1L water samples (Mendez et al., 2004). Rapid bacteriophage detection methods involving enrichment steps followed by latex agglutination or bioluminescence (Love and Sobsey, 2007) and molecular approaches have also been developed and recently reviewed by Jofre et al. (2011).

Specific methods for the production of the large-volume, high-titer purified bacteriophage suspensions that are necessary for many types of environmental fate and transport studies were, until very recently, difficult to find in the refereed literature. Given that system chemistry and other surface-related characteristics of phage particles, may substantially contribute to observations of their environmental fate and transport behavior in many types of porous media filtration systems used for water treatment (Pieper et al., 1997; Harvey and Ryan, 2004; Cheng et al., 2007), it is critical to consider the impacts of the propagation/purification protocol on those factors. In response to this need, a selected sequence of rapid, reliable, and cost-effective procedures to propagate and purify high-titer bacteriophage suspensions has recently been proposed (Mesquita et al., 2010). This methodology emphasizes the most important factors required to ensure maximum bacteriophage yields, minimum change on phage particles surface characteristics, and low dissolved organic carbon (DOC) concentration in the final suspensions.

Many of the methods routinely used to quantify microscopic discrete particles such as bacteriophages are known to yield highly variable results arising from sampling error and variations in analytical recovery (i.e., losses during sample processing and errors in counting); thereby leading to considerable uncertainty in particle concentration or $\log_{10}$-reduction estimates (Emelko et al., 2008; 2010; Schmidt et al., 2010). For example, sampling error is substantially greater than analytical error when organisms are present in relatively low concentrations; in these cases, improved sampling (i.e., resulting in counts of approximately 10 or more organisms in a sample or, in some cases, several replicates) substantially contributes to reducing uncertainty. In contrast, when organisms are present in higher and homogeneous concentrations, uncertainty in concentration estimates can be reduced by decreasing analytical errors (Emelko et al., 2008; 2010). Emelko et al. (2010) demonstrated that uncertainty in concentration and removal estimates derived from microbial enumeration data can be addressed when these errors are properly considered and quantified. The development and use of such quantitative approaches is an essential component of strategies (e.g., the monitoring of surrogate parameters/pathogens, experimental design, and data analysis) for better evaluating microorganism transport and fate in source and treated drinking waters.

5. Bacteriophages contribution to predicting pathogen transport in filtration porous media

In the last two centuries a large number of field studies have evaluated the transport of bacteriophages in the subsurface (especially through the vadose zone) at different field sites around the world (Rossi, 1994; Collins et al., 2006; Pieper et al., 1997; Bales et al., 1997; Dowd et al., 1998; Rossi et al., 1998; Sinton et al., 1997; Ryan et al., 1999; Auckenthaler et al., 2002; McKay et al., 2000; Schijven and Hassanizadeh, 2000; Schijven, 2001; Harvey and Harms, 2002; Ryan et al., 2002; Harvey and Ryan, 2004; Blanford et al., 2005; Harvey et al., 2007; Ferguson et al., 2007). PRD-1, MS-2 and ΦX174 have also been extensively used at controlled laboratory conditions to elucidate physicochemical effects on virus transport through a variety of porous media (Bales et al., 1991; Bales et al., 1993; Schulze-Makuche et al., 2003; Zhuang and Jin, 2003; Han et al., 2006; Sadeghi et al., 2011).

Based on existing data, major environmental factors affecting enteric viruses and phage survival and transport through soil, porous media and in groundwater have been identified

(Table 5). Due to the complexity of interactive factors controlling survival and transport there is great variability among study outcomes, however. It is, at present, generally accepted that the main processes for viral removal in water filtration through porous media

Factors	Findings
1.	**Temperature** - a major controlling factor for virus inactivation usually with greater inactivation at temperatures above 20°C. This may be due to more rapid denaturation of viral capsid proteins or potential degradation of extracellular enzymes with increased temperature
2.	**Native Microbial activity** - Inactivation rates have often been reported to be lower in the absence of groundwater bacteria possibly because bacterial enzymes and protozoa may destroy viral capsid protein. However, other studies have found the opposite to be true.
3.	**Moisture content** - Different viruses and phage (MS2 and PRD-1) have been reported to have different inactivation rates in groundwater, saturated, unsaturated and dry soils. Migration seems to increase under saturated flow conditions.
4.	**Nutrients** - addition when native organisms are present seems to determine decreased viral inactivation. Possibly because the nutrients offered protection from inactivation by enzymatic attack or acted as alternate nutrient sources for the native bacteria
5.	**Aerobic and anaerobic condition** - Anaerobic conditions have been shown to slow down poliovirus and coxsackievirus inactivation. It has been suggested this is potentially an interactive factor with the impact of native microorganisms since low oxygen will minimize negative microbial activity.,
6.	**pH** - most enteroviruses are stable over a pH range of 3 to 9, survival may be prolonged at near neutral; low pH favors virus attachment and high pH detachment from soil particles
7.	**Salt species and concentration** - some viruses are protected from inactivation by certain cations: the reverse is also true. Generally increasing the concentration of ionic salts and cation valences enhances virus attachment.
8.	**Association with soil and other particles** – in many cases viral survival is prolonged by attachment to soil, although the opposite has also been observed. Usually virus transport through the soil is slowed or prevented by association with particles. However, attachment to solid surfaces appears to be virus-type-dependent
9.	**Soil properties** – effects on survival are probably related to the degree of virus attachment: greater virus migration is usually observed in coarse-textured soils, while there is a high degree of virus retention by the clay fraction of soil.
10.	**Virus type** – particle-structure may be a deciding factor in attachment/detachment and inactivation by physical, chemical and biological factors.
11.	**Organic matter** (OM) – may protect virus from inactivation or reversibly retard virus infectivity. Soluble OM seems to compete with virus particles for attachment sites on soil.
12.	**Hydraulic conditions** – increasing hydraulic loads and flow rates usually increase virus transport.

Table 5. Major factors determining viral survival and transport in the subsurface and in groundwater (adapted from: Azadpour-Keeley et al., 2003; John and Rose, 2005).

are physio-chemical attachment/detachment and inactivation (Keswick and Gerba, 1980; Yates et al., 1987; Bales et al, 1991; 1997; Gitis et al., 2002; Tufenkji and Emelko, 2011). Virus attachment and inactivation depend on the type virus, as well as on the physico-chemical properties of the water and soil or filtration media grain (Schijven and Hassanizadeh, 2000; Tufenkji and Emelko, 2011). Physical and physico-chemical processes such as advection, dispersion, diffusion, and physico-chemical filtration all contribute to attenuation of virus concentrations (Schijven and Hassanizadeh 2000; Tufenkji and Emelko, 2011). Various physico-chemical forces may be involved in the attachment of viruses to soil or filtration media particles including, hydrogen bonding, electrostatic attraction and repulsion, Van der Waals forces and covalent ionic interaction (Murray and Parks; 1980). Straining (i.e. physical blocking of movement) may come into play in some environments as well (Bradford et al., 2006).

The unsaturated or vadose zone (i.e. the layer between the land surface and the groundwater table) where much of the subsurface contamination originates, passes through, or can be eliminated before it contaminates surface and subsurface water resources has gained particular attention in recent years. In unsaturated conditions, additional and more complex mechanisms are involved in pathogen transport such as: variability in ionic strength , pH and water content, particle capture at the water-gas interface, particle capture at the solid-water-gas interface, and preferential flow or retention in the immobilization zone (Sen, 2011). Biological processes such as growth and decay, active attachment or detachment, survival, random mobility and chemotaxis are also believed to strongly affect virus transport in saturated and unsaturated porous media (Sen, 2011). Less information is available regarding the fate of pathogenic protozoa in the vadose zone (Harvey et al., 1995; Harvey et al., 2002; Hancock et al., 1998; Brush et al., 1999; Harter et al., 2000; Darnault et al., 2004; Davies et al., 2005), however, the physico-chemical processes that affect virus fate and transport also apply to protozoan cysts and oocysts during soil transport, albeit to a different extent (Schijven and Hassanizadeh, 2000).

The growing database of information concerning phage attachment, inactivation and transport behavior in porous media has led to their use as viral surrogates in mathematical models used to describe viral transport within physically or geochemically heterogeneous granular media at environmentally-relevant field scales (Rehmann et al., 1999; Schijven and Hassanizadeh, 2000; Schijven et al., 2000; Bhatacharjee et al., 2002; Schijven et al., 2010). As they continue to improve, such models may become useful tools in decision making related to in public health protection because they may ultimately be incorporated into quantitative microbial risk assessment to: (1) access groundwater vulnerability, especially of highly vulnerable geological settings (i.e. fractured rock aquifers, cross-connecting bore holes, or leaking well cases in sandstone and shale aquifers) in combination with significant sources of contamination (i.e. wastewater treatment plants, septic tanks and animal manure), (2) simulate the transport of viruses from a contamination source at or near the surface to a groundwater abstraction well, and (3) evaluate set back distances from abstraction wells from potential contamination sources for source protection (Schijven et al. 2010).

6. Conclusions and recommendations for future research

Considerable progress has been made in understanding how suitable bacteriophages are as surrogates for pathogenic enteric viruses. As a result, they have become invaluable tools in environmental research and are often successfully used in a variety of applications, namely:

- The use of somatic and F-specific coliphages as indices of water contamination by sewage and as process indicators in the evaluation of drinking water and the efficacy of drinking water treatment processes.
- The use of F-specific bacteriophages as indices and models of human enteric viruses in contaminated water, shellfish and agricultural products and in microbial source tracking.
- The use of particular somatic and F-specific bacteriophages to improve the understanding of the multiple physical, chemical and biological processes affecting biocolloid transport in saturated and unsaturated subsurface environments.
- The use of bacteriophages of B. *fragilis* as indicators of human fecal contamination and in microbial source tracking.

Additional research efforts are needed in the following areas:

- Use of more sensitive and reliable methodologies (i.e. standardized cultural procedures, molecular and other techniques) to minimize the variance between reported and actual numbers of bacteriophages in field and laboratory studies and allow the development of more complete and reliable databases.
- Use of more consistent experimental procedures to reduce variability among researchers' findings. Standardized protocols are required for the preparation (propagation, concentration and purification) of bacteriophages to be used in laboratory and field scale studies, as well the use of phage from well known sources such as the American Type Culture Collection (ATCC) or the Canadian Felix d'Herelle Reference Center for Bacterial Viruses to avoid differences in the viruses themselves.
- Evaluation of the complex interactions of native groundwater organisms with introduced enteric microbes (including enteric bacteriophage) and the environmental factors that influence them.
- Evaluation of the impact of viral structure and surface properties on attachment/detachment and inactivation of virus particles in various environments.
- Improved understanding of the transport and survival of both bacteriophages and pathogenic enteric viruses in surface water and the subsurface is needed; not only at laboratory scale to clarify the generic mechanisms involved, but also at field scale at settings with specific environmental conditions (water matrixes, flow regimes, hydrogeological and filtration media characteristics, etc.) in an attempt to clarify conflicting evidence previously reported on the extent of inactivation and immobilization of viruses by some physico-chemical and biological factors.
- Development of sound databases reflecting the occurrence, persistence and transport of viral particles in natural environments and water treatment systems that can be used to improve mathematical models of microbial fate and transport.
- Development of microbial fate and transport models taking into account the many factors affecting virus fate and transport under various conditions applicable to: improve viral contamination control in specific environments, ensure compliance with current water quality regulations, help in the selection and control of treatment processes and ultimately improve public health protection.
- Further investigation of the usefulness of bacteriophages for source tracking purposes. Taking advantage of the stringent host specificity of some phage groups and the speed, high specificity and sensitivity of molecular detection methods in order to better

characterize sources of contamination in aquatic environments so that appropriate and cost-effective water quality remediation plans can be developed.

In the future, the progress of such applications will reveal the true potential of bacteriophages as viral pathogen surrogates in water and water treatment.

7. References

Abbaszadegan, M; Monteiro, P.; Nwachuku, N.; Alum A. & Ryu H. (2008). Removal of adenoviruses, calicivirus, and bacteriophages by conventional drinking water treatment. *Journal of Environmental Science and Health,* 43 (A): 171-177.

Aboytes R.; DiGiovanni G.D.; Abrams F.A.; Rheinecker C.; McElroy W.; Shaw N. & LeChevallier M.W. (2004). Detection of infectious *Cryptosporidium* in filtered drinking water. *Journal of the American Water Works Association,* 96: 88-98.

Anonymous (2001). Loi sur la qualité de l'environement : règlement sur la qualité de l'eau potable c. Q-2, r.18.1.1. *Gazette Officielle du Québec,* 24: 3561.

Anonymous (2005). Draft Guidelines for Drinking Water Quality Management in New Zealand, Volume 3 - Micro-organisms and Chemical and Physical Determinands.

APHA, EWWA & WEF (2005). Section 9224 C- Male-specific coliphage assay using Escherichia coli. In *American Public Health Association, Standard Methods for the Examination of Water and Wastewater,* 21th ed. APHA, Washington, DC. pp. 9-77-9-78.

Ashbolt, N. J., Grabow, W.O.K. & Snozzi, M. (2001). Indicators of microbial water quality. In WHO *Water Quality Guidelines, Standards and Health.* Fewtrell L. and Bartram J. (eds.). IWA Publishers, London UK.

Auckenthaler, A., Raso, G. & Huggenberger, P. (2002). Particle transport in a karst aquifer: natural and artificial tracer experiments with bacteria, bacteriophages and microspheres. *Water Science and Technology,* 46 (3): 131–138.

AWWA (2006). *Waterborne Pathogens.* AWWA Manual M48. 2nd edition. Published by the American Water Works Association.

Azadpour-Keeley A., Faulkner B.R. & Chen J-S. (2003). Movement and longevity of viruses in the subsurface. US-EPA.

Betancourt W. Q. & Rose J. B, (2004). Drinking water treatment processes for removal of Cryptosporidium and Giardia. *Veterinary Parasitology* 126(1-2): 219-234.

Bales R.C., Hinkle, S.R., Kroeger, T.W., Stocking K. & Gerba C.P. (1991). Bacteriophage adsorption during transport through porous media: chemical perturbations and reversibility. *Environmental Science and Technology,* 25: 2088-2095.

Bales R.C.; Li S.; Maguire K.M.; Yahya M.T. & Gerba C.P. (1993). MS-2 and poliovirus transport in porous media: hydrophobic effects and chemical perturbations. *Water Resources Research,.* 29, 957–963.

Bales R.C., Li S.M. Yeh T.C.J., Lenczewski M.E. & Gerba C.P. (1997). Bacteriophage and microsphere transport in saturated porous media: forced gradient experiment at Borden, Ontario. *Water Resources Research,* 33(4): 639-648.

Bateman K., Coombs, P. Harrison H., Milodowsky A.E., Noy D., Van C.H. Wagner D. & West J.M. (2006). Microbial transport and microbial indicators of mass transport through geological media – A literature survey. In A Geological Background and Planning for any Area. British Geological Survey. Internal report IR 06/029. Natural Environment Research Council. Keyworth, Nothingham, UK.

Bhattacharjee S., Ryan J.N., & Elimelech M. (2002). Virus transport in physically and geochemically heterogeneous subsurface porous media. *Journal of Contaminant Hydrology*, 57: 161-187.

Blanford, W.J., Brusseau M.L., Yeh T.C.J., Gerba C.P. & Harvey R. (2005). Influence of water chemistry and travel distance on bacteriophage PRD-1 transport in a sandy aquifer. *Water Research*, 39: 2345-2357.

Borchardt M.A., Bertz P.D., Spencer S., & Battigelli D.A. (2003). Incidence of enteric viruses in groundwater from household wells in Wisconsin. *Applied Environmental Microbiology*, 69 (2), 1172-1180.

Bradford S.A., Tadassa Y.F., & Jin Y. (2006). Transport of coliphage in the presence and absence of manure suspension. *Journal of Environmental Quality*, 35(5):1692-701.

Brush C. F., Ghiorse W.C., Anuish L.J., Parlange J.-Y. & Grimes H.G.(1998). Transport of Cryptosporidium parvum oocysts through saturated columns. *Journal of Environmental Quality*, 28: 809-815.

CDC (2003). Guideline for Environmental Infection Control in Health-Care Facilities, Communicable Disease Control Advisory Committee. USA.

Cheng L., Chetochine A.S., Pepper I. L. & Brusseau M. L. (2007). Influence of DOC on MS2 bacteriophage transport in a sandy soil. *Water Air and Soil Pollution*, 178 (1-4):315-322.

Collins, K.E., Cronin, A.A., Rueedi, J., Pedley, S., Joyce, E., Humble, P.J. & Tellam, J.H. (2006). Fate and transport of bacteriophage in UK aquifers as surrogates for pathogenic viruses. *Engineering Geology*, 85 (1/2), 33-38.

Chung H. & Sobsey M.D. (1993). Comparative survival of indicator viruses and enteric viruses in seawater and sediment. *Water Science and Technology*, 27, 425-429.

Darnault, C. J. G., Steenhuis T. S., Garnier P., Kim Y.-J., Jenkins M., Ghiorse W. C., Baveye P. C., & Parlange J.-Y. (2004). Preferential flow and transport of Cryptosporidium parvum oocysts through the vadose zone: experiments and modeling. *Vadose Zone Journal*, 3: 262-270.

Davies, C. M., Ferguson C. M., Kaucner C., Altavilla N. & Deere D. A., (2005). Fate and transport of surface water pathogens in watersheds. Cooperative Research Centre for Water Quality and Treatment (Australia) and AWWA. AWWA Research Foundation. Denver, CO, USA, 261pp.

Davies-Colley, R.J., Craggs, R.J., Park, J., Sukias, J.P.S. Nagels J.W. & Stott, R. (2005). Virus removal in a pilot-scale 'advanced' pond system as indicated by somatic and F-RNA bacteriophages. *Water Science and Technology*, 51 (12), 107-110

Dowd S.E., Pillai S.D., Wang S. & Corapcioglu M.Y (1998). Delineating the specific influence of virus isoelectric point and size on virus adsorption and transport through sandy soils. *Applied Environmental Microbiology*, 64(2): 405–410.

Durán A.E., Muniesa M., Mocé-Llivina L., Campos C., Jofre J. & Lucena F. (2003). Usefulness of different groups of bacteriophages as model microorganisms for evaluating chlorination. *Journal of Applied Microbiology*, 2003, 95, 29–37

Ebdon J., Muniesa, M. & Taylor, H. (2007). The application of a recently isolated strain of Bacteroides (GB-124) to identify human sources of fecal pollution in a temperate river catchment. *Water Research*, 41: 3683-3690.

Emelko, M.B., Schmidt, P.J., & Reilly, P.M. (2010). Microbiological Data: Enabling Quantitative Rigor and Judicious Interpretation. *Environmental Science and Technology*, 44:461-472.

Emelko, M.B., Schmidt, P.J., & Roberson, J.A. (2008). Quantification of Uncertainty in Microbial Data: Reporting and Regulatory Implications. *Journal of the American Water Works Association*, 100(3)94-104.

Erickson M. C. & Ortega Y.R. (2006). Inactivation of protozoan parasites in food, water and environmental systems. *Journal of Food Protection,* 69: 2786-2808.

EU (2000). Directive (2000/60/EC) of the European of the Parliament and the Council of 23 de October 2000 establishing a framework for Community action in the field of water policy. *Journal of the European Union.*

EU (2006). Directive (2006/7/EC) of the European of the Parliament and the Council of 15 February 2006 concerning the management of bathing water quality and repealing Directive 76/160/EEC. *Journal of the European Union.*

Ferguson C.M, Davies C.M., Kaucner C., Krogh M., Rodehutskors M.K., Deere D.A. & Ashbold N. J. (2007). Field scale quantification of microbial transport from bovine faeces under simulated rainfall events. *Journal of Water and Health,* 5(1): 83-95.

Fong T-T. & Lipp E.K. (2005). Enteric Viruses of Humans and Animals in Aquatic Environments: Health Risks, Detection, and Potential Water Quality Assessment Tools. *Microbiology and Molecular Biology Reviews,* 69(2): 357-371.

Gerba C.P. (1987). Phage as indicators of fecal pollution, in Goyal S.M., Gerba C.P. and Bitton G. (eds.), *Phage Ecology.* Wiley. New York, USA, pp197-210.

Gerba C. P. (2006). Bacteriophage as pollution indicators. In Calender R. (ed.) *The Bacteriophages.* Second edition. Oxford University Press. New York, USA.

Gerba C .P. & Bitton G. (1984). Microbial pollutants: their survival and transport pattern to groundwater, p. 65-68 In G . Bitton and Gerba C .P. (eds.). *Groundwater pollution microbiology,* John Wiley & Sons Inc., New York, USA.

Gerba C .P. & Rose J .B. (1990). Viruses in source and drinking water, p. 380-396. In McFeters G.A . (ed.), *Drinking water microbiology: progress and recent development,* Springer-Verlag, New York, USA.

Gitis V., Adin A., Nasser A., Gun J. & Lev O. (2002). Fluorescent dye labeled bacteriophages – as new tracer for the investigation of viral transport in porous media : 1. Introduction and characterization. *Water Research,* 36(17): 4227-4234

Goyal S.M. 1983. Indicators of viruses in Berg, G. (ed), *Viral Pollution of the Environment,* CRC Press Inc., Boca Raton, FL, USA.

Goyal S.M. & Gerba C.P. (1979). Comparative adsorption of human enteroviruses, simian rotavirus and selected bacteriophages to soils. *Applied Environmental Microbiology,* 38:241-247.

Grabow,1996. Waterborne diseases: Update on water quality assessment and control. *Water S.A.,* 22: 193-202.

Guzmán C., Mocé-Levina L., Lucena F. & Jofre J. (2008). Evaluation of *Escherichia coli* host strain (CB390) for simultaneous detection of somatic and F-specific coliphages. *Applied Environmental Microbiology,* 74: 531-534.

Han, J., Jin Y. & Willson C.W. (2006). Effect of ionic strength and surface hydrophobicity on MS-2 and ΦX174 transport through saturated and unsaturated columns. *Environmental Science and Technology,* 40: 1547-1555.

Hancock, C.M., Rose, J.B. & Callahan, C.M. 1998. *Crypto* and *Giardia* in US groundwater. *Journal of the American Water Works Association,* 90 (3): 58-61.

Harter, T., Wagner, S. & Atwill, E.R. 2000. Colloid transport and filtration of *Cryptosporidium parvum* in sandy soils and aquifer sediments. *Enviromental Science and Technology,* 34(1): 62-70.

Harvey R.W., Kinner N.E., Bunn A., MacDonald D. & Metge D.W. (1995). Transport behavior of groundwater protozoa and protozoan-sized microspheres in sandy aquifer sediments. *Applied Environmental Microbiology,* 61:209-17.

Harvey R.W. & Harms H. (2002). Tracers in groundwater: use of microorganisms and microspheres, in Bitton, G. (ed), *Encyclopedia of Environmental Microbiology*, John Wiley & Sons, New York, USA, pp 3194-3202.

Harvey R.W., Mayberry N., Kinner N.E., Metge D.W. & Novarino F. (2002). Effect of growth conditions and staining procedure upon the subsurface transport and attachment behaviors of a groundwater protist. *Applied Environmental Microbiology*, 68:1872–1881.

Harvey R.W. & Ryan J.N. (2004). Use of PRD1 bacteriophage in groundwater viral transport, inactivation, and attachments studies: *FEMS Microbiology Ecology*, 49 (1), 3-16.

Harvey R.W., Harms H. & Landkamer L. (2007). Transport of microorganisms in the terrestrial subsurface: In situ and laboratory methods, in Hurst C.J., Crawford R.L., Garland J.L., Lipson D.A., Mills A.L., and Stetzenback L.D. (eds.), *Manual of Environmental Microbiology*. 3rd Edition, Washington, ASM Press, pp. 872-897.

Havelaar, A.H. (1986). *F-specific Bacteriophages as Model Viruses in Water Treatment Processes*. Ph.D. thesis. University of Utrecht, Utrecht, Netherlands.

Havelaar, A.H., Furuse K. & Hogeboom W.M. (1986). Bacteriophages and indicator bacteria in human and animal feces. *Journal of Applied Bacteriology*, 60: 255-262.

Havelaar A.H. (1987). Bacteriophages as model organisms in water treatment. *Microbiological Sciences*, 4(12): 362-364.

Havelaar A.H., Pot-Hogeboom W.M., Furuse K., Pot R. & Hormann M. P. (1990). F-specific RNA bacteriophages and sensitive host strains in faeces and wastewater of human and animal origin. *Journal Applied Microbiology*, 69(1): 30-37.

Havelaar A.H. (1993). Bacteriophages as models of enteric viruses in water treatment processes. *American Society for Microbiology News*, 59: 614-619.

Havelaar A.H., Vanophen M. & Drost Y.C. (1993). F-specific RNA bacteriophages are adequate model organisms for enteric viruses in freshwater. *Applied Environmental Microbiology*, 59 (9): 2956-2962.

Health Canada (2004). *Guidelines for Canadian Drinking Water Quality: Supporting Documentation*. Protozoa: *Giardia* and *Cryptosporidium*. Enteric viruses. Prepared by the Federal-Provincial- Territorial Committee on Drinking Water. Ottawa, Ontario, Canada.

Hsu F.C., Shieh Y.S., Van Duin J., Beekwilder M.J. & Sobsey M.D. (1995). Genotyping male specific RNA coliphages by hydridization with oligonucleotide probes. *Appied Environmental Microbiology*, 61 (11): 3960-3966.

Hsu F.-C., Chang A., Amante A., Shieh Y.-S.C ., Wait D. & Sobsey M.C. (1996). Distinguishing human from animal feacal contamination in water by typing male-specific RNA coliphages. *AWWA Water Technology Conference* 1996.

IAWPRC Study Group on Health related water microbiology (1991). Bacteriophages as model viruses in water quality control. *Water Research*, 25: 529-545.

ISO (1995). ISO 10705-1: *Water quality: Detection and enumeration of bacteriophages*. Part 1: Enumeration of F-specific RNA bacteriophages. International Organization for Standardization, Geneva, Switzerland.

ISO (2000). ISO 10705-2 *Water quality: Detection and enumeration of bacteriophages*. Part 2: Enumeration of somatic coliphages. International Organization for Standardization, Geneva, Switzerland.

ISO (2001). ISO 10705-4 *Water quality: Detection and enumeration of bacteriophages*. Part 4: Enumeration of bacteriophages infecting *Bacteroides fragilis*. International Organization for Standardization, Geneva, Switzerland.

Jofre J. (2009). Is the replication of coliphages in water environments significant? *Journal of Applied Microbiology*, 106(4):1059-69.

Jofre J., Stewart J. R. & Grabow W. G. (2011). Phage Methods, in Hagedorn C. Blanch A.R and Harwood V.J. (eds.), *Microbial Source Tracking: Methods, Applications, and Case Studies*, 1st edition. Springer, NY, USA.

John D.E. & Rose J.B. (2005). Review of factors affecting microbial survival in groundwater. *Enviromental Science and Technology,*. 39 (19): 7345-7356.

Karanis P., Papadopoulou C., Kimura A., Economou E., Kourenti C., & Sakkas H. (2002). *Cryptosporidium* and *Giardia* in natural, drinking and recreational water of northwestern Greece. *Acta Hydrochimica et Hydrobiologica*, 30: 49-58.

Keswick, B.H. & Gerba C.P. (1980). Viruses in groundwater. *Environmental Science and Technology*, 14: 1290-1297.

Keswick, B.H., Gerba C.P., Secor S.L. & Cech I. (1982). Survival of enteric viruses and indicator bacteria in groundwater. *Journal of Environmental Science and Health*, A17: 903-912.

LeChevallier M.W., Norton W.D. & Lee R.G. (1991). Giardia and Cryptosporidium spp. in filtered drinking water supplies. *Applied Environmental Microbiology*, 57(9): 2617-262.

Leclerc H., Edberg S., Pierzo V. & Delattre J.M. (2000). Bacteriophages as indicators of enteric viruses and public health risk in groundwaters. *Journal of Applied Microbiology*, 88 (5): 5–21.

Love D. & Sobsey M.D. (2007). Simple and rapid F+ coliphage culture, latex agglutination, and typing assay to detect and source track fecal contamination. Applied Environmental Microbiology, 73 (13): 4110-4118.

Lucena F. & Jofre J. (2010). Potential use of bacteriophages as indicators of water quality and wastewater treatment processes, in Sabour P. M and Griffiths M.W. (eds), *Bacteriophages in the Control of Food and Waterborne Pathogen*, ASM Press, New York.

Lucena, F., Ribas, F., Duran, A.E., Skraber, S., Gantzer, C., Campos,C., Morón, A., Calderón, E. & Jofre, J. (2006). Occurrence of bacterial indicators and bacteriophages infecting enteric bacteria in groundwater in different geographical areas. *Journal of Applied Microbiology*, 101 (1): 96-102.

McGrath S. & van Sinderen D. (eds) (2007). *In* Bacteriophage: Genetics and Molecular Biology . First edition. Caister Academic Press, Cork, Ireland.

McKay L.D., Sanford W.E. & Strong J.M. (2000). Field-scale migration of colloidal tracers in a fractured shale saprolite: *Ground Water*, 38: 139-147.

Melnick J .L. & Gerba C .P. (1980). Viruses in water and soil. *Public Health Rev.* 9: 185-213.

Méndez, J., Audicana A., Isern A., Llaneza J., Moreno B., Tarancón M. L., Jofre J. & F. Lucena. 2004. Standardized evaluation of the performance of a simple membrane filtration-elution method to concentrate bacteriophages from drinking water. *Journal of Virological Methods*, 117: 19-25.

Mesquita M.M.F., Stimson J., Chae G-T, Tufenkji N., Ptacek C.J., Blowes D.W. & Emelko M.B. (2010). Optimal preparation and purification of PRD1-like bacteriophages for use in environmental fate and transport studies. *Water Research*, 44(4): 1114-1125.

Metcalf T.G. (1978). Indicators for viruses in natural waters, in Mitchell D.J. (ed.), *Water Pollution Microbiology*, Vol. 2, Wiley Interscience Inc., NY, USA.

Moce-Llivina, L., Lucena, F. & Jofre, J. (2005). Enteroviruses and bacteriophages in bathing waters. *Applied Environmental Microbiology*, 71 (11), 6838-6844.

Montemayor M., Costan A., Lucena F., Jofre J. Munoz J., Dalmau E., Mujeriego R. & Sala (2008). The combined performance of UV light and chlorine during reclaimed water disinfection. *Water Science and Technology*, 57(6): 35-940.

Morinigo M.A., Wheeler D., Berry C., Jones C. Munoz M.A, Cornax R., Borrego J.J. (1992). Evaluation of different bacteriophage groups as faecal indicators in contaminated natural waters in Southern England. *Water Research*, 26(3): 267.

Murray J.P. & Parks G.A. (1980). Particulates in Water, in Leckie M.C., Kavanaugh M. and Leckie J.O. (eds), *Advances in Chemistry Series 189*, American Chemical Society, Washington DC. USA.

Noonan M.J. & McNabb J.F. (1979). Contamination of Canterbury groundwater by viruses, in Noonan M. (ed), *The Quality and Movement of Groundwater in Alluvial Aquifers of New Zealand*, Technical Publication No 2., Department of Agricultural Microbiology, Lincoln College, Canterbury, New Zealand.

Payment P. & Franco E. (1993). *Clostridium perfringens* and somatic coliphages as indicators of the efficiency of drinking water treatment for viruses and protozoan cysts. *Applied Environmental Microbiology*, 59(8): 2418-2424.

Payment P., Trudel, M. & Plante R. (1985). Elimination of viruses and indicator bacteria each step of treatment during preparation of drinking water at seven water treatment plants. *Applied Environmental Microbiology*, 49(6): 1418–1428.

Payment P., Berte A., Prevost M., Menard B. & Barbeau B. (2000). Occurrence of pathogenic microorganisms in the Saint Lawrence River (Canada) and comparison of health risks for populations using it as their source of drinking water. *Canadian Journal of Microbiology*, 46: 565-576.

Payment P. & Locas A. (2011). Pathogens in water: value and limits of correlation with microbial indicators. *Ground Water* 49(1): 4-11.

Persson F., Langmark, J., Heinicke, G., Hedbergb, T., Tobiasonc, J., Stenstro, T. & Hermanssona, M. (2005). Characterization of the behavior of particles in biofilters for pre-treatment of drinking water. *Water Research*, 39 (16): 3791-3800.

Pieper A.P., Ryan J.N., Harvey R.W., Gary L.A., Illangasekore T.H. & Metge D.W. (1997). Transport and recovery of bacteriophage PRD1 in a sand and gravel aquifer: effect of sewage derived organic matter. *Environmental science and Technology*, 31: 1163-1170.

Pina S., Puig M., Lucena F., Jofre J., & Girones R.. (1998). Viral pollution in the environment and in shellfish: human adenovirus detection by PCR as an index of human viruses. *Applied Environmental Microbiology*, 64: 3376-3382.

Powelson D.K., Simpson G.R. & Gerba C.P. (1990). Virus transport and survival in saturated and unsaturated flow through soil columns. *Journal of Environmental Quality*, 19: 396-401

Powelson, D. K. & Gerba C. P. (1994). Virus removal from sewage effluents during saturated and unsaturated flow through soil columns. *Water Research*, 28: 2175-2181.

Rose, J.B. 1997. Environmental ecology of *Cryptosporidium* and public health concerns. *Annual Review of Public Health*, 18: 135-161.

Rehmann L.L.C., Welty C. & Harvey R. W. (1999). Stochastic analysis of virus transport in aquifers . *Water Resources Research*, 35(7): 1987-2006.

Rice E.W., Fox K.R., Miltner R.J., Lytle D.A. & Johnson C.H. (1996). Evaluating plant performance with endospores. *Journal of the American Water Works Association*, 88(9): 122-130.

Rossi, P. (1994). *Advances in Biological Tracer Techniques for Hydrology and Hydrogeology using Bacteriophages: Optimization of the Methods and Investigation of the Behaviour of Bacterial Viruses in Surface Waters and in Porous and Fractured Aquifers*. Ph.D Thesis. University of Neufchâtel, Neufchâtel, Switzerland.

Rossi P., Doerfliger N., Kennedy K., Müller I. & Aragno M. (1998). Bacteriophages as surface and ground water tracers. *Hydrology and Earth System Sciences,* 2: 101-110.

Ryan J.N. & Elimelech M. (1996). Colloid Mobilization and Transport in Groundwater. *Colloids and Surfaces* 107: 1-56.

Ryan J.N., Elimelech M., Ard R.A., Harvey R.W. & Johnson P.R. (1999). Bacteriophage PRD1 and silica colloid transport and recovery in an iron oxide-coated sand aquifer. *Environmental Science and Technology,* 33: 63-73.

Ryan J.N., Harvey R.W., Metge D, .Elimelech M., Navigato T. & Pieper A.P. (2002). Field and laboratory investigations of inactivation of viruses (PRD1 and MS2) attached to iron oxide-coated quartz sand. *Environmental Science and Technology,* 36 (11): 2403-2413.

Runge J. & Mann J. (2008). State of the industry report 2008: Charting the course ahead. *Journal of The American water Works Association,* 100(10): 61-74.

Sadeghi G., Schijven J.F., Behrends T., Hassanizadeh S.M., Gerritse J. & Kleingeld P.J. (2011). Systematic study of effects of pH and ionic strength on attachment of phage PRD1. *Ground Water,* 49 (1): 12-19.

Schijven, J.F. (2001). *Virus Removal from Groundwater by Soil Passage - Modeling, Field and Laboratory Experiments*. Ph.D. thesis. University of Delft. Delft, the Netherlands.

Schijven J.F. & Hassanizadeh M.S (2000). Removal of Viruses by Soil Passage: Overview of Modeling, Processes and Parameters. *Critical Reviews in Environmental Science and Technology,* 30(1): 49-127.

Schijven J.F., Hassanizadeh M.S., Dowd S.E. & Pillai S.D. (2000). Modeling Virus Adsorption in Batch and Column Experiments. *Quantitative Microbioogy,* 2 (1): 5-20.

Schijven, J. F. Hassanizadeh, S. M. & de Roda Husman, A. M. (2010). Vulnerability of unconfined aquifers to virus contamination. *Water Research,* 44 (4): 1170-1181.

Schmidt, P.J., Reilly, P.M., & Emelko, M.B. (2010). Quantification of Methodological Recovery in Particle and Microorganism Enumeration Methods. *Environmental Science and Technology,* 44:51705-1712.

Schulze-Makuch D., Guan H. & Pillai S.D. (2003). Effects of pH and geological medium on bacteriophage MS-2 transport in a model aquifer. *Geomicrobiology,* 20: 73-84.

Scott T.M., Rose J.B., Jenkins T.M., Farrah S.R. & Lukasik J. (2002). Microbial source tracking: current methodology and future directions. *Applied Environmental Microbiology,* 68(12): 5796-5803.

Sen T. K. (2011). Processes in pathogenic biocolloidal contaminants transport in saturated and unsaturated porous media: A review. *Water, Air and Soil Pollution,* 216: 239-256.

Sinton, L.W., Finlay R.K., Pang L. & Scott D.M. (1997). Transport of bacteria and bacteriophages in irrigated effluent into and through an alluvial gravel aquifer. *Water Air and Soil Pollution,* 98: 17-42.

Sobsey M.D., Shields P.A., Hauchman F.H., Hazard R.L. & Caton III L.W. (1986). Survival and transport of hepatitis A virus in soils, groundwater and wastewater. *Water Science and Technology,* 18: 97-106.

Sobsey M. D., Schwab K. J. & Hanzel T.R. (1990). A simple membrane filter method to concentrate and enumerate male-specific RNA coliphages. *J. AWWA* 82: 52-59.

Stetler R.E. (1984). Coliphages as indicators of enteroviruses. *Applied Environmental Microbiology,* 48(3): 668-670.

Simpson , J.M.; Santo Domingo J.W. & Reasoner D.J. (2002). Microbial source tracking: state of the science. *Environmental Science and Technology,*. 36(24): 5279-5288.

Tartera C. & Jofre J. (1987). Bacteriophages active against *Bacteroides fragilis* in sewage-polluted waters. *Appl. Environ. Microbiol.* 53(7): 1632-1637.

Tartera C., Lucena F. & Jofre J. (1989). Human origin of Bacteroides fragilis bacteriophages present in the environment. *Applied Environmental Microbiology,* 55(10): 2696–2701.

Thurston-Enriquez J. A., Haas C. N., Jacangelo J., Riley K., & Gerba C. P. (2003). Inactivation of feline calicivirus and adenovirus type 40 by UV radiation. *Applied Environmental Microbiology,* 69: 577-582.

Tufenkji, N. and Emelko, M.B. (2011) Groundwater Pollution: Impacts on Human Health: Fate and Transport of Microbial Contaminants. In Encyclopedia of Environmental Health, J. Nriagu, Ed., Elsevier Publishing Inc.

UNICEF, WHO (2009). *Diarrhoea: Why children are still dying and what can be done.* The United Nations Children's Fund (UNICEF)/ World Health Organization (WHO).

US-EPA (2001a). Method 1601. *Male-specific (F+) and somatic coliphage in water by two-step enrichment procedure.* US-Environmental Protection Agency, Washington, DC.

US-EPA (2001b). Method 1602. *Male-specific (F+) and somatic coliphage in water by single agar layer (SAL) procedure.* US-Environmental Protection Agency, Washington, DC.

US-EPA (1989a, 2002). *Surface Water Treatment Rules* (SWTR). US-Environmental Protection Agency. Washington DC.

US-EPA (2003). *Draft Ultraviolet Disinfection Guidance Manual.* US-EPA report No 815-D-03-007. US-Environmental Protection Agency. Washington DC.

US-EPA (2006). *Ground Water Rule* (GWR). US-Environmental Protection Agency 215-F-06-003. Washington DC.

WHO (2004). Water, sanitation and hygiene links to health: Facts and figures. World Health Organization, Geneva, Switzerland.

WHO (2011). Guidelines for drinking-water quality. 4th edition. World Health Organization, Geneva, Switzerland.

Vaughan J.M., Landry E.F., Thomas M.Z. (1983). Entrainment of viruses from septic tank leach fields through a shallow sandy soil aquifer. *Applied Environmental Microbiolog,*. 45:1474-1480.

Yates M.V., Gerba C.P. & Kelley L.M. (1985). Virus persistence in groundwater. *Applied Environmental Microbiology,* 49 (4):778-781.

Yates M.V., Yates S.R., Wagner J. & Gerba C.P. (1987). Modeling virus survival and transport in the subsurface. *Journal of Contaminants Hydrology,* 1: 329-345.

Zhuang J. & Jin Y. (2003). Virus retention and transport through Al-oxide coated sand columns: effects of ionic strength and composition. *Journal of Contaminants Hydrology,* 60(3-4): 193 – 209.

Part 3

Bacteriophages as Tools
and Biological Control Agents

7

Application of Therapeutic Phages in Medicine

Sanjay Chhibber and Seema Kumari

Department of Microbiology, Basic Medical Sciences Building,
Panjab University, Chandigarh,
India

1. Introduction

For more than half a century, the doctors and clinicians have been relying primarily on antibiotics to treat infectious diseases caused by pathogenic bacteria. However, the emergence of bacterial resistance to antibiotics following widespread clinical, veterinary, and animal or agricultural usage has made antibiotics less and less effective (Fischetti, 2008; Perisien et al., 2008). These days scientists are now facing the threat of superbugs, i.e. pathogenic bacteria resistant to most or all available antibiotics (Livemore, 2004; Fischetti, 2006). During the last 30 years, no new classes of antibiotics have been found, even with the help of modern biotechnology such as genetic engineering. Pharmaceutical companies have mainly focused on the development of new products derived from the known classes of antibiotics (Carlton, 1999; Sulakvelidze et al., 2001) which is a cause of major concern. Thus, exploring alternative approaches to develop antibacterial products is also a worthwhile task, and re-examining the potential of promising older methods might be of value. One of the possible replacements for antibiotics is the use of bacteriophages or simply phages as antimicrobial agents (Shasha et al., 2004; Vinodkumar et al., 2008). Phage therapy involves the use of lytic phages for the treatment of bacterial infections, especially those caused by antibiotic resistant bacteria. In general, there are two major types of phages, lytic and lysogenic. Only the lytic phages (also known as virulent phages) are a good choice for developing therapeutic phage preparations (Sandeep, 2006; Borysowski and Gorski, 2008). The bactericidal ability of phages has been used to treat human infections for years as a complement or alternative to antibiotic therapy (Alisky et al., 1998; Matsuzaki et al., 2005; Kysela & Turner, 2007). Bacteriophages, nature's tiniest viruses and it is estimated that there are about 10^{31} phages on earth making viruses the most abundant life form on earth (Ashelford et al., 2000; Hendrix, 2002; Dabrowska et al., 2005). Bacteriophages not only help in the treatments of bacterial infections in animals and human beings but also used in birds, fishes, plants, food material and biofilm eradication (Flaherty et al., 2000; Goode et al., 2003; Leverentz et al., 2003; Park & Nakai, 2003; Curtin & Donlan, 2006).

2. Benefits of phage therapy over antibiotics

Phages appear to be better therapeutic agents as they have several advantages over traditional antibiotics (Pirisi, 2000; Sulakvelidze et al., 2001; Matsuzaki et al., 2005). Majority of them are summarized in the Table given below.

Bacteriophages	Antibiotics
Phages are highly effective in killing their targeted bacteria i.e., their action is bactericidal	Some antibiotics are bacteriostatic, i.e., they inhibit the growth of bacteria, rather than killing them (e.g., chloramphenicol).
Production is simple and cheap.	Production is complex and expensive.
Phages are an 'intelligent' drug. They multiply at the site of the infection until there are no more bacteria. Then they are excreted.	They are metabolized and eliminated from the body and do not necessarily concentrate at the site of infection.
The pharmacokinetics of bacteriophage therapy is such that the initial dose increases exponentially if the susceptible bacterial host is available. In such cases, there is no need to administer the phages repeatedly.	Repeated doses of antibiotic is required to cure the bacterial disease.
The high selectivity/specificity of bacteriophages permits the targeting of specific pathogens, without affecting desirable bacterial flora which means that phages are unlikely to affect the "colonization pressure" of the patients	Antibiotics demonstrate bactericidal or bacteriostatic effects not only on the cause of bacterial disease, but on all microorganisms present in the body including the host normal microflora.. Thus their non-selective action affects the patient's microbial balance, which may lead to various side effects.
Because of phages specificity, their use is not likely to select for phage resistance in other (non-target) bacterial species	The broad spectrum activity of antibiotics may select for resistant mutants of many pathogenic bacterial species.
Humans are exposed to phages throughout life, and well tolerate them. No serious side effects have been described.	Multiple side effects, including intestinal disorders, allergies, and secondary infections (e.g., yeast infections) have been reported.
Phage-resistant bacteria remain susceptible to other phages having a similar host range.	Resistance to antibiotics is not limited to targeted bacteria.
Phages are found throughout nature. This means that it is easy to find new phages when bacteria become resistant to them. Selecting a new phage (e.g., against phage-resistant bacteria) is a rapid process and frequently can be accomplished in days.	Developing a new antibiotic (against antibiotic resistant bacteria) is a time consuming process and may take several years to accomplish.
Phages may be considered as good alternative for patients allergic to antibiotics.	If patient is allergic to antibiotic, treatment is very difficult

Table 1. Comparison of phages and antibiotics regarding their prophylactic and therapeutic use.

There are also some disadvantages with the phage therapy approach. These include:

- The problem which requires attention is the rapid clearance of phage by the spleen, liver and other filtering organs of reticuloendothelial system (Carlton, 1999). This can be taken care by doing serial passage in mice (Merril et al., 1996) so as to obtain a phage mutant capable of evading the reticuloendothelial system and therefore capable of long circulation in the blood. The minor variations in their coat proteins enable some variants to be less easily recognized by the RES organs, allowing them in the circulation for longer periods than the "average" wild-type phage.
- This therapy can not be used for intracellular bacteria as the host is not available for interaction.
- Theoretically development of neutralizing antibodies against phages could be an obstacle to the use phage therapy in recurrent infections. This needs to be confirmed experimentally. However, in the immunocompromised host where the immune system is depressed such as chronic infections, the phage therapy may work in this situation (Skurnik & Strauch, 2006).
- The shelf life of phages varies and needs to be tested and monitored.
- Phages are more difficult to administer than antibiotics. A physician needs special training in order to correctly prescribe and use phages.

3. Safety of the therapeutic phage preparation

During the long history of using phages as therapeutic agents through Eastern Europe and the former Soviet Union, there has been no report of serious complications associated with their use (Sulakvelidze & Morris, 2001). Phages are extremely common in environment and regularly consumed in foods (Bergh et al. 1989). In fact humans are exposed to phages from birth itself and therefore these constitute the normal microflora of the human body. They have been commonly found in human gastrointestinal tract, skin and mouth, where they are harboured in saliva and dental plaques (Bachrach et al., 2003). Phages are also abundant in environment including saltwater, freshwater, soil, plants and animals and they have been shown to be unintentional contents of some vaccines and sera commercially available in United States (Merril et al., 1972; Geier et al., 1975; Milch & Fornosi, 1975). Phages have high specificity for specific bacterial strains, a characteristic which requires careful targetting (Merril et al. 2003; Bradbury 2004). Therefore, phage therapy can be used to lyse specific pathogens without disturbing normal bacterial flora and phages pose no risk to anything other than their specific bacterial host (Lorch, 1999; Sulakvelidze et al., 2001; Duckworth & Gulig, 2002).

From a clinical standpoint, phage therapy appears to be very safe. Efficacy of natural phages against antibiotic-resistant *Streptococci, Escherichia, Pseudomonas, Proteus, Salmonella, Shigella, Serratia, Klebsiella* (Kumari et al., 2010), *Enterobacter, Campylobacter, Yersinia, Acinetobacter* and *Brucella* are being evaluated by researchers (Matsuzaki et al., 2005). However, in the last few years, modified phages are being explored increasingly, due to the limitations of phage therapy using lytic phages. The safety concerns regarding spontaneously propagating live microorganisms and the inconsistency of phage therapy results in the treatment of bacterial infections specifically induced scientists to explore more controllable phages (Skurnik et al., 2007). Phages can be modified to be an excellent therapeutic agent by directed mutation of the phage genome, recombination of phage genomes, artificial selection of phages *in vivo*, chimeric phages and other rational designs which confer new properties on the phages. These new modified phages have been shown to successfully overcome challenges to earlier phage therapy (Moradpour & Ghasemian, 2011).

As with antibiotic therapy and other methods of countering bacterial infections, endotoxins (lipopolysaccharide) are released by the gram negative bacteria as a component of outer membrane. This can cause symptoms of fever, or in extreme cases, toxic shock (Herxheimer reaction) (Theil, 2004). To address the endotoxin release issue, recombinant phage derived from *P. aeruginosa* filamentous phage Pf3 was constructed by genetic modifications and the results showed that this filamentous phages could be used as effective anti-infection agent (Hagen & Blasi, 2003; Hagens et al. 2004). This phage had the benefit of minimizing the release of membrane associated endotoxins during phage therapy (Parisien et al., 2008). In order not to compromise on the issue of the safe use of therapeutic phage preparation, rigorous characterizations of each phage to be used therapeutically should be done, in particular, especially looking for potentially harmful genes in their genome (Payne & Jensen, 2000; Carlton et al., 2005; Hanlon, 2007; Mattey & Spencer, 2008).

4. Clinical application of bacteriophages

4.1 Whole phage as antimicrobial agents

4.1.1 Phage therapy in Humans

However, although d'Hérelle carried out the first human therapeutic phage trial, the first article documenting phage therapy was on research conducted in Belgium by Bruynoghe and Maisin in 1921. They reported that phages when injected in six patients targeted staphylococcus near the base of cutaneous boils (furuncles and carbuncles), resulted in improvement within 48 hours and reduction in pain, swelling and fever. Merabishvili and workers (2009) used phage cocktail, consisting of exclusively lytic bacteriophages for the treatment of *Pseudomonas aeruginosa* and *Staphylococcus aureus* infections in burn wound patients in the Burn Centre of the Queen Astrid Military Hospital in Brussels, Belgium. The first controlled clinical trial of a therapeutic bacteriophage preparation (Biophage-PA) showed efficacy and safety in chronic otitis because of drug resistant P. aeruginosa in UCL Ear Institute and Royal National Throat, Nose and Ear Hospital, London, UK (Wright et al., 2009). Several clinical trials on phage therapy in humans were reported with the majority coming from researchers in Eastern Europe and the former Soviet Union (Abdul-Hassan et al., 1990; Sulakvelidze et al., 2001). One of the most extensive studies evaluating the application of therapeutic phages for prophylaxis of infectious diseases was conducted in Tbilisi, Georgia, during 1963 and 1964 and involved phages against bacterial dysentery (Babalova et al., 1968). The most detailed English language reports on phage therapy in humans were by Slopek and co workers who published a number of papers on the effectiveness of phages against infections caused by several bacterial pathogens, including multidrug-resistant mutants (Slopek et al., 1983, 1984, 1985; Kucharewicz-Krukowska et al., 1987; Weber-Dabrowska et al., 1987). Phages have been reported to be effective in treating various bacterial diseases such as cerebrospinal meningitis in a newborn (Stroj et al., 1999), skin infections caused by *Pseudomonas, Staphylococcus, Klebsiella, Proteus, E. coli* (Cislo et al., 1987), recurrent subphrenic and subhepatic abscesses (Kwarcinski et al., 1987), Staphylococcal lung infections

((Ioseliani et al., 1980; Kaczkowski et al., 1990), *Pseudomonas aeruginosa* infections in cystic fibrosis patients (Shabalova et al., 1995), eye infections (Proskurov, 1970), neonatal sepsis (Pavlenishvili & Tsertsvadze, 1993), urinary tract infections (Perepanova et al., 1995), and cancer (Weber-Dabrowska et al., 2001). Abdul-Hassan et al. (1990) reported on the treatment of 30 cases of burn-wound associated antibiotic-resistant *Pseudomonas aeruginosa* sepsis. Bandages soaked with 10^{10} phages/ml were applied three times daily. Half of the cases were found to be improved. Markoishvili et al., (2002) reported the use of PhagoBioDerm, the phage impregnated polymer, to treat infected venous stasis skin ulcers. To patients that had failed to respond to other treatment approaches, PhagoBioDerm was applied to ulcers both alone and, where appropriate, in combination with other treatment strategies. Complete healing of ulcers was observed in 70% of the patients. Mushtaq et al., (2005) reported that a bacteriophage encoded enzyme, endosialidase E (endo E) selectively degrades the linear homopolymeric a-2, 8-linked N acetylneuraminic acid capsule associated with the capacity of *E. coli* K1 strain to cause severe infection in the newborn infant. In one of the study, PhagoBioDerm (a wound-healing preparation consisting of a biodegradable polymer impregnated with ciprofloxacin and bacteriophages) was used in three Georgian lumberjacks from the village of Lia who were exposed to a strontium-90 source from two Soviet-era radiothermal generators they found near their village. In addition to systemic effects, two of them developed severe local radiation injuries which subsequently became infected with *Staphylococcus aureus*. Approximately 1 month after hospitalization, treatment with phage bioderm was initiated. Purulent drainage stopped within 2–7 days. Clinical improvement was associated with rapid (7 days) elimination of the *S. aureus* resistant to many antibiotics (including ciprofloxacin), but susceptible to the bacteriophages contained in the PhagoBioDerm preparation (Jikia et al., 2005). Leszczynski and co workers (2006) described the use of oral phage therapy for targeting Methicillin Resistant Staphylococcus aureus (MPSA) in a nurse who was a carrier. She had MRSA colonized in her gastrointestinal tract and also had a urinary tract infection. The result of phage therapy was complete elimination of culturable MRSA (Leszczynski et al., 2006).

4.1.2 Animal trials

In Britain, Smith and Huggins (1982, 1983) carried out a series of excellent, well-controlled studies on the use of phages in systemic *E. coli* infections in mice and then in diarrheic disease in young calves and pigs. Bogovazova et al., (1991) studied the effectiveness of specific phage therapy in non inbred white mice, caused by intraperitoneal injection of *K. pneumoniae* K25053 into the animals. Soothill, (1994) examined the ability of bacteriophage to prevent the rejection of skin grafts of experimentally infected guinea pigs. His findings demonstrated that the phage-treated grafts were protected in six of seven cases, while untreated grafts failed uniformly, suggesting that phage might be useful for the prevention of *P. aeruginosa* infections in patients with burn wounds. Phage therapy has been successfully used to remove *E. coli* 0157:H7 from livestock (Barrow et al., 1998; Kudva et al., 1999; Tanji et al., 2004). One of the most successful studies was carried out by Biswas and coworkers (2002). These workers suggested that a single i.p. injection of 3×10^8 PFU of the phage strain, administered 45 minutes after the bacterial challenge (vancomycin-resistant *Enterococcus faecium* (VRE) was sufficient to rescue 100% of the animals. Even when treatment was delayed to the point where all animals were moribund, approximately 50% of them were rescued by a single injection of the phage. The protective effect of bacteriophage was assessed against experimental *S. aureus* infection in mice. Subsequent intraperitoneal administration of purified ØMR11 (MOI of 0.1) suppressed *S. aureus*-induced lethality. This lifesaving effect coincided with the rapid appearance of ØMR11 in the circulation, which remained at substantial levels until the bacteria were eradicated (Matsuzaki et al. 2003). Benedict & Flamiano, (2004) evaluated the use of bacteriophages as therapy for Escherichia coli-induced bacteremia in mice. This experimental study showed clearly that a single dose of crude phage lysates administered by i.p. injection was enough to rescue bacteremic mice back to normal health after having been challenged with a lethal concentration of *E. coli*. Vinodkumar and co-workers (2005) studied the ability of bacterial viruses to rescue septicemic mice with multidrug resistant (MDR) Klebsiella pneumoniae isolated from neonatal septicemia. A single i.p. injection of 3×10^8 PFU of the phage strain administered 45 minutes after the bacterial challenge rescued 100% of the animals. Wills and colleagues (2005) also demonstrated the efficacy of bacteriophage therapy against *S. aureus* in a rabbit abscess model. 2×10^9 PFU of staphylococcal phage prevented abscess formation in rabbits when it was injected simultaneously with S. aureus (8×10^7 CFU) into the same subcutaneous site. Phage multiplied in the tissues. The sewerage-derived bacteriophage reduced the abscess area and the count of *S. aureus* in the abscess was lowered in a bacteriophage dose dependent way (Will et al., 2005). Marza et al. (2006) reported the treatment of a dog with chronic bilateral otitis external that had consistently grown P. aeruginosa. This infection had failed to be resolved after repeated courses of topical and systemic antibiotics. After inoculation with 400 PFU of bacteriophage into the auditory canal there was a marked improvement in the clinical signs, 27 hours after treatment. Wang et al., (2006) examined the effectiveness of phages in the treatment of imipenem resistant *Pseudomonas aeruginosa* (IMPR-Pa) infection in an experimental mouse model. A single i.p. inoculation of the phage strain ØA392 (MOI > 0.01) at up to 60 min after the bacterial challenge was sufficient to rescue 100% of the animals. The workers demonstrated that the ability of the phage to rescue bacteremic animals was due to the functional capabilities of the phage and not to a non-specific immune effect. McVay and co-workers (2007) examined the efficacy of phage therapy in treating fatal *Pseudomonas aeruginosa* infections in mouse burn wound model. The results showed that a single dose of the *Pseudomonas aeruginosa* phage cocktail could significantly decrease the mortality of thermally injured, *Pseudomonas aeruginosa*-infected mice (from 6% survival without treatment to 22 to 87% survival with treatment) and that the route of administration was particularly important to the efficacy of the treatment, with the i.p. route providing the most significant (87%) protection. Watanabe et al. (2007) examined the efficacy of bacteriophage by using a gut-derived sepsis model caused by *Pseudomonas aeruginosa* in mice. Oral administration of a newly isolated lytic phage strain (KPP10) significantly protected mice against mortality with survival rates, 66.7% for the phage-treated group as compared to 0% survival in saline treated control group. Mice treated with phage also had significantly lower numbers of viable *Pseudomonas aeruginosa* cells and lower level of inflammatory cytokines (tumor necrosis factor alpha TNF-α, interleukin-1ß [IL-1ß], and IL-6) in their blood and different organs such as liver and spleen.

In recent years the phage therapy has received lot of attention due to an increase in the prevalence of antibiotic resistant strains in clinical settings. A numbers of recent experimental studies have proved the efficacy of phages in treating different infections. Chhibber & co workers (2008) had reported the therapeutic potential of phage SS in treating

Klebsiella pneumoniae induced respiratory infection in mice. A single intraperitoneal injection of (MOI of 200) phage (SS) administered immediately after i.n. challenge was sufficient to rescue 100% of animals from *K. pneumoniae*-mediated respiratory infections. The use of lytic bacteriophages to rescue septicemic mice with multidrug-resistant (MDR) *Pseudomonas aeruginosa* infection was evaluated (Vinodkumar et al., 2008). A single i.p. injection of 10^9 PFU of the phage strain, administered 45 min after the bacterial challenge (10^7), was sufficient to rescue 100% of the animals. Malik & Chhibber (2009) investigated the protective effect of *K. pneumoniae*–specific bacteriophage KØ1 isolated from the environment in a mouse model of burn wound infection caused by *K. pneumoniae*. A substantial decrease in the bacterial load of blood, peritoneal lavage, and lung tissue was noted following treatment with the bacteriophage preparation. Recently in other studies, workers have successfully employed well characterized phages to treat burn wound infection induced by *Klebsiella pneumoniae* in mice. In this study, a single dose of phages, intraperitoneally (i.p.) at an MOI of 1.0, resulted in significant decrease in mortality, and this dose was found to be sufficient to completely cure *K. pneumoniae* infection in the burn wound model. Maximum decrease in bacterial counts in different organs was observed at 72 hours post infection (Kumari et al., 2009). Kumari and co- workers (2010) evaluated the therapeutic potential of a well characterized phage Kpn5 in treating burn wound infection in mice as a single topical application of this phage was able to rescue mice from infection caused by *K. pneumoniae* B5055 in comparison to multiple applications of honey and Aloe vera gel (Kumari et al., 2010). Recently, Kumari and co-workers (2011) evaluated the efficacy of silver nitrate and gentamicin in the treatment of burn wound infection and compared it with phage therapy using an isolated and well-characterized *Klebsiella* -specific phage, Kpn5. Phage Kpn5 mixed in hydrogel was applied topically at an MOI of 200 on the burn wound site. The efficacy of these antimicrobial agents was assessed on the basis of percentage survival of infected mice following treatment. The results showed that a single dose of phage Kpn5 resulted in a significant reduction in mortality ($P<0.001$) as compared to daily application of silver nitrate and gentamicin (Kumari et al., 2011).

4.1.3 Phages in the eradication of biofilms

Biofilms are densely packed communities of microorganisms growing on a range of biotic and abiotic surfaces and surround themselves with secreted extracellular polymer (EPS). Many bacterial species form biofilms and it is an important bacterial survival strategy. Biofilm formation is thought to begin when bacteria sense environmental conditions that trigger the transition to life on a surface. The structural and physiological complexity of biofilms has led to the idea that they are coordinated and cooperative groups, analogous to multicellular organisms (Passerini et al., 1992). In humans biofilms are responsible for many pathologies, most of them associated with the use of medical devices. A major problem of biofilms is their inherent tolerance to host defences and antibiotic therapies. Therefore there is an urgent need to develop alternative ways to prevent and control biofilm-associated clinical infections (Azeredo & Sutherland, 2008). Bacteriophages have been suggested as effective antibiofilm agents (Donlan, 2009). Use of indwelling catheters was often compromised as a result of biofilm formation. Curtin and Donlan (2006) investigated if hydrogel-coated catheters pretreated with coagulase negative bacteriophage would reduce *Staphylococcus epidermidis* biofilm formation. In our laboratory, efficacy of bacteriophage was assessed alone or in combination with amoxicillin, for the eradication of biofilm produced by *Klebsiella pneumoniae* B5055 (Bedi et al., 2009). Similarly Verma *et. al.* (2009) also evaluated the efficacy of lytic bacteriophage KPO1K2 alone or in combination with another antibiotic, ciprofloxacin for eradicating the biofilm of *Klebsiella pneumoniae in vitro* (Verma et al., 2009). Despite the efficacy of antibiotics as well as bacteriophages in the treatment of bacterial infections, their role in treatment of biofilm associated infections is still under consideration especially in case of older biofilms. The ability of bacteriophage and their associated polysaccharide depolymerases was investigated to control enteric biofilm formation. The action of combined treatments of disinfectant and phage enzyme as a potentially effective biofilm control strategy was evaluated and the results showed that the combination of phage enzyme and disinfectant was found to be more effective than either of these when used alone (Tait et al., 2002). Since age of biofilm is a decisive factor in determining the outcome of antibiotic treatment, in one recent study, biofilm of *K. pneumoniae* was grown for extended periods and treated with ciprofloxacin and/or depolymerase producing lytic bacteriophage (KPO1K2). The reduction in bacterial numbers of older biofilm was greater after application of the two agents in combination as ciprofloxacin alone could not reduce bacterial biomass significantly in older biofilms (Verma et al., 2010).

4.2 Phage products or phage lysins

With the increasing worldwide prevalence of antibiotic resistant bacteria, bacteriophage endolysins represent a very promising novel alternative class of antibacterial in the fight against infectious disease. Pathogenic bacteria are increasingly becoming resistant to antibiotics. For nearly a century, scientists have attempted to treat bacterial infections with whole phages. Vincent Fischetti (1940) was the first, however, to focus on the deadly weapons, the potent and specific enzymes called lysins produced by these viruses. These lysins create lethal holes in bacterial cell walls. Fischetti has identified lysins that can kill a wide range of Gram-positive pathogenic bacteria, and have proven their effectiveness in both preventing and treating infections in mice, an important step towards their potential application in human disease (Fischetti, 2008). As an alternative to "classic" bacteriophage therapy, in which whole viable phage particles are used, one can also apply bacteriophage-encoded lysis-inducing proteins, either as recombinant proteins or as lead structures for the development of novel antibiotics. Phage endolysins, or lysins, are enzymes that damage the cell walls' integrity by hydrolyzing the four major bonds in its peptidoglycan component (Loessner et al., 1997; Lopez et al., 2004). A number of studies have shown the enormous potential of the use of phage endolysins, rather than the intact phage, as potential therapeutics. The great majority of human infections such as viral or bacterial start at a mucous membrane site (upper and lower respiratory, intestinal, urogenital, and ocular) which are the reservoir for many pathogenic bacteria found in the environment (i.e., pneumococci, staphylococci, streptococci), many of which are reported to be resistant to antibiotics (Young, 1994). Therefore, various animal models of mucosal colonization were used to test the efficacy of phage lysins to kill organisms on these surfaces. An oral colonization model was developed for prevention and elimination of upper respiratory colonization of mice by group A streptococci by using a purified C1 phage lysin C1 (Nelson et al., 2001). Phage lytic enzymes have recently been proposed for the reduction of nasopharyngeal carriage of *S. pneumoniae* (Loeffler et al., 2001, 2003). In both these cases, when the animals were colonized with their respective bacteria and treated with a small amount of lysin specific for the colonizing organism, the animals were found to be free of colonizing bacteria two to five hours after lysin treatment. Group B streptococci are the leading cause of neonatal meningitis and sepsis all over the world. A vaginal model for group B streptococci was established to remove colonization of the vagina and oropharynx of mice with a phage lysin (named PlyGBS). A single dose of PlyGBS significantly reduced bacterial colonization in both the vagina and oropharynx (Cheng et al., 2005). These results support the idea that such enzymes may be used in specific high-risk populations to control the reservoir of pathogenic bacteria and therefore control the disease. These phage enzymes are so efficient in killing pathogenic bacteria that they may be considered as valuable tools in controlling biowarfare bacteria. To determine the feasibility of this approach, Schuch and co workers (2002) identified a lytic enzyme PlyG from the gamma phage that is specific for *Bacillus anthracis*. This approach may be used in post-exposure cases of anthrax, in which individuals can be treated intravenously with PlyG to control the bacilli entering the blood after germination because higher doses of phage lysin or multiple doses will result in nearly 100% protection. Recently, antimicrobial therapy of recombinant Cpl-1, a phage lysin specific for *Streptococcus pneumoniae* was reported to be effective in experimental pneumococcal meningitis using infant Wistar rats (Grandgirard et al., 2008).

5. Phage application in food industry

Food contamination is a serious issue because it results in foodborne diseases. Food contamination can be microbial or environmental, with the former being more common. Meat and poultry can become contaminated during slaughter through cross-contamination from intestinal fecal matter. Similarly, fresh fruits and vegetables can be contaminated if they are washed using water contaminated with animal manure or human sewage. During food processing, contamination is also possible from infected food handlers. Food contamination usually causes abdominal discomfort and pain, and diarrhea, but symptoms vary depending on the type of infection. At the present time, the leading causes of death due to foodborne bacterial pathogens are *Listeria* and *Salmonella*, followed closely by other foodborne pathogens such as *Escherichia coli* (*E. coli* O157:H7, in particular) and *Campylobacter jejunii*. Bacteriophages may provide a natural, non-toxic, safe, and effective means for significantly reducing or eliminating contamination of foods with specific pathogenic bacteria, thereby eliminating the risk, or significantly reducing the magnitude and severity, of foodborne illness caused by the consumption of foods contaminated with those bacteria (Meadet et al., 1999; Atturbury et al., 2003). The effectiveness of phage administration for the control of fish diseases and for food disinfection has also been documented. Nakai and co-workers (1999) and some other workers succeeded in saving the lives of cultured fish challenged by *Lactococcus garvieae* and *Pseudomonas plecoglossicida*, which are fish pathogens (Nakai et al., 1999; Nakai & Park, 2002; Park & Nakai, 2003). The need for control of pathogens during the manufacture of food is reflected by the incidence of foodborne bacterial infections. The use of phage or phage products in food production has recently become an option for the food industry as a novel method for biocontrol of unwanted pathogens, enhancing the safety of especially fresh and ready-to-eat food products (Hagens & Loessner, 2010). Phages were also shown to be effective for the elimination of food poisoning pathogens such as *Listeria monocytogenes* (Leverentz et al., 2003), *Campylobacter jejuni* (Atterbury et al., 2003) and *Salmonella* spp. (Leverentz et al., 2001; Goode et al., 2003) from the surface of foods. The bacterial spot pathogen of tomato plants, *Xanthomonas campestris* pv. *vesicatoria* was successfully controlled with bacteriophage (Flaherty et al., 2000).

6. Phages as antibacterial nanomedicines

Nowadays, apart from phage therapy, phages are also being used for phage display, DNA vaccine delivery, therapeutic gene delivery and bacterial typing Recently whole bacteriophage was constructed by fusing immunogenic peptides to modified coat proteins, which was found to be highly efficient DNA vaccine delivery vehicle (phage-display vaccination). Similarly the other approach has been incorporation of a eukaryotic promoter-driven vaccine gene within the phage genome (phage DNA vaccination) (Clark & March, 2006; Gao et al., 2010). Bacteriophages (phages) have been used for about two decades as tools for the discovery of specific target-binding proteins and peptides, and for almost a decade as tools for vaccine development. Drug-carrying phage represents a versatile therapeutic nanoparticle which because of tailoring of its coat can be equipped with a targeting moiety, and its massive drug-carrying capacity may become an important general targeting drug-delivery platform. In comparison to particulate drug-carrying devices, such as liposomes or virus-like particles, the arrangement of drug that is conjugated in high density on the external surface of the targeted particle is unique. A dense coating of the phage with aminoglycosides and other drugs might produce advantages that have been regarded as challenges in the application of phages as therapeutic agent. Most important issue in this field is the immunogenicity of bacteriophages on *in vivo* administration. This problem can be tackled as it has been shown that drug-carrying phages are hardly recognized by commercial antiphage antibodies and generate significantly lower antiphage antibody titers when used to vaccinate mice (in comparison to 'naked' phages). Filamentous bacteriophages are the workhorse of antibody engineering and are gaining increasing importance in nanobiotechnology because of its nanometric dimentions (Yacoby et al., 2007). Vaks and Benhar (2011) described a new application in the area of antibacterial nanomedicines where antibody targeted, chloramphenicol drug loaded filamentous phage (M13) was used for inhibiting the growth of *Staphylococcus aureus* bacteria. Systemic administration of chemotherapeutic agents, in addition to its anti-tumor benefits, results in indiscriminate drug distribution and severe toxicity. Therefore to solve this problem, Bar and co workers (2008) used targeted anti-cancer therapy in the form of targeted drug-carrying phage nanoparticles. The bacteriophages are also being currently evaluated for their biosensor potential. In a recent study it has been proposed to develop a unique and innovative biosensor based on induced luminescence of captured Biowarfare bacterial agents and organic light emitting diode (OLED) technology. The system would use array of bacteriophage engineered to express fluorescent protein in infected Biowarfare agents (Gooding, 2006). The specificity of the phage provides capture of only targets of interest, while the infection of the bacteria and natural replication of the expressed protein will provide the detection signal. Using novel OLED arrays, a phage array chip can be constructed similar to DNA chips for multianalyte detection.

7. Conclusion

Phage therapy for eliminating multidrug resistant bacteria is gaining importance. The abundance of phages in the environment makes it a relatively simple task to isolate phages against any given pathogen which can be characterized using a series of known protocols. The timescale and costs for the development of a new phage(s) for therapy will be a fraction of those for introducing a new antibiotic. Currently, many pathogenic bacteria have acquired multiple drug resistance, which is a serious clinical problem. Phages, when properly selected, offer the most cost-effective alternative to antibiotics. These have proved to be efficient in bacterial elimination on single application and recently accepted for food treatment as well to counter food contamination during storage. Phages should be essentially free of contaminating bacterial toxin and also capable of evading the clearance by reticulendothelial system. Although some problems remain to be solved, many experts are of the opinion that phage therapy will find a niche in modern Western medicine in the future. Phage lytic enzymes have a broad application in the treatment of bacterial diseases. Whenever there is a need to kill bacteria, phage enzymes may be freely utilized. They may be used not only to control pathogenic bacteria on human mucous membranes, but may find application in the food industry to control disease causing bacteria. Phage lytic enzymes have yet to be exploited. Because of the serious problems of resistant bacteria in hospitals, day care centers, and nursing homes, particularly *staphylococci* and *pneumococci*, such enzymes may be of immediate benefit in these environments.

8. References

Abdul-Hassan, H.S.; El-Tahan, K.; Massoud, B. & Gomaa, R. (1990). Bacteriophage therapy of Pseudomonas burn wound sepsis. *Annals Mediterranean Burn Club*, Vol.3, pp 262-4, ISSN 1592-9566

Alisky, J.; Iczkowski, K.; Rapoport, A. & Troitsky, N. (1998). Bacteriophages show promise as antimicrobial agents. *Journal of Infection*, Vol.36, pp 5-15, ISSN 0163-4453

Ashelford, K.E.; Norris, S.J.; Fry, J.C.; Bailey, M.J. & Day, M.J. (2000). Seasonal population dynamics and interactions of competing bacteriophages and their host in the rhizosphere. *Applied Environmental Microbiology*, Vol.66, pp ISSN 4193 – 4199 1098-533

Atterbury, R.J.; Connerton, P.L.; Dodd, C.E.; Rees, C.E. & Connerton, I.F., (2003). Isolation and characterization of Campylobacter bacteriophages from retail poultry. *Applied Environmental Microbiology*, Vol.69, pp 4511–4518, ISSN 1462-2920

Azeredo, J. & Sutherland, I.W. (2008). The use of phages for the removal of infectious biofilms. *Current Pharmaceutical Biotechnology* Vol.9, No.4, pp 261-6, ISSN 1389-2010

Babalova, E.G.; Katsitadze, K.T.; Sakvarelidze, L.A.; Imnaishvili, N. S.; Sharashidze, T.G.; Badashvili, V. A.; Kiknadze, G. P.; Meipariani, A. N.; Gendzekhadze, N. D.; Machavariani, E. V.; Gogoberidze, K. L.; Gozalov, E. I. & Dekanosidze, N.G. (1968). Preventive value of dried dysentery bacteriophage. *Zhurnal Mikrobiologii Epidemiologii Immunobiologii*, Vol.2, pp 143–145, ISSN 0372-9311

Bachrach, G.; Leizerovici-Zigmond, M.; Zlotkin, A.; Naor, R. & Steinberg, D. (2003). Bacteriophage isolation from human saliva. *Letters in Applied Microbiology*, Vol.36, pp 50-53, ISSN 1365-2672

Bar, H.; Yacoby, I.& Benhar, I. (2008). Killing cancer cells by targeted drug-carrying phage nanomedicines. *BMC Biotechnology*, Vol.8, pp 37, ISSN 1472-6750

Barrow, P.; Lovell, M. & Berchieri, A. (1998). Use of lytic bacteriophage for control of experimental *Escherichia coli* septicemia and meningitis in chickens and calves. *Clinical and Diagnostic Laboratory Immunology*, Vol.5, pp 294 – 298, ISSN 1071-4138

Bedi, M.S.; Verma, V. & Chhibber, S. (2009.) Amoxicillin and specific bacteriophage can be used together for eradication of biofilm of *Klebsiella pneumoniae* B5055. *World Journal of Microbiology and Biotechnology*, Vol.25, pp 1145–1151, ISSN 0959-3993

Benedict, L.R.N.; Flamiano, R.S. (2004). Use of bacteriophages as therapy for *Escherichia coli*-induced bacteremia in mouse models. *Philippines Journal of Microbiology and Infectious Diseases*, Vol.33, No.2, pp 47 – 51, ISSN 0115-0324

Bergh, O.; Borsgeim, G.; Bratbak, S. & Heldal, M. (1989). High abundance of viruses found in aquatic environments. *Nature*, Vol.340, pp 467-468, ISSN 0028-0836

Biswas, B., Adhya, S., Washart, P., Paul, B., Trostel, A., Powell, B., Carlton, R., Merril, C. 2002. Bacteriophage therapy rescues mice bacteremic from a clinical isolate of vancomycin-resistant Enterococcus faecium. *Infection and Immunity*. Vol. 70, pp 204-210, ISSN 1098- 5522

Bogovazova, G.G.; Voroshilova, N.N.;& Bondarenko, V.M. (1991). The efficacy of *Klebsiella pneumoniae* bacteriophage in the therapy of experimental Klebsiella infection. *Zhurnal Mikrobiologii Epidemiologii Immunobiologii*, Vol.4, pp 5 – 8, ISSN 0372-9311

Borysowski, J. & Gorski, A. (2008). Is phage therapy acceptable in the immunocompromised host?. *International Journal of Infectious diseases*, Vol.12, pp 466 – 471, ISSN 1201-9712

Bradbury, J. (2004). "My enemy's enemy is my friend": using phages to fight bacteria. *Lancet*, Vol.363, pp 624 – 625, ISSN 0140-6736

Brussow, H. & Hendrix, R.W. (2002). Phage genomics: small is beautiful. *Cell*, Vol.108 pp 13 -16, ISSN 0092-8674

Bruynoghe, R. &, Maisin J. (1921). Essais de thérapeutique au moyen du bactériophage du Staphylocoque. *Comptes Rendus des Séances et Mémoires de la Société de Biologie*, Vol.85, pp 1120-1, ISSN 0037-9026

Carlton, R. (1999). Phage therapy: past history and future prospects. *Archivum Immunologiae et Therapiae Experimentalis*, Vol.47, No.5, pp 267-274, ISSN 0004-069X

Carlton, R.M.; Noordman, W.H.; Biswas B. et al., (2005). Bacteriophage P100 for control of Listeria monocytogenes in foods: genome sequence, bioinformatic analyses, oral toxicity study, and application. *Regulatory Toxicology and Pharmacology*, Vol.43, pp 301 – 312, ISSN: 0273-2300

Cheng, Q.; Nelson, D.; Zhu, S. & Fischetti, V.A. (2005). Removal of group B streptococci colonizing the vagina and oropharynx of mice with a bacteriophage lytic enzyme. *Antimicrobial Agents and Chemotherapy*, Vol.49, pp 111–117, ISSN 1098-6596

Chhibber, S.; Kaur, S.; & Kumari, S. (2008). Therapeutic potential of bacteriophage in treating *Klebsiella pneumoniae* B5055-mediated lobar pneumonia in mice. *Journal of Medical Microbiology*, Vol.57, pp 1508 -1513, ISSN 0022-2615

Cislo, M.; Dabrowski, M.; Weber-Dabrowska, B. & Woyton, A. (1987). Bacteriophage treatment of suppurative skin infections. *Archivum Immunologiae et Therapiae Experimentalis*, Vol.2, pp 175–183, ISSN 0004-069X

Clark, J.R. & March, J.B. (2006). Bacteriophages and biotechnology: vaccines, gene therapy and antibacterials. *Trends in Biotechnology*, Vol.24, pp 212–218, ISSN 0167-7799

Curtin J. J. & Donlan, R.M. (2006). Using bacteriophages to reduce formation of catheter-associated biofilms by *Staphylococcus epidermidis*. *Antimicrobial Agents and Chemotherapy*, Vol.50, No.4, pp 1268–1275, ISSN 1098-6596

Dabrowska, K.; Swita a-Jelen, K., Opolski, A.; Weber-Dabrowska, B. & Gorski, A. (2005). Bacteriophage penetration in vertebrates. *Journal of Applied Microbiology*, Vol.98, pp 7-13, ISSN 1365-2672

Donlan, R.M. (2009). Preventing biofilms of clinically relevant organisms using bacteriophage. *Trends in Microbiology*, Vol.17, pp 66–72, ISSN 0966-842X

Duckworth, D. & Gulig, P. (2002). Bacteriophage: potential treatment for bacterial infections. *Biodrugs*, Vol 16, pp 57 – 62, ISSN 1179-190X

Fischetti, V.A. (2008). Bacteriophage lysins as effective antibacterials. *Current Opinion in Microbiology*, Vol.11, pp 393 – 400, ISSN 1369-5274

Fischetti, V.A.: Nelson, D. & Schuch, R. (2006). Reinventing phage therapy: are the parts greater than the sum? *Nature Biotechnology*, Vol.24, pp 1508 -1511, ISSN 1087-0156

Flaherty, J. E.; Jones, J.B.; Harbaugh, B .K.; Somodi, G .C. & Jackson, L.E. (2000). Control of bacterial spot on tomato in the greenhouse and field with h-mutant bacteriophages. *Bioscience,* Vol. 35, pp 882–888, ISSN 0006-3568

Gao, J.; Wang, Y.; Liu, Z. & Wang, Z. (2010). Phage display and its application in vaccine design. *Annals of Microbiology*, Vol. 60, pp 13-19, ISSN 1590-4261

Geier, M.R.; Attallah, A.F. & Merril, C.R. (1975). Characterization of *Escherichia coli* Bacterial viruses in commercial sera. *In Vitro,* Vol.11, pp 55 – 78, ISSN 1054-5476.

Goode, D.; Allen, V.M. & Barrow, P.A. (2003). Reduction of Experimental Salmonella and Campylobacter Contamination of Chicken Skin by Application of Lytic Bacteriophages. *Applied and Environmental Microbiology*, Vol.69, pp 5032-5036, ISSN 0099-2240.

Gooding, J.J. (2006). Biosensor technology for detecting biological warfare agents: Recent progress and future trends. *Analytica Chemica Acta*, Vol.559, pp 137-151, ISSN 0003-2670.

Grandgirard, D.; Loeffler, J.M.; Fischetti, V.A. & Leib, S.L. (2008). Phage lytic enzyme cpl-1 for antibacterial therapy in experimental pneumococcal meningitis. *Journal of Infectious Diseases*, Vol.197, pp 1519–1522, ISSN 1537-6613

Hagens, S. & Blasi, U. (2003). Genetically modified filamentous phage as bactericidal agents: a pilot study. *Letters in Applied Microbiology*, Vol.37, pp 318 – 323, ISSN 1472-765X

Hagens, S. & Loessner, M.J. (2010). Bacteriophage for Biocontrol of Foodborne Pathogens: Calculations and Considerations. *Current Pharmaceutical Biotechnology*, Vol.11, pp 58-68, ISSN 1389-2010

Hagens, S.; Habel, A.; von Ahsen, U.; von Gabain, A. & Blasi, U. (2004). Therapy of experimental *Pseudomonas* infections with a nonreplicating genetically modified phage. *Antimicrobial Agents and Chemotheapy*, Vol.48, pp 3817 – 3822, ISSN 1098-6596.

Hanlon, G.W. (2007). Bacteriophages: an appraisal of their role in the treatment of bacterial infections. *International Journal of Antimicrobial Agents*, Vol.30, pp 118–28, ISSN 0924 -8579

Hendrix, R.W. (2002). Bacteriophages: evolution of the majority. *Theoretical Population Biology*, Vol.61, pp 471 – 480, ISSN 0040-5809

Ho, K. (2001). Bacteriophage therapy for bacterial infections. Rekindling a memory from the pre-antibiotics era. *Perspective in Biology and Medicine* , Vol.44, pp 1 -16, ISSN 1529-8795

Inal, J. (2003). Phage therapy: a reappraisal of bacteriophages as antibiotic. *Archivum Immunologiae et Therapiae Experimentalis*, Vol.51, pp 237-244, ISSN 0004-069X

Ioseliani, G.D.; Meladze, G.D.; Chkhetiia, N.S.; Mebuke, M.G. & Kiknadze, N.I. (1980). Use of bacteriophage and antibiotics for prevention of acute postoperative empyema in chronic suppurative lung diseases. *Grudnaia khirurgiia*, Vol.6, pp 63 - 67, ISSN 0017-4866

Jamalludeen, N.; Johnson, R.P.; Shewen, P.E. & Gyles. C. L. (2009). Evaluation of bacteriophages for prevention and treatment of diarrhea due to experimental enterotoxigenic *Escherichia coli* O149 infection of pigs. *Veterinary Microbiology*, 136, pp 135–141, ISSN 0378-1135

Jikia, D.; Chkhaidze, N.; Imedashvili, E.; Mgaloblishvili, I.; Tsitlanadze, G.; Katsarava, R.; Morris, J.G. & Sulakvelidze, A. (2005). The use of a novel biodegradable preparation capable of the sustained release of bacteriophages and ciprofloxacin, in the complex treatment of multidrug-resistant *Staphylococcus aureus*-infected local radiation injuries caused by exposure to Sr90. *Clinical and Experimental Dermatology*, Vol.30, pp 23-26, ISSN 1365-2230.

Kaczkowski, H.; Weber-Dabrowska, B.; Dabrowski, M.; Zdrojewicz, Z. & Cwioro, F. (1990). Use of bacteriophages in the treatment of chronic bacterial diseases. *Wiadomosci Lekarskie*, Vol.43, pp 136–141, ISSN 0860-8865

Kropinski, A.M. (2006). Phage therapy – everything old is new again. *Canadian Journal of Infectious Diseases and Medical Microbiology*, Vol.17, pp 297 - 306, ISSN 1712-9532

Kucharewicz-Krukowska, A. & Slopek S. (1987). Immunogenic effect of bacteriophage in patients subjected to phage therapy. *Archivum Immunologiae et Therapiae Experimentalis*, Vol 35, No. 5, pp 553–61 ISSN 0004-069X

Kudva, I.T.; Jelacic, S.; Tarr, P.I.; Youderian, P. & Hovde, C.J. (1999). Biocontrol of *E.coli* O157 with O157-specific bacteriophages. *Applied and Environmental Microbiology*, Vol.65, pp 3767 - 3773, ISSN 1098-5336

Kumari, S.; Harjai, K. & Chhibber, S. (2009). Efficacy of bacteriophage treatment in murine burn wound infection induced by *Klebsiella pneumoniae*. *Journal of Microbiology and Biotechnology*, Vol.19, No. 6, pp 622 - 628, ISSN 1738-8872

Kumari, S.; Harjai, K. & Chhibber, S. (2010). Topical treatment of *Klebsiella pneumoniae* B5055 induced burn wound infection in mice using natural products. *Journal of Infection in Developing Countries*, Vol.4, No.6, pp 367-377, ISSN 1972-2680

Kumari, S.; Harjai, K.& Chhibber, S. (2011). Bacteriophage versus antimicrobial agents for the treatment of murine burn wound infection caused by *Klebsiella pneumoniae* B5055. *Journal of Medical Microbiology*, 60, pp 205-210. 0022-2615

Kwarcinski, W.; Lazarkiewicz, B.; Weber-Dabrowska, B.; Rudnicki, J.; Kaminski, K. & Sciebura, M. (1994). Bacteriophage therapy in the treatment of recurrent subphrenic and subhepatic abscess with jejunal fistula after stomach resection. *Polski tygodnik lekarski*, Vol.49, pp 535, ISSN 0032-3756

Kysela, D.T. & Turner, P.E. (2007). Optimal bacteriophage mutation rates for phage therapy. *Journal of Theoretical Biology*, Vol. 249, pp 411–421, ISSN 0022-5193

Leszczynski, P.; Weber-Dabrowska, B.; Kohutnicka, M.; Luczak, M. & Gorski, A. (2006). Successful eradication of methicillin-resistant *Staphylococcus aureus* (MRSA)

intestinal carrier status in a healthcare worker--case report. *Folia Microbiolica*, Vol.51, pp 236-8, ISSN 0015-5632

Leverentz, B.; Conway, W. S.; Camp, M. J.; Janisiewicz, W. J.; Abuladze, T., Yang, M.; Saftner, R. & Sulakvelidze, A. (2003). Biocontrol of Listeria monocytogenes on Fresh-Cut Produce by Treatment with Lytic Bacteriophages and a Bacteriocin. *Applied and Environmental Microbiology*, Vol.69, pp 4519-4526, ISSN 1098-5336

Leverentz, B.; Conway, W.S.; Alavidze, Z.; Janisiewicz, W.J.; Fuchs, Y.; Camp, M.J.; Chighladze, E. & Sulakvelidze, A. (2001). Examination of bacteriophage as a biocontrol method for *Salmonella* on fresh-cut fruit: a model study. *Journal of Food Protection*, Vol.64, pp 1116 – 1121, ISSN 0362-028X

Livermore, D.H. (2004). The need for new antibiotics. *Clinical Microbiology and Infection*, Vol.10 (Suppl 4), pp 1 – 9, ISSN 1469-0691

Loeffler, J.M.; Djurkovic, S. & Fischetti, V.A. (2003). Phage Lytic Enzyme Cpl-1 as a Novel Antimicrobial for Pneumococcal Bacteremia. *Infection and Immunity* pp 6199–6204, ISSN 1098-5522

Loeffler, J.M.; Nelson, D. & Fischetti, V.A. (2001). Rapid killing of Streptococcus pneumoniae with a bacteriophage cell wall hydrolase. *Science*, Vol.294, pp 2170–2172, ISSN 1095-9203.

Loessner, M.J.; Maier, S.K.; Daubek-Puza, H.; Wendlinger, G. & Scherer, S. (1997). Three *Bacillus cereus* bacteriophage endolysins are unrelated but reveal high homology to cell wall hydrolases from different bacilli. *Journal of Bacteriology*, Vol.179, pp 2845–2851, ISSN 1098-5530

Lopez, R.; Garcia, E. & Garcia, P. (2004). Enzymes for anti-infective therapy: phage lysins. *Drug Discovery Today*, Vol.1, No.4, pp 469 – 474, ISSN 1740-6773

Lorch, A. (1999). "Bacteriophages: An alternative to antibiotics?" *Biotechnology and Development Monitor*, Vol.39, pp 14 -17, ISSN 0924-9877

Malik, R. & Chhibber, S. (2009). Protection with bacteriophage KØ1 against fatal Klebsiella pneumoniae–induced burn wound infection in mice. *Journal of Microbiology Immunology and Infection*, Vol.42, pp 134-140, ISSN 1684-1182

Markoishvili, K.; Tsitlanadze, G.; Katsarava, R.; Morris, G. & Sulakvelidze, A. (2002). A novel sustained-release matrix based on biodegradable poly (esteramide)s and impregnated with bacteriophages and an antibiotic shows promise in management of infected venous stasis ulcers and other poorly healing wounds. *Internation Journal of Dermatology*, Vol.41, pp 453 – 458, ISSN 0011-9059

Marza, J.; Soothill, J.; Boydell, P. & Collyns, T. (2006). Multiplication of therapeutically administered bacteriophages in *Pseudomonas aeruginosa* infected patients. *Burns*, Vol.32, pp 644 – 646, ISSN 0305-4179

Mathur, M.D.; Bidhani, S. & Mehndiratta, P.L. (2003). Bacteriophage therapy: an alternative to conventional antibiotics. *Journal of Association of Physicians of India*, Vol.51, pp 593 – 596, ISSN 0004-5772

Matsuzaki, S.; Rashel, M.; Uchiyma, J.; Ujihara, T.; Kuroda, M.; Ikeuchi, M.; Fujieda, M.; Wakiguchi, J. & Imai, S. (2005). Bacteriophage therapy: a revitalized therapy against bacterial infectious diseases. *Journal of Infection and Chemotherapy*, Vol.11, pp 211 – 219, ISSN 1437-7780

Matsuzaki, S.; Yasuda, M.; Nishikawa, H.; Kuroda, M.; Ujihara, T.; Shuin, T.; Shen, Y.; Jin, Z Fujimoto, S.; Nasimuzzan, M.D.; Wakiguchi, H.; Sugihara, S.; Sugiura, T.; Koda, S.; Muraoka, A. & Imai. S. (2003). Experimental protection of mice against lethal *Staphylococcus aureus* infection by novel bacteriophage ΦMR11. *Journal of Infectious Diseases*, Vol.187, pp 613 – 624, ISSN 0022-1899

Mattey, M. & Spencer, J. (2008). Bacteriophage therapy - cooked goose or Phoenix rising? *Current Opinion in Biotechnology*, Vol.19, pp 1 – 5, ISSN 0958-1669

McVay, C.; Velasquez, S.M.; & Fralick, J.A. (2007). Phage therapy of *Pseudomonas aeruginosa* infection in a mouse burn wound model. *Antimicrobial Agents and Chemotherapy*, Vol.51, No.6, pp 1934 -1938, ISSN 1098-6596

Mead, P.S.; Slutsker, L.; Dietz, V.; McCaig, L. F.; Bresee, J. S.; Shapiro, C.; Griffin, P. M. & Tauxe, R.V. (1999). Food-related illness and death in the United States. *Emerging Infectious Diseases*, Vol.5, No.5, pp 607-25, ISSN 1080-6059

Merabishvili, M.; Pirnay, J.P.;; Verbeken, G.; Chanishvili, N.; Tediashvili, M.; Lashkhi, N.; Glonti, T.; Krylov, V. et al. (2009) Quality-controlled small-scale production of a well- defined bacteriophage cocktail for use in human clinical trials. PLoS ONE Vol.4, pp e4944, ISSN 1932-6203

Merril, C.R.; Friedman, T.B., Attallah, A.F.M.; Geier, M.R.; Krell, K. & Yarkin, R. (1972). Isolation of bacteriophages from commercial sera. *In Vitro*, Vol.8, pp 91 – 93, ISSN 1071-2690.

Merril, C.R.; Biswas, B.; Carlon, R.; Jensen, N.C.; Creed, G.J.; Zullo, S.; & Adhya, S. (1996). Long-circulating bacteriophage as antibacterial agents. *Proceedings of the National Academy of Sciences of the United States of America* Vol.93, pp 3188 – 3192, ISSN 0027-8424

Merril, C.R.; Scholl, D. & Adhya, S.L. (2003). The prospect for bacteriophage therapy in Western medicine. *Nature Reviews Drug Discovery*, Vol.2, pp 489–497, ISSN 1474-1776

Milch, H. & Fornosi, F. (1975). Bacteriophage contamination in live poliovirus vaccine. *Journal of Biological Standardization*, Vol. 3, pp 307 – 310, ISSN 0092-1157

Moradpour, Z. & Ghasemian, A. (2011). Modified phages: Novel antimicrobial agents to combat infectious diseases. *Biotechnology Advances*, Vol.29, pp 732–738, ISSN 0734-9750

Mushtaq, N.; Redpath, M.B.; Luzia, J.P. & Taylor, P.W. (2005). Treatment of experimental *Escherichia coli* infection with recombinant bacteriophage derived capsule depolymerase. *Journal of Antimicrobial Chemotherapy*, Vol.56, pp 160 -165, ISSN 0305-7453

Nakai, T. & Park, S.C. (2002). Bacteriophage therapy of infectious disease in aquaculture. *Research in Microbiology*, Vol.153, pp 13 – 8, ISSN 0923-2508

Nakai, T.; Sugimoto, R.; Park, K. H.; Matsuoka, S.; Mori, K.; Nishioka, T. & Maruyama, K. (1999). Protective effects of bacteriophage on experimental *Lactococcus graviae* infection in yellow tail. *Diseases of Aquatic Organisms*, Vol.37, pp 33-41, ISSN 0177-5103

Nelson, D.; Loomis, L. & Fischetti, V.A. (2001). Prevention and elimination of upper respiratory colonization of mice by group A streptococci by using a bacteriophage lytic enzyme. *Proceedings of the National Academy of Sciences of the United States of America, Vol.*98, pp 4107–4112, ISSN 0027-8424

Oliveira, A.; Sereno, R. & Azeredo, J. (2010). *In vivo* efficiency evaluation of a phage cocktail in controlling severe colibacillosis in confined conditions and experimental poultry houses. *Veterinary Microbiology*, Vol.146, No.(3-4), pp 303-308, ISSN 0378-1135.

Parisien, A.; Allain, B.; Zhang, J.; Mandeville, R. & Lan, C.Q. (2008). Novel alternatives to antibiotics: bacteriophages, bacterial cell wall hydrolases, and antimicrobial peptides. *Journal of Applied Microbiology*, Vol.104, pp 1 -13, ISSN 1365-2672

Park, S.C., Nakai, T. (2003). Bacteriophage control of Pseudomonas plecoglossicida infection in ayu Plecoglossus altivelis. *Diseases of Aquatic Organisms*, Vol.53, pp 33–39, ISSN 0177-5103

Passerini, L.; Lam, K.; Costerton J.W. & King, E.G. (1992). Biofilms on indwelling vascular catheters. *Critical Care Medicine, Vol.* 20, pp 665-673, ISSN 1530-0293

Pavlenishvili, & Tsertsvadze, T. (1993). Bacteriophagotherapy and enterosorbtion in treatment of sepsis of newborns caused by gram-negative bacteria. *Pren Neon Infection*, Vol.11, pp 104, ISSN 0975-5241

Payne, R.J.H. & Jansen, V.A.A. (2000). Phage therapy: the peculiar kinetics of self replicating pharmaceuticals. *Clinical Pharmacology and Therapeutics*, Vol.68, pp 225-230, ISSN 0009-9236

Perepanova, T.S.; Darbeeva, O.S.; Kotliarova, G.A.; . Kondrat'eva, E.M.; Maiskaia, L.M.; Malysheva, V.F.; Baiguzina, F.A. & Grishkova, N.V. (1995). The efficacy of bacteriophage preparations in treating inflammatory urologic diseases. Urologica e Nefrologica, Vol.5, pp 14-17, ISSN 0393-2249

Pirisi, A. (2000). Phage therapy - advantages over antibiotics? *Lancet*, Vol.356, pp 1418, ISSN 0140-6736

Platt, R.; Reynolds, D.L. & Phillips, G.J. (2003). Development of a novel method of lytic phage delivery by use of a bacteriophage P22 site-specific recombination system". *FEMS Microbiology Letters*, Vol.223, pp 259-265, ISSN 1574-6968

Proskurov, V.A. (1970). Use of staphylococcal bacteriophage for therapeutic and preventive purposes. *Zhurnal Mikrobiologii Epidemiologii Immunobiologii*, Vol.2, pp 104-107, ISSN 0372-9311

Rohwer, F. (2003). Global phage diversity. *Cell,* Vol.113, pp 141, ISSN 0092-8674

Sandeep, K. (2006). Bacteriophage precision drug against bacterial infections. *Current Science*, Vol.90, pp 361 – 363, ISSN 0011-3891

Schuch, R.; Nelson, D. & Fischetti, V.A. (2002). A bacteriolytic agent that detects and kills *Bacillus anthracis. Nature*, Vol 418, pp 884–889, ISSN 0028-0836

Shabalova, I.A.; Karpanov, N.I.; Krylov, V.N.; Sharibjanova, T.O. & Akhverdijan, V.Z. (1995). *Pseudomonas aeruginosa* bacteriophage in treatment of *P. aeruginosa* infection in cystic fibrosis patients, p. 443. *In* Proceedings of IX International Cystic Fibrosis Congress. International Cystic Fibrosis Association, Zurich, Switzerland.

Shasha, S.M.; Sharon, N. & Inbar, M. (2004). Bacteriophages as antibacterial agents. *Harefuah*, Vol.143, pp 121–125, ISSN 0017-7768

Skurnik, M. & Strauch, E. (2006). Phage therapy: facts and fiction. *International Journal of Medical Microbiology*, Vol.296, pp 5–14, ISSN 1438-4221

Skurnik, M.; Pajunen, M. & Kiljunen, S. (2007). Biotechnological challenges of phage therapy. *Biotechnology Letters*, Vol.29, pp 995 – 1003, ISSN 0141-5492

Slopek, S.; Durlakowa, I.; Weber-Dabrowska, B.; Kucharewicz-Krukowska, A.; Dabrowski, M. & Bisikiewicz, R. (1983). Results of bacteriophage treatment of suppurative bacterial infections. I. General evaluation of the results. *Archivum Immunologiae et Therapiae Experimentalis*, Vol.31, pp 267–291, ISSN 0004-069X

Slopek, S.; Kucharewicz-Krukowska, A.; Weber-Dabrowska, B. & Dabrowski M. (1985). Results of bacteriophage treatment of suppurative bacterial infections. V. Evaluation of the results obtained in children. *Archivum Immunologiae et Therapiae Experimentalis*, Vol.33, pp 241–259, ISSN 0004-069X

Slopek, S.; Durlakowa, I.; Weber-Dabrowska, B.; Dabrowski, M. & Kucharewicz-Krukowska, A. (1984). Results of bacteriophage treatment of suppurative bacterial infections. III. Detailed evaluation of the results obtained in a further 150 cases. *Archivum Immunologiae et Therapiae Experimentalis*, Vol.32, pp 317–335, ISSN 0004-069X

Smith H.W. & Huggins M.B. (1983). Effectiveness of phages in treating experimental *Escherichia coli* diarrhea in calves, piglets and lambs. *Jouranl of General Microbiology*, Vol.129, pp 2659 – 2675, ISSN 0022-1287

Smith, H.W. & Huggins, M.B. (1982). Successful treatment of experimental *Escherichia coli* infections in mice using phages: its general superiority over antibiotics *Jouranl of General Microbiology*, Vol.128, pp 307 – 318, ISSN 0022-1287

Soothill J. (1994). Bacteriophage prevents destruction of skin grafts by *Pseudomonas aeruginosa*. *Burns*, Vol.20, pp 209 – 211, ISSN 0305-4179

Stroj, L.; Weber-Dabrowska, B.; Partyka, K.; Mulczyk, M. & Wojcik, M. (1999). Successful treatment with bacteriophage in purulent cerebrospinal meningitis in a newborn. *Neurologia i neurochirurgia polska*, Vol.3, pp 693 – 698, ISSN 0028-3843

Sulakvelidze, A. & Morris, J.G. (2001). Bacteriophages as therapeutic agents. *Annals of Medicine*, Vol.33, pp 507 – 509, ISSN 1365-2060

Sulakvelidze, A.; Alavidze, Z. & Morris, J. (2001). Bacteriophage therapy. *Antimicrobial Agents and Chemotherapy*, Vol.45, pp 649 – 659, ISSN 1098-6596

Tait, K.; Skillman, L.C. & Sutherland, I.W. (2002). The Efficacy of Bacteriophage as a method of biofilm eradication. *Biofouling*, Vol.18, No. 4, pp 305–311, ISSN 0892-7014

Tanji, Y.; Shimada, T.; Fukudomi, H.; Miyanaga, K.; Nakai, Y. & Unno, H. (2005). Therapeutic use of phage cocktail for controlling *Escherichia coli* O157:H7 in gastrointestinal tract of mice. *Journal of Bioscience and Bioengineering*, Vol.100, pp 280–287, ISSN 1389-1723

Tanji, Y.; Shimada, T.; Yoichi, M.; Miyanaga, K.; Hori, K. & Unno, H. (2004). Toward rational control of *Escherichia coli* O157:H7 by a phage cocktail. *Applied Microbiology and Biotechnology*, Vol.64, pp 270 – 274, ISSN 0175-7598

Teuber, M. (2001). Veterinary use and antibiotic resistance. *Current Opinion in Microbiology*, Vol.4, pp 493–499, ISSN 1369-5274

Thacker, P.D. (2003). Set a microbe to kill a microbe: drug resistance renews interest in phage therapy. *Journal of American Medical Association*, Vol.290, pp 3183 – 3185, ISSN 1538- 3598

Thiel, K. (2004). Old dogma, new tricks – 21st century phage therapy. *Nature Biotechnology*, Vol.22, pp 31-36, ISSN 1087-0156

Vaks, L. & Benhar, I. (2011). Antibacterial application of engineered bacteriophage nanomedicines: antibody-targeted, chloramphenicol prodrug loaded bacteriophages for inhibiting the growth of *Staphylococcus aureus* bacteria. *Methods in Molecular Biology*, Vol .72, pp 187-206, ISSN 1064-3745

Verma, V.; Harjai, K. & Chhibber, S. (2009). Restricting ciprofloxacin induced resistant variant formation in biofilm of *Klebsiella pneumoniae* B5055 by complementary bacteriophage treatment. *Journal of Antimicrobial Chemotherapy*, Vol.64, pp 1212–1218, 0305-7453,

Verma, V.; Harjai, K. & Chhibber, S. (2010). Structural changes induced by a lytic bacteriophage make ciprofloxacin effective against older biofilm of *Klebsiella pneumoniae. Biofouling*, Vol.26, No. 6, pp 729-737, ISSN 0892-7014

Vinodkumar, C.S.; Kalsurmath, S. & Neelagund Y.F. (2008). Utility of lytic bacteriophage in the treatment of multidrug-resistant *Pseudomonas aeruginosa* septicemia in mice. *Indian Journal of Pathology and Microbiology*, Vol.5, No. 3, pp 360 – 366, ISSN 0974-5130, ISSN 0377-4929

Vinodkumar, C.S.; Neelagund, Y.F. & Kalsurmath, S. (2005). Bacteriophage in the treatment of experimental septicemic mice from a clinical isolate of multidrug resistant *Klebsiella pneumoniae. Journal of Communicable Diseases*, Vol.37, No. 1, pp 18 – 29, ISSN 0019-5138

Wang, J.; Hu, B.; Xu, M.; Yan, Q.; Liu, S.; Zhu, X.; Sun, Z.; Reed, E.; Ding, L.; Gong, J.; Li, G.Q. & Hu, J. (2006). Use of bacteriophage in the treatment of experimental animal bacteremia from imipenem-resistant *Pseudomonas aeruginosa. International Journal of Molecular Medicine*, Vol. 17, pp 309 - 317, ISSN 1107-3756.

Watanabe, R.; Matsumoto, T.; Sano, G.; Ishii, Y.; Tateda, K.; Sumiyama, Y.; Uchiyama, J.; Sakurai, S.; Matsuzaki, S.; Imai, S. & Yamaguchi. K. (2007). Efficacy of bacteriophage therapy against gut-derived sepsis caused by *Pseudomonas aeruginosa* in mice. *Antimicrobial Agents and Chemotherapy*, Vol 51, No. 2, pp 446 – 452, ISSN 1098-6596

Weber-Dabrowska, B., Dabrowski, M. & Slopek, S. (1987). Studies on bacteriophage penetration in patients subjected to phage therapy. *Archivum Immunologiae et Therapiae Experimentalis*, Vol.35, No.5, pp 563–68, ISSN 0004-069X

Weber-Dabrowska, B.; Mulczyk, M. & Górski, A. (2001). Bacteriophage therapy for infections in cancer patients. *Clinical and Applied Immunology Reviews*, Vol.1, pp 131–134, ISSN 1529-1049

Wills, Q.; Kerrigan, C. & Soothill, J. (2005). Experimental bacteriophage protection against *Staphylococcus aureus* abscesses in a rabbit model. *Antimicrobial Agents and Chemotherapy*, Vol.49, pp 1220 – 1221, ISSN 1098-6596

Wright, A.; Hawkins, C.H.; Anggard, E.E.& Harper, D.R. (2009). A controlled Quality-controlled small-scale production of a well- defined bacteriophage cocktail for use in human clinical trials. clinical trial of a therapeutic bacteriophage preparation in chronic otitis due to antibi- otic-resistant Pseudomonas aeruginosa; a preliminary report of efficacy. Clinical Otolaryngology, Vol.34, pp 349–357, ISSN 1749-4486

Yacoby, I.; Bar, H. & Benhar, I. (2007). Targeted drug-carrying bacteriophages as antibacterial nanomedicines. *Antimicrobial Agents and Chemotherapy*, Vol.51, No. 6, pp 2156–63, ISSN 1098-6596

Young, R. (1992). Bacteriophage lysis: mechanism and regulation. *Microbiology and Molecular Biology Reviews*, Vol. 56, pp 430–481, ISSN 1092-2172.

8

Successes and Failures of Bacteriophage Treatment of Enterobacteriaceae Infections in the Gastrointestinal Tract of Domestic Animals

L.R. Bielke, G. Tellez and B.M. Hargis
Department of Poultry Science, Division of Agriculture,
University of Arkansas,
USA

1. Introduction

Bacteriophages are numerous in the ecosystem and play a central role in bacterial ecology (Ashelford et al., 2000; Brabban et al., 2005; Breitbart et al., 2003; Danovaro et al., 2001; Fuhrman, 1999). Bacteriophages have frequently been isolated from various environmental sources and the gastrointestinal tract of animals (Adams et al., 1966; Bielke et al., 2007a; Breitbart et al., 2003; Callaway et al., 2003, 2006; Dhillon et al., 1976; Filho et al., 2007; Higgins et al., 2008; Klieve & Bauchop, 1988; Klieve & Swain, 1993; Kudva et al., 1999; Orpin & Munn, 1973; Raya et al., 2006; Smith & Huggins, 1982). Breitbart and co-workers (2003) found bacteriophages to be the second most abundant uncultured biological group in their analysis of human feces and Furhman (1999) suggests that bacteriophages could be responsible for as much as 50% of bacterial death in surface waters. It has been suggested that bacteriophages in cattle help maintain microbial diversity and balance, allowing the ecology of the gut, particularly the rumen, to adapt to changes in feed and water intake (Klieve and Swain 1993; Swain et al., 1996). Bacteriophages lytic for *E. coli* and *Salmonella* were isolated from cattle feedlots with no correlation between presence of *E. coli* O157:H7 or *Salmonella* and bacteriophages against the specific pathogen (Callaway et al., 2006). *Salmonella* targeted bacteriophages were isolated from *Salmonella*-positive poultry farms, with bacteriophages found at only one *Salmonella*-negative farm. A total of seven bacteriophages were isolated from farms that were *Salmonella*-positive, and two bacteriophages from the single *Salmonella*-negative chicken house (Higgins et al., 2008). This might suggest that as environmental *Salmonella* increased, a near-simultaneous increase in bacteriophages may have also occurred. The hypothesis corresponds with other reports where it was found that bacteriophages within treated animals remained in the animal for the duration of the infection, but once the bacterial host was no longer present, the presence of bacteriophages also rapidly dropped (Barrow et al., 1998; Calloway et al., 2003; Hurley et al., 2008; Smith and Huggins, 1987).

Because bacteriophages are a natural component of gastrointestinal microbial populations, they are presumably a potentially effective control strategy against bacterial pathogens. However, *in vivo* attempts have yielded mixed results (Bach et al., 2003; Bielke et al., 2007b,

Hurley et al., 2008; Kudva et al., 1999; O'Flynn et al., 2004; Higgins et al., 2007; Smith and Huggins, 1983, 1987; Toro et al., 2005). This chapter will review both successes and failures in research aimed to reduce enterobacterial infections of the gastrointestinal tract. During the last approximately 60 years, there have been sporadic published reports of efficacy in treating Enterobacteriaceae infections systemically and within the gastrointestinal tract. While a number of reports have rather consistently indicated that systemic or tissue-associated infections were treatable by parenteral administration of appropriate bacteriophage cocktails, reports of successful treatment of enteric Enterobacteriaceae are much more sporadic, and are interspersed with a number of reports of failed attempts for enteric treatment. The present chapter will discuss selected successes and failures and describe the possible differences in these studies and the potential for development of more effective strategies.

Bacteriophages can be regarded as natural enemies of bacteria, and therefore are logical candidates to evaluate as agents for the control of bacterial pathogens. Bacteriophages can be selected to kill bacterial pathogen target cells, and not affect desired bacteria such as starter cultures, commensals in the gastrointestinal tract or on skin, or accompanying bacterial flora in the environment. Bacteriophages harbor the potential for precise targeting of bacterial contamination, without compromising the viability of beneficial microorganisms in the habitat. Additionally, since bacteriophages are generally composed entirely of proteins and nucleic acids, the eventual breakdown products consist exclusively of amino acids and nucleotides, and, unlike antibiotics and antiseptic agents, their introduction into and distribution within a given environment may be seen as a natural process. With respect to their potential application for the biocontrol of pathogens, it should be considered that bacteriophages are the most abundant self-replicating units in our environment, and are present in significant numbers in water and foods of various origins, and most surfaces in our environment (Sulakvelidze and Barrow, 2005). A test of the safety of bacteriophages when administered orally to human volunteers revealed no adverse side effects (Bruttin and Brüssow, 2005). Very low levels of bacteriophage were found in the serum, suggesting low passage from the intestinal lumen to the blood flow, liver enzymes were not affected by bacteriophage ingestion, and no antibodies to the bacteriophages were detected. Mai and co-workers (2010) noted that treatment of mice with anti-*Listeria* bacteriophages did not significantly affect gastrointestinal microflora diversity. Additionally, Carlton et al. (2005) reported no adverse effects in rats after five continuous days of oral bacteriophage administration, suggesting that bacteriophages can in fact be regarded as safe.

Prior to the discovery of antibiotics, bacteriophages were researched as bacterial control agents (For a review, see Alisky et al., 1998). However, a lack of understanding of mechanisms resulted in therapeutic difficulties and resulted in poor experimental results. When treating bacterial infections, the goal is to take advantage of the lytic cycle of bacteriophages, rather than the lysogenic cycle in which bacteria are not killed (Figure 1). With an increase in bacterial pathogens that are resistant to traditional antibiotics, the scientific community has developed a renewed interest in using bacteriophages and they are currently being investigated in numerous laboratories and companies as alternative treatments for a variety of problems. Indeed, for some specific applications, bacteriophage therapy holds significant promise, and there is growing evidence that bacteriophage may be effective for some applications, with the caveat that these viruses are incredibly specific by definition, and selection of product for specific applications may be of critical importance.

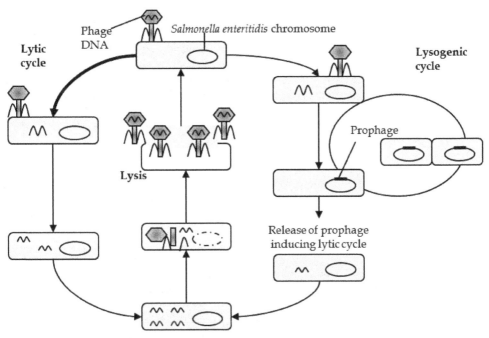

Figure provided by Dr. Jack Higgins.

Fig. 1. **Life Cycle of Bacteriophages.** Bacteriophages are capable of entering either a lytic cycle, in which the bacteriophage inserts its DNA into the host cell and replicates to make multiple copies of the bacteriophage virus before lysing the cell wall, and releasing daughter viruses to repeat the cycle. Bacteriophages can also enter a lysogenic cycle. Instead of using the cell to replicate large numbers of virus particles, the bacteriophage genome can remain in the host's genome and replicate through binary fission of the cell. Some lysogenic bacteriophages are capable of entering into the lytic cycle, while others will remain lysogenic.

2. Successes

The bacteriocidal effects of bacteriophages have long been studied for their usefulness in treating gastrointestinal infections. Early studies originating from the former Soviet Union, Eastern Europe, and Western Asia suggested bacteriophages could prevent and treat *Vibrio cholera* infections (Dubos et al., 1943; Dutta, 1963; Marčuk et al., 1971; Sayamov, 1963). In the 1980s, Slopek and co-workers (1983a-b, 1984, 1985a-c, 1987) published numerous papers showing the promising results of treating septic patients with bacteriophages. While the validity of these studies has been questioned, in part due to relaxed scientific rigor in these regions during the time when these studies were completed (Alisky et al, 1998; Merril et al., 2003) and are not often cited by bacteriophage researchers in recent years, they have served as an inspiration for continued research into the possibility that bacteriophages can cure gastrointestinal diseases in humans and animals.

Some bacteriophage research has also focused on the treatment of animals to cure a variety of diseases. In relatively recent years, Smith and Huggins (1982) compared the efficacy of bacteriophages with that of antibiotics in treating both generalized and cerebral infections in mice. They isolated anti-K1 bacteriophages that were able to lyse K1+ *E. coli*. When administered by intramuscular injection at the same time as, or eight hours after, infection with *E. coli*. These bacteriophages were able to cure infection, even when used at a low titer. The same effects were seen with intracerebrally infected mice treated with bacteriophages 16 hours after infection. The bacteriophages were more effective than numerous types of antibiotics at curing mice. Smith and Huggins (1983) also successfully used bacteriophage therapy to treat calves, pigs, and lambs that had been infected with *E. coli*. They selected a bacteriophage that would lyse *E. coli* and also selected a second bacteriophage that would lyse *E. coli* cultures that had become resistant to the first bacteriophage. Key to the success of this selection method was the idea that, by selecting a bacteriophage that affected the K antigens of *E. coli*, resistance would require a modification to an important component of virulence for the cell. Resistant *E. coli* strains had different colony morphology on agar plates and were K-negative. Treatment consisted of two bacteriophages, one that resulted in a K-negative strain as resistance was developed, and a second to lyse the K-negative cells. The combination of bacteriophages to combat resistance was better able to prevent death in calves with diarrhea than a single bacteriophage or no bacteriophage treatment. Sheng et al. (2006) followed a similar method to select bacteriophages against *E. coli* O157:H7. They selected bacteriophage that attached to LPS on the cell surface so that resistant cells had to change LPS expression with the idea that it would decrease the pathogenicity of the bacteria. Resistant cultures had rough colony morphology instead of the usual smooth mucoid texture of many typical pathogenic Enterobacteriaceae. When bacteriophage KH1, selected to attach to LPS, was administered alone it did not reduce *E. coli* O157:H7 recovery in sheep. However, when combined with another bacteriophage, recovery of *E. coli* O157:H7 was reduced. In 1987, Smith and Huggins used bacteriophages to treat calves with *E. coli*-caused diarrhea. They selected their bacteriophages by administering *E. coli* to a calf followed by a bacteriophage cocktail. Bacteriophages able to survive the gastrointestinal tract were collected in the feces 24 hours post-administration. These bacteriophages were used to treat subsequent calves. Calves given bacteriophages within 24 hours of the onset of diarrhea recovered within 20 hours. Also, sick calves placed on bedding that had been sprayed with bacteriophages recovered from diarrhea. Smith and Huggins noted that during the period of disease, bacteriophages continued to persist in the feces, but after recovery, bacteriophage numbers dropped dramatically.

Biswas et al. (2002) successfully cured vancomycin-resistant *Enterococcus faecium*-infected mice with bacteriophage therapy. Mice were treated with bacteriophages just 45 minutes after infection with bacteria. Treatment at a multiplicity of infection (MOI) level of 0.3 to 3.0 was able to cure all of the infected mice. However, lower multiplicity of infection ratios (MOI) of 0.03 to 0.003 resulted in just 60% and 40% survival of mice, respectively. They also noted that bacteriophage treatment could be delayed for up to five hours after infection. However, if treatment was delayed for 18 or 24 hours, only 50% recovery was seen. In other studies, preparations of the appropriate bacteriophage have been able to protect mice and guinea pigs against systemic infections with strains of *Pseudomonas aeruginosa* and *Acinetobacter baumanii* and have been able to inhibit the rejection of skin grafts caused by *P. aeruginosa* infections (Soothill 1992, 1994). Interestingly, a very low MOI of 10^{-6}, one bacteriophage to one million bacteria, was able to protect mice against infection with *P. aeruginosa* (Soothill, 1992).

Successes and Failures of Bacteriophage Treatment of Enterobacteriaceae Infections in the Gastrointestinal
Tract of Domestic Animals

137

While each of these studies was successful at curing the infected animals, treatment was given simultaneously with the bacteria, or just a few hours after infection. These would not be practical conditions for treating disease under real world conditions because it is often not known that an animal is sick until clinical signs are noticed several days after the infection begins. Furthermore, key to the success of these experiments, is knowledge that the selected host is susceptible to the bacteriophage (or combination of bacteriophages) which was administered. As discussed below, multiple researchers have focused on improving food safety by treating animals pre-harvest with bacteriophages targeted against foodborne pathogens.

One such pathogen has been *Escherichia coli* O157:H7 in ruminants. Much research has investigated the ability of bacteriophages to treat *E. coli* O157:H7 infections in sheep and cattle. Enterohemorrhagic *E. coli*, such as *E. coli* O157:H7, cause severe enterohemorrhagic enteritis, renal uremic syndrome, and are capable of causing death, especially among young children, immune-suppressed individuals, and the elderly (Gyles, 2007; Rangel et al., 2005). Cattle and sheep are the primary sources of waterborne and foodborne cases, and with an infectious dose of approximately 100 cells, control of this pathogen is of utmost importance to meat processors (Besser et al., 1999; Chapman et al., 2001; Grauke et al., 2002; Wells et al., 1991; Zhao et al., 1995). Like other *E. coli*, the primary site of infection for cattle and sheep is the hindgut. Grauke et al. (2002) noted a correlation between positive fecal samples and isolation from the rumen and duodenum while Naylor et al. (2003) found the primary site of infection in the recto-anal junction. Other reports have confirmed the rectum and cecum as primary sites of infection in cattle (Buchko et al., 2000; Dean-Nystrom et al., 1999). These findings suggest that oral bacteriophage treatment, with bacteriophages selected for anaerobic activity, may affect *E. coli* O157:H7 colonization in sheep and cattle. In fact, multiple studies have focused on the application of anaerobically active bacteriophages to ruminants for the control of *E. coli* O157:H7, as described below.

Bacteriophage CEV1, isolated from sheep with short and transient *E. coli* O157:H7 infections, was found to be lytic against 20 pathogenic strains of *E. coli* and had both aerobic and anaerobic *in vitro* activity. Oral administration of CEV1 to infected sheep, three days post-infection, showed reduced levels of *E. coli* O157:H7 in the ruminal, cecal, and rectal contents two days after bacteriophage treatment (Raya et al., 2006). Similarly, rectal administration of bacteriophage KH1 and SH1 to cattle infected with *E. coli* O157:H7 was able to reduce the levels of recovered pathogen. The same bacteriophages eliminated detectable levels of *E. coli* O157:H7 in mice (Sheng et al., 2006). In 2008 Calloway et al. isolated bacteriophages from cattle feces and noted an effectiveness of these bacteriophages at reducing *E. coli* O157:H7 levels throughout the intestine of sheep. Bacteriophages were selected *in vitro* for their ability to lyse *E. coli* O157:H7, and eight different bacteriophages were included in the culture to combat resistance by the bacteria. While each of these studies effectively reduced the levels of *E. coli* O157:H7 in sheep and cattle, the researchers noted that bacteriophages may not prove a long-term treatment and application should be considered immediately prior to processing for maximum effectiveness. The principle reason for this is that as with antimicrobial chemicals, serial applications have often led to selection of bacteriophage-resistant bacteria.

Like *E. coli* O157:H7, researchers have attempted to reduce multiple serovars of *Salmonella* in poultry, a frequent source of cases of *Salmonella* in humans. *Salmonella enterica* serovars continue to be among the most important foodborne pathogens worldwide due to the considerable human rates of illness reported and the wide range of hosts that are colonized by members of this genus, which serve as vectors and reservoirs for spreading these agents to animal and human populations (CDC 2005, 2006a,b, 2007, 2008a,b). Furthermore, public concern for the appearance of resistant strains to many antibiotics, particularly among zoonotic pathogens such as common *Salmonella* isolates, is also challenging several sectors of agriculture to find alternative means of control (Boyle et al., 2007). In the United States, it is estimated that 1.4 million humans contract salmonellosis and that the annual cost of this illness, including lost productivity, is $3 billion annually (WHO, 2006). In the year 2004, surveillance data indicated that the greatest number of foodborne illnesses was caused by *Salmonella*, comprising 42% of all laboratory diagnoses (FoodNet, 2005). Because many of these illnesses are associated with poultry and poultry products (Bean & Griffin, 1990; Persson & Jendteg, 1992), the reduction of microbial contamination during the production of poultry is important. Further considerations for bacteriophage treatment to control *Salmonella* in poultry are issues associated with antiobiotic treatment. Poultry harboring *Salmonella* infection can be treated with antibiotics with some success (Goodnough & Johnson, 1991; Muirhead, 1994). However, Manning and co-workers (1994, 1992) reported increased *Salmonella* colonization when chickens were treated with selected antibiotics, possibly due to reduction of normal bacterial flora in the gastrointestinal tract that serve as a natural barrier to *Salmonella* infection. Additionally, Kobland et al. (1987) and Gast et al. (1988) have recovered antibiotic resistant *Salmonella* from experimentally challenged birds treated with antibiotics. Recently, the United States Food and Drug Administration (FDA) has banned the use of enrofloxacin in poultry production because of concerns regarding an increase in resistant *Campylobacter* infections in humans (FDA, 2005). Recently, a ban on antibiotics and coccidiostats was put in place by European Parliament Council Directive 1831/2003. The regulation stated that antibiotics, other than coccidiostats and histomonostats, had to be removed from feed by the end of 2005, and that anticoccidial substances would be prohibited by 2013. After these dates, medical substances in animal feeds will supposedly be limited to therapeutic use by veterinary prescription (European Parliamant and of the Council, 2003). Thus, it is increasingly important that effective and inexpensive methods or products to treat bacterial infections in food production animals be developed.

Recently, Toro et al. (2005) reported using a combination of bacteriophages and competitive exclusion to treat *Salmonella*-infected chickens. They were able to reduce recovery of *Salmonella* Typhimurium (ST) from the ceca of chickens. In the successful experiments, chickens were challenged with ST during the course of treatment with bacteriophages. However, treatment with bacteriophages was not better than treatment with a competitive exclusion product. And, combination of competitive exclusion and bacteriophages did not further reduce ST recovery. Similarly, Filho and co-workers (2007) reported that administration of bacteriophage cocktails could temporarily reduce the incidence of *Salmonella* recovery in broiler chickens, but by 48 hours post-treatment there was no difference between treated and non-treated controls. Additionally, combining the bacteriophage cocktail with a probiotic culture had no effect on *Salmonella* recovery when compared to bacteriophages alone. The pattern was also noted by Higgins et al. (2007) when *Salmonella* was reduced to zero recovery, but 48 hours after administration recovery from bacteriophage-treated birds increased to higher than non-treated control groups.

In addition to control of paratyphoid *Salmonella*, much research has been completed on the study of bacteriophages in poultry for other diseases. The first reports of bacteriophage therapeutics in poultry were made by d'Herelle in 1922 when he reported successful treatment of at least 19 barnyards affected by fowl typhoid. Bacteriophages administered were selected specifically to target the pathogens involved. The pathogens causing the typhoid were either one or a combination of five different bacterial species. The data reported in these studies is anecdotal, with the statement "The sick recovered and the epizootic stopped at once" comprising all the results for those studies. However, treatment of a highly pathogenic and systemic host-adapted *Salmonella*, in this case, might be more effective as even a temporary reduction in pathogen levels could buy a critical amount of time for acquired immunity and eventual recovery of infected birds from the disease. This could be distinctly different than the case with the essentially non-pathogenic (for poultry) common paratyphoid isolates that are commonly implicated in food borne disease of humans. Supporting this hypothesis, other attempts to alleviate bacterial disease in poultry have focused on non-paratyphoid *Salmonella*. In 1926, Pyle isolated bacteriophages specific for *Salmonella pullorum* from the feces of birds affected with the disease. He demonstrated that even after 120 passages, strains of bacteriophage that were not initially lytic did not become lytic. Additionally he observed that two bacteriophage strains which were initially lytic became more lytic over 60 serial passages. Two experiments employing the lytic bacteriophages ensued. *Salmonella pullorum* and bacteriophages were injected simultaneously into the pectoral muscle of chickens in one treatment group, with the treated group receiving bacteriophage eight hours post-challenge. When compared to controls that received no bacteriophage treatment, mortality was delayed in both treatment groups. In the second experiment, the bacteriophages and *S. pullorum* were administered or the bacteriophages were administered eight hours post-challenge in the drinking water. Again, when compared to controls which did not receive bacteriophage treatment, the onset of mortality was delayed.

Berchieri et al. (1991) treated birds infected with ST with bacteriophages and found that the levels of ST could be reduced by several $\log_{10}$, and mortality associated with this unusually pathogenic ST was reduced significantly. However, ST was not eliminated, and returned to original levels within six hours of treatment. Also, the bacteriophages did not persist in the gastrointestinal tract for as long as the *Salmonella* was present. In fact, bacteriophages persisted only as long as they were added to the feed. In order to be effective, bacteriophages had to be administered in large numbers, and soon after infection with ST. Similar to reports below (Hurley et al., 2008), the bacteriophages may have been killing the bacteria via lysis from without and, instead of infecting and replicating within the cell, the bacteriophages may have been killing the ST by an excess of penetration from the bacteriophages. This may explain the decline in bacteriophage numbers despite the presence of a host for replication.

Multiple researchers have investigated the possibility of curing *E. coli* infections in poultry. In 1998 Barrow et al. prevented morbidity and mortality in chickens using bacteriophages lytic for *E. coli*. When chickens were challenged intramuscularly with *E. coli* and simultaneously treated with $10^6 - 10^8$ pfu of bacteriophages, mortality was reduced by 100%. This study also demonstrated that bacteriophages can cross the blood brain barrier, and furthermore can amplify in both the brain and the blood.

Huff et al. (2003) used bacteriophages to treat airsacculitis caused by *E. coli* in chickens. Marked efficacy was achieved when administering bacteriophage with the bacterial challenge inoculum, by injection in the thoracic air sac. However, drinking water administration of the same bacteriophages was ineffective at preventing the manifestation of the disease syndrome. This indicated that it is important to deliver bacteriophages directly to the site of infection. It was also shown that an aerosol treatment of bacteriophages, followed by an *E. coli* challenge on the same day, the next day, or three days later reduced morbidity and mortality associated with respiratory infection (Huff et al., 2002a). Thus, the study demonstrated a prophylactic ability of bacteriophages in the respiratory tract. However, given the evidence that bacteriophages do not typically remain in an environment without an appropriate host (Ashelford et al., 2000; Fiorentin et al., 2005; Hurley et al., 2008; Oot et al., 2005), prophylaxis could be difficult without continued administration or by knowing an animal had been exposed.

In summary, there are few current reports of efficacy of bacteriophage treatment in chickens other than when treatment was administered at the same time as bacterial challenge or via injection. Outside of experimentally controlled situations, it is not usually possible to treat a disease at the same time as the challenge. Also, it is not practical to treat commercial poultry flocks by individual injection, though highly valuable breeder flocks might warrant the time and money involved. However, such of these limited successes do not necessarily translate into effective enteric treatments. Host-associated pressure against pathogen infections may predispose systemic bacteriophage therapy toward success. In these cases, where bacteriophages are used to treat systemic or tissue-associated infections, an acute efficacy of merely reducing the infection load by 90% or more, could greatly reduce mortality and reduce the duration and magnitude of disease by allowing time for acquired immunity in the animal host. In the intestinal lumen, host pressures against the infection may not be as severe and many Enterobacteriaceae are capable of free living status within the gut without eliciting robust acquired immune responses from the infected animal. In these cases, a temporary reduction in enteric colonization may not be as likely to be curative, as discussed below.

3. Failures

As the history of published successful bacteriophage treatments of enteric disease is reviewed, it is readily evident that such reports, while often dramatic in effect, are relatively sporadic during the last approximately 60 years. Given that experimental failures frequently are not published, as the cause of failure can often not be ascertained, the authors suspect that history is replete with unpublished examples of failures to treat enteric Enterobacteriaceae infections. Still, some reports of failures, or incomplete successes, have been documented and are described below.

Bacteriophage KH1, shown to lyse 12 of 16 *E. coli* O157 strains tested, originally showed promise as results of *in vitro* tests demonstrated an ability to lyse bacterial cultures, by plaque formation, at both 37 °C and 4 °C (Kudva et al., 1999). However, when administered to *E. coli* O157:H7 infected sheep, bacteriophage KH1 did not effectively reduce levels of recovered pathogen, despite continued recovery of bacteriophages from the feces for eight days post-treatment (Sheng et al., 2006). Aerobic, instead of anaerobic, selection may have played a key role in the ability of this bacteriophage to effectively eliminate intestinal carriage of *E. coli* O157:H7. Multiple researchers have suggested anaerobic environments can affect bacteriophage activity (Bach et al., 2003; Kudva et al., 1999; Raya et al., 2006; Tanji et al., 2005). When Bach and co-workers (2003) tested the effects of bacteriophage DC22 on *E. coli* O157:H7 in an *in vitro* fermentation system prior to treating infected sheep, bacteriophage DC22 only decreased microbial levels in the artificial ruminant set up at high multiplicities of infection (MOI), and failure of the bacteriophage to replicate and increase PFU over the course of 120 hours suggests that the bacteriophages may have reduced *E. coli* by lysis from without rather than by infecting, replicating within, and lysing the cells. Subsequent *in vivo* studies in sheep did not result in decreased shedding of *E. coli* O157:H7, and bacteriophages were found in the feces for only two days post-treatment. Similarly, Tanji et al., (2005) administered a bacteriophage with promising *in vitro* test results to *E. coli* O157:H7 mice with little success. These studies reinforce the need to understand how *in vitro* conditions relate to the *in vivo* infection parameters and show the need for an appreciation for the ecosystem where the bacteriophage will be used.

In silico modeling was used by Hurley et al. (2008) to predict parameters for treating *Salmonella*-infected chickens with bacteriophage SP6 in an attempt to better comprehend the biological factors of the luminal ecosystem, *Salmonella*, and bacteriophages, and how they interact within the gastrointestinal tract. Among the factors considered were varying growth rates, feed and water intake, and *Salmonella* resistance to the bacteriophages. The results of these *in silico* test results were considered when an *in vivo* challenge was designed. However, after bacteriophage treatment *Salmonella* was detected at levels that did not differ from control groups not treated with bacteriophages. In fact, bacteriophage infection may have selected for resistant bacteria because half of the *Salmonella* isolates from a treated group were resistant to bacteriophage SP6 on day 29, one day after the second dose of bacteriophage treatment. Moreover, many of the *Salmonella* cultured from other samples of bacteriophage-treated birds showed at least a partial resistance to bacteriophages, with only partially clear plaques forming on soft agar overlays when, prior to bacteriophage treatment, the *Salmonella* isolate was susceptible and clear plaques routinely formed on soft agar overlays. The authors also noted a steady decrease in bacteriophage excretion, despite continued high levels of *Salmonella* recovery within the cecum. This data is similar to the results of Fiorentin et al. (2005), where *Salmonella* continued to be detected 21 days after inoculation, but bacteriophage levels had declined to undetectable levels. In another related study, bacteriophage treatment resulted in higher levels of *Salmonella* recovery in turkeys 48 hours post-treatment after an initial decrease in *Salmonella* at 6, 12, and 24 hour post-treatment time points (Higgins et al., 2007). These bacteriophages were selected for ability to survive low pH, to simulate passage through the ventriculus of poultry, and were administered with $Mg(OH)_2$ to aid adhesion of bacteriophages to the cell walls of bacteria (Eisenstark, 1967). The authors also noted that bacteriophage resistance was common in all cultures.

Our laboratory and others have demonstrated that resistance to bacteriophages selected against *Salmonella* isolates quickly occurs, often in a single passage (Bastias et al., 2010; Hurley et al., 2008; Fiorentin et al., 2005). When bacteriophage cocktails of 71 different bacteriophages selected for treatment of experimental *Salmonella* Enteritidis infections in chickens, a brief reduction in enteric colonization was noted during the first 24 hours, but rebound levels were similar to controls within 48 hours, even with repeated or continuous dosage of the bacteriophage cocktail (Higgins et al., 2007). Because of the demonstrated temporary reduction in enteric colonization in these studies, effective bacteriophages were demonstrably able to pass to the lower gastrointestinal tract. As continued treatments failed to maintain this reduction, development of resistance by the enteric *Salmonella* Enteritidis is the most likely explanation.

In order to potentially deliver higher levels of bacteriophage, several attempts to protect the bacteriophage cocktail through the upper gastrointestinal tract were made in our laboratory. Pre-treatment of infected poultry with antacid preparations designed to reduce the acidity of the proventriculus (the true stomach of birds) were successful in increasing the number of administered bacteriophage that successfully passed into the intestinal tract, but this treatment did not improve the outcome of bacteriophage treatment of *Salmonella* Enteritidis infection (Higgins et al., 2007).

An alternative approach is to select for alternative non-pathogenic bacteriophage hosts which could potentially "carry" bacteriophage through the gastrointestinal tract and, with continuous dietary administration of the non-infected alternative host bacterium, provide a means of amplification within the gut of the host (Bielke et al., 2007a). Bielke and co-workers (2007b) demonstrated that non-pathogenic alternative hosts can be selected for some bacteriophages that were originally isolated using a *Salmonella* Enteritidis target. This approach, which has potential utility for amplification of large numbers of phage without the necessity to thoroughly separate bacteriophage from a pathogenic target host, was also used to create a potential "Trojan Horse" model for protecting the bacteriophages through the upper gastrointestinal tract, thus potentially providing a vehicle for enteric amplification of those surviving bacteriophages. In these studies, neither the Trojan Horse approach, nor the continuous feeding of the alternative host bacteria as a source of enteric amplification, were effective in producing even more than a transient reduction in enteric *Salmonella* infections.

Through these failures, many investigators have concluded that the escape of even a minority of target bacteria within the enteric ecosystem allows for almost immediate selection of resistant target bacteria and rebound to pre-treatment levels of infection may even exceed the levels of non-treated controls in some cases.

4. Potential strategies to overcome failures

Bacteriophage resistance is an important component of therapy to overcome before bacteriophages can really be a viable antimicrobial for infection. The generation time for bacteria is typically short enough that mutants with bacteriophage resistance can emerge within hours (Higgins et al., 2007; Lowbury and Hood, 1953). One possible strategy to overcome this problem is administration of multiple bacteriophage isolates for treatment. Smith and Huggins (1983) selected a bacteriophage against *E. coli* K$^+$, and then subsequently selected a bacteriophage against a resistant strain of *E. coli* K$^+$. The combination of these two bacteriophages reportedly cured calves, pigs, and lambs of intestinal colibacillosis. Despite this success, resistance is difficult, if not impossible, to predict and combining the correct cocktail of bacteriophages to overcome resistance would be a blind guess in most cases. This, combined with the highly selective nature of individual bacteriophage isolates and even cocktails, as described above, is discouraging from the perspective of enteric therapeutic development, especially for very low level or opportunistic pathogens.

The most success is likely to come from treating points in the system that are continually bombarded with bacteria that have not been previously subjected to the bacteriophages being used for treatment. Also important for this system is keeping exposure of the bacteria to bacteriophages to a minimal amount of time. If the bacteriophages interact with the bacteria for long periods of time, the bacteria will become resistant as repeatedly demonstrated in the above discussion. Food and meat processing facilities are an excellent example. As live animals enter a slaughter/processing facility, the bacteria have not likely been exposed to the bacteriophages used to treat the infection, thus greatly increases the chances of success. Similar potential exists for hatchery applications, and applications to food products, as discussed below.

Higgins and co-workers (2005) successfully treated turkey carcasses at a processing facility with bacteriophages specific to the *Salmonella* to which they were infected. This process was effective when either an autogenous bacteriophage treatment targeted to the specific *Salmonella* strain infecting the turkeys was used, or a cocktail of nine wide host-range *Salmonella*-targeting bacteriophage were used. Similarly, a bacteriophage treatment for cattle carcass contamination has been effective at reducing the *E. coli* 0157:H7 load at processing has been developed and commercially licensed in the United States. These successes avoid development of bacteriophage resistance by applying treatment at a single point during production, in an environment where proliferation of the target organism is extremely limited. In this way, since the target organism is never intentionally exposed twice to the same treatment, resistance is unlikely to ever increase beyond the naturally-occurring resistance to the bacteriophage (or cocktail) used.

One of the most well documented successes of published treatment of enteric Enterobacteriaceae infections with bacteriophages was the study of Smith and Huggins (1983) as described above. It is notable that in this successful study, the bacteriophage cocktail used was a combination of two bacteriophages, but the second was isolated using the target organism which was resistant to the first bacteriophage. This approach of selecting for bacteriophage isolates using target bacteria that are resistant to sequential bacteriophage treatments was not used in the work of Higgins et al. (2007), or in several other published studies. Higgins and co-workers (2007) used a collection of bacteriophages, independently isolated from different sources and with several different plaque morphologies, suggesting that a number of different bacteriophages were employed – and failed to persistently reduce enteric colonization. Similarly, application of a bacteriophage combination failed to reduce *S.* Enteritidis PT4 infections in broilers (Fiorentin et al., 2005). However, some cocktails have been successful. A combination of three bacteriophages isolated from feces of patients infected with *E. coli* O157:H7 were applied to contaminated processed beef for reduction of the pathogen (O'Flynn et al., 2004). In 2006, Sheng et al. reported that a combination of two bacteriophages worked better at reducing *E. coli* O157:H7 in ruminants than each bacteriophage administered solely. Perhaps, with a defined method to select for bacteriophages that have become resistant to bacteriophages, a combination can overcome the resistance issue. However, the resistance acquired by the pathogen would have to be predictable and consistent. Smith and Huggins' (1983) method to first apply a bacteriophage specific for an antigen on the cell surface was successful, and may prove to be a procedure that could consistently overcome the issue of resistance in bacteriophage therapy. However, the ability to simultaneously target a broad range of wild-type isolates under field conditions has not been explored.

It is possible that one of the most notable exceptions to the many failures to treat enteric Enterobacteraceae infections during recent years, that of Smith and Huggins (1983), provides a singular clue as to the potential for enhancing the likelihood of enteric Enterobacteriaceae efficacy. It is possible that selection of multiple bacteriophages for the same target cell phenotype results in selection of bacteriophages that are effective through identical mechanisms of adhesion, penetration, replication, and release. When new bacteriophages are isolated for efficacy against sequentially resistant isolates of the target bacteria, and these are combined for administration as a cocktail, the ability of the target cell to shift phenotype may be severely limited, resulting in a much larger proportion of target cell reduction, thereby increasing the probability of elimination or cure. Multiple researchers have noticed a change in colony morphologies of *E. coli* that had become resistant to bacteriophages selected to adhere to key-components of pathogenicity of the organism, such as lipopolysaccharides (O'Flynn et al., 2004; Sheng et al., 2006). This change in morphology may relate to a decreased ability to cause infection because the bacteria may be inhibited as a result of the bacteriophage resistance.

Another consideration for bacteriophage treatment could be the application of bacteriophages to foods post-processing. Multiple researchers have noticed a successful reduction of foodborne pathogens on meats, and fruits (Bielke et al., 2007c; Higgins et al., 2005; Leverentz et al., 2001, 2003; O'Flynn et al., 2004). Treating processed poultry carcasses with different bacteriophages was able to eliminate *S.* Enteritidis (Bielke et al., 2007c) or field isolates of *Salmonella* to below detection limits (Higgins et al., 2005). A mixture of three different bacteriophages reduced the levels of *E. coli* O157:H7 detected on processed beef, though the bacteriophage treatment was not as effective at temperatures below 30 °C, making them ineffective at refrigeration temperatures (O'Flynn et al., 2004). Leverentz et al. (2001, 2003) successfully reduced the levels of *Salmonella* and *Listeria monocytogenes* on selected fruits with bacteriophage application. A loss of effectiveness was seen on cut fruits, possibly due to low pH of the fruit flesh. While this research appears promising, the studies do not report the long-term susceptibility of the contaminating bacteria to the bacteriophages. Perhaps, selection of a cocktail of bacteriophages to combat resistance, and for the ability to lyse bacteria at refrigeration temperatures could result in successful reduction of these foodborne pathogens.

5. Conclusions

While bacteriophage treatment of enteric infections has had some success, failures do occur and the system has not yet been perfected. Chemical antibiotics are often effective against multiple species of bacteria and do not require specific selection to treat infections. Unlike bacteriophages, which tend to be at least somewhat host specific and may not even kill bacterial isolates within the same species. Still, with the rise of antibiotic resistance, bacteriophages may be able to offer a line of defense in situations for which antibiotics are not available, or are not effective. For example, with the restriction or elimination of antibiotics usage in food animals, researchers have been investigating the possibility of bacteriophages to control foodborne pathogens. With the realization that resistance of pathogenic isolates of bacteria to bacteriophages can, and do, emerge, perhaps the best application would be to apply the bacteriophages immediately prior to slaughter. Thus, the pathogen could be effectively reduced in the gastrointestinal tract, and subsequently reduce the risk of contamination during processing. With the frequency of reported failures, and the assumption that many are not reported, bacteriophage therapeutics has not been perfected well enough for widespread application. In addition to resistance, safety, specificity, and long-term effectiveness must be demonstrated, and although several products have been licensed in the United States and elsewhere, procedures for demonstration of these characteristics are not well established, providing an additional regulatory burden for commercialization.

Clearly, widespread bacteriophage treatments with Enterobacteriaceae within the gastrointestinal tract have not been adopted for any animal species during the last 60 years and successful research in this area has been modest and sporadic. Nevertheless, the occasional reports by reputable scientists in solid journals must indicate that there is potential for improved therapeutic efficacy of bacteriophages for this purpose. With the diminution of new antimicrobial pharmaceuticals and the widespread resistance among many pathogenic enteric Enterobacteriaceaes, a breakthrough in this area is sorely needed.

6. References

Adams, J.C.; Gazaway, M.D.; Brailsford, M.D.; Hartman, P.A. & Jacobson, N.L. (1966) Isolation of bacteriophages from the bovine rumen. *Experientia* Vol. 22, No. 11, pp. 717-718.

Alisky, J.; Iczkowski, K.; Rapoport, A. & Troitsky, N. (1998) Bacteriophage show promise as antimicrobial agents. *Journal of Infection* Vol. 36, pp. 5-15.

Ashelford, K.E.; Norris, S.J.; Fry, J.C.; Bailey, M.J. & Day, M.J. (2000) Seasonal population dynamics and interactions of competing bacteriophages and their host in the rhizosphere. *Applied Environmental Microbiology* Vol. 66, pp. 4193–4199.

Bach, S.A.; McAllister, T.A.; Veira, D.M.; Gannon, V.P.J. & Holley, R.A. (2003) Effect of bacteriophage DC22 on *Escherichia coli* O157:H7 in an artificial rumen system (Rusitec) and inoculated sheep. *Animal Research* Vol. 52, pp. 89-101.

Barrow, P.; Lovell, M. & Berchieri, Jr, A. (1998) Use of Lytic Bacteriophages for Control of Experimental *Escherichia coli* Septicemia and Meningitis in chickens and calves. *Clinical Diagnostics and Laboratory Immunolgy* Vol. 5, pp. 294-298.

Bastias, R.; Higuera, G.; Sierralta, W. & Espejo, R.T. (2010) A new group of cosmopolitan bacteriophages induce a carrier state in the pandemic strain of *Vibrio parahaemolyticus*. *Environmental Microbiology* Vol. 12, pp. 990-1000.

Bean, N.H. and P.M. Griffin, (1990). Food-borne disease outbreaks in the United States, 1973-1987: Pathogens and trends. *Journal of Food Protection*, 53: 804.

Berchieri, A, MA Lovell, and PA Barrow, 1991. The activity in the chicken alimentary tract of bacteriophages lytic for *Salmonella typhimurium*. *Research Microbiology* 142: 541-549

Besser, R.E.; Griffin, P.M. & Slutsker, L. (1999) *Escherichia coli* O157:H7 gastroenteritis and hemolytic uremic syndrome: an emerging infectious disease. *Annual Reviews in Medicine* Vol. 50, pp. 355-367.

Bielke, L.R.; Higgins, S.E.; Donoghue, A.M.; Kral, T.; Donoghue, D.J.; Hargis, B.M. & Tellez, G.I. (2007a) Evaluation of alternative host bacteria as vehicles for oral administration of bacteriophages. *Internation Journal Poultry Science* Vol. 6, pp. 758-761.

Bielke, L.; Higgins, S.; Donoghue, A.; Donoghue, D; & Hargis, B.M. (2007b) *Salmonella* host range of bacteriophages that infect multiple genera. *Poultry Science* Vol. 86, pp. 2536-2540.

Bielke, L.R.; Higgins, S.E.; Donoghue, A.M.; Donoghue, D.J.; Hargis, B.M. & Tellez, G. (2007c) Use of wide-host-range bacteriophages to reduce *Salmonella* on poultry products. *International Journal of Poultry Science* Vol. 6, pp. 754-757.

Biswas, B.; Adhya, S.; Washart, P.; Paul, B.; Trostel, A.N.; Powell, B.; Carlton, R. & Merril, C.R. (2002) Bacteriophage therapy rescues mice bacteremic from a clinical isolate of vancomycin-resistant *Enterococcus faecium*. *Infection and Immunolology* Vol. 70, pp. 204-210.

Boyle, E. C.; Bishop, J. L.; Grassl, G. A.; & Finlay, B. B. (2007). *Salmonella*: From pathogenesis to therapeutics. *Journal of Bacteriology* Vol. 189, No. 5, pp. 1489-1495.

Brabban, A.D.; Nelson, D.A.; Kutter, E.; Edrington, T.S. & Callaway, T.R. (2004) Approaches to controlling *Escherichia coli* O157:H7, a food-borne pathogen and an emerging environmental hazard. *Environmental Practice* Vol. 6, pp. 208-229.

Breitbart, M.; Hewson, I.; Felts, B.; Mahaffy, J.M.; Nulton, J.; Salamon, P. & Rohwer, F. (2003) Metagenomic analyses of an uncultured viral community from human feces. *Journal of Bacteriology* Vol. 185, pp. 6220-6223.

Bruttin A. & Brüssow, H. (2005) Human volunteers receiving *Escherichia coli* phage T4 orally: a safety test of phage therapy. *Antimicrobial Agents and Chemotherapy* Vol. 49, Iss. 7, pp. 2874-2878

Buchko, S.J.; Holley, R.A.; Olson, W.O.; Gannon, V.P. & Veira, D.M. (2000) The effect of different grain diets on fecal shedding of *Escherichia coli* O157:H7 by steers. *Journal of Food Protection* Vol. 63, pp. 1467-1474.

Callaway, T.R.; Elder, R.O.; Keen, J.E.; Anderson, R.C. & Nisbet, D.J. (2003) Forage feeding to reduce preharvest *Escherichia coli* populations in cattle: a review. *Journal of Dairy Science* Vol. 86, pp. 852-860.

Callaway, T.R.; Edrington, T.S.; Brabban, A.D.; Keen, J.E.; Anderson, R.C.; Rossman, M.L.; Engler, M.J.; Genovese, K.J.; Gwartney, B.L.; Reagan, J.O.; Poole, T.L.; Harvey, R.B.; Kutter, E.M. & Nisbet, D.J. (2006) Fecal prevalence of *Escherichia coli* O157, *Salmonella*, *Listeria*, and bacteriophage infecting *E. coli* O157:H7 in feedlot cattle in the southern plains region of the United States. *Foodborne Pathogens and Disease* Vol. 3, Iss. 3, pp. 234-244.

Callaway, T.R.; Edrington, T.S.; Brabban, A.D.; Anderson, R.C.; Rossman, M.L.; Engler, M.J.; Carr, M.A.; Genovese, K.J.; Keen, J.E.; Looper, M.L.; Kutter, E.M. & Nisbet, D.J. (2008) Bacteriophages isolated from feedlot cattle can reduce *Escherichia coli* O157:H7 populations in ruminant gastrointestinal tracts. *Foodborne Pathogens and Disease* Vol. 5, Iss. 2, pp. 183-191.

Carlton, R.M.; Noordman, W.H.; Biswas, B.; deMeester, E.D. & Loessner, M.J. (2005) Bacteriophage P100 for control of *Listeria monocytogenes* in foods: genome sequence, bioinformatics analyses, oral toxicity study, and application. *Regulatory Toxicology and Pharmacology* Vol. 43, Iss. 3, pp. 301-312.

CDC (2005) Outbreak of multidrug-resistant *Salmonella* Typhimurium associated with rodents purchased at retail pet stores – United States, December 2003-October 2004. *Morbidity and Mortality Weekly Report* Vol. 57, No. 17, pp. 429-433.

CDC, (2007a) Three outbreaks of salmonellosis associated with baby poultry from three hatcheries – United States, 2006. *Morbidity and Mortality Weekly Report* Vol. 56, No. 12, pp. 273-276.

CDC (2007b) *Salmonella* Oranienburg infections associated with fruit salad served in health-care facilities – Northeastern United States and Canada, 2006. *Morbidity and Mortality Weekly Report* Vol. 56, No. 39, pp. 1025-1028.

CDC (2008a) Multistate outbreak of human *Salmonella* infections associated with exposure to turtles – United States, 2007-2008. *Morbidity and Mortality Weekly Report* Vol. 57, No. 3, pp. 69-72.

CDC (2008b) Multistate outbreak of human *Salmonella* infections caused by contaminated dry dog food --- United States, 2006 – 2007. *Morbidity and Mortality Weekly Report MMWR* Vol. 57, No. 19, pp. 521-524

Chapman, P.A.; Cerdan Malo, A.T.; Ellin, M.; Ashton, R. & M.A. Harkin (2001) *Escherichia coli* O157 in cattle and sheep at slaughter, on beef and lamb carcasses and in raw beef and lamb products in South Yorkshire, UK. *International Journal of Food Microbiology* Vol. 64, pp. 139–150.

Danovaro, R.; Dell'Anno, A.; Trucco, A.; Serresi, M. & Vanucci, S. (2001) Determination of virus abundance in marine sediments. *Applied Environmental Microbiology* Vol. 67, pp. 1384–1387.

Dean-Nystrom, E.A.; Bosworth, B.T.; & Moon, H.W. (1999) Pathogenesis of *Escherichia coli* O157:H7 in weaned calves. *Advances in Experimental Medicine and Biology.* Vol. 473, pp. 173–177.

Dhillon, T.S.; Dhillon, E.K.; Chau, H.C.; Li, W.K. & Tsang, A.H. (1976) Studies on bacteriophage distribution: virulent and temperate bacteriophage content of mammalian feces. *Applied Environmental Microbiology* Vol. 32, pp. 68–74.

Dubos, R.J.; Straus, J.H. & Pierce, C. (1943) The multiplication of bacteriophage *in vivo* and its protective effect against an experimental infection with *Shigella dysenteriae.* *Journal of Experimental Medince* Vol. 78, pp. 161–168.

Dutta, N.K. (1963) An experimental study on the usefulness of bacteriophage in the prophylaxis and treatment of cholera. *Bulletein of Organization mond Santé Bulletein World Health Org* Vol. 28, pp. 357-360.

Eisenstark, A. (1967) Bacteriophage Techniques. In: Methods in Virology, pp: 449-524.

European Parliamant and of the Council (2003) http://eur-lex.europa.eu/LexUriServ/LexUriServ.do?uri=CELEX:32003R1831:EN:HTML.

Filho, R.L.A.; Higgins, J.P.; Higgins, S.E.; Gaona, G.; Wolfenden, A.D.; Tellez, G. & Hargis, B.M. (2007) Ability of bacteriophages isolated from different sources to reduce *Salmonella enterica* serovar Enteritidis *in vitro* and *in vivo*. *Poultry Science* Vol. 86, pp. 1904-1909.

Fiorentin, L.; Vieira, N.D. & Barioni, W. (2005) Oral treatment with bacteriophages reduces the concentration of *Salmonella* Enteritidis PT4 in caecal contents of broilers. *Avian Pathology* Vol. 34, pp.258–263.

Food and Drug Administration, U.S. (2005) FDA Announces Final Decision About Veterinary Medicine. P05-48, Docket No. 2000N-1571.

FoodNet (2005) Preliminary FoodNet Data on the Incidence of Infection with Pathogens Transmitted Commonly Through Food---10 Sites, United States, 2004. *Morbidity and Mortality Weekly Report* Vol. 54, pp. 352-356.

Fuhrman, J.A. (1999) Marine viruses and their biogeochemical and ecological effects. *Nature* Vol. 399, pp. 541–548.

Gast, R.K.; Stephens, J.F. & Foster, D.F. (1988) Effect of kanamycin administration to poultry on the proliferation of drug-resistant *Salmonella*. *Poultry Sciecne* Vol. 67, pp. 699.

Goodnough, M.C. & Johnson, E.A. (1991) Control of *Salmonella enteritidis* infections in poultry by polymyxin B and trimethoprim. *Appl. Enviro. Micro* Vol. 57, pp. 785.

Grauke, L.J.; Kudva, I.T.; Yoon, J.W.; Hunt, C.W.; Williams, C.J. & Hovde, C.J. (2002) Gastrointestinal tract location of *Escherichia coli* O157:H7 in ruminants. *Appl Environ Microbiol* Vol. 68, pp. 2269-2277.

Gyles, C.L. (2007) Shiga toxin-producing *Escherichia coli*: an overview. *Journal of Animal Science* Vol. 85E, pp. 45-62.

Higgins, J.P.; Higgins, S.E.; Guenther, K.L.; Newberry, L.A.; Huff, W.E. & Hargis, B.M. (2005) Use of a specific bacteriophage treatment to reduce *Salmonella* in poultry. *Poult Sci* Vol. 84, pp. 1141-1145.

Higgins, J.P.; Andreatti Filho, R.L.; Higgins, S.E.; Wolfenden, A.D.; Tellez, G. & Hargis, B.M. (2008) Evaluation of *Salmonella*-lytic properties of bacteriophages isolated from commercial broiler houses. *Avian Diseases* Vol. 52, pp. 139-142.

Higgins, S.E.; Higgins, J.P.; Bielke, L.R. & Hargis, B.M. (2007) Selection and application of bacteriophages for treating *Salmonella enteritidis* infections in poultry. *International Journal of Poultry Science* Vol. 6, pp. 163-168.

Huff, W.E.; Huff, G.R.; Rath, N.C.; Balog, J.M. & Donoghue, A.M. (2002a) Prevention of *Escherichia coli* infection in broiler chickens with a bacteriophage aerosol spray. *Poult Sci* Vol. 81, pp. 1486-1491.

Huff, W.E.; Huff, G.R.; Rath, N.C.; Balog, J.M.; Xie, H.; Moore, Jr, P.A.; & Donoghue, A.M. (2002b) Prevention of *Escherichia coli* respiratory infection in broiler chickens with bacteriophage (SPR02). *Poultry Science* Vol. 81, pp. 437-441.

Huff, W.E.; Huff, G.R.; Rath, N.C.; Balog, J.M. & Donoghue, A.M. (2003a). Evaluation of aerosol spray and intramuscular injection of bacteriophage to treat an *Escherichia coli* respiratory infection. *Poultry Science* 82: 1108-1112

Huff, W.E.; Huff, G.R.; Rath, N.C.; Balog, J.M. & Donoghue, A.M. (2003b) Bacteriophage treatment of severe *Escherichia coli* respiratory infection in broiler chickens. *Avian Disease* 47: 1399-1405

Hurley, A.; Maurer, J.J. & Lee, M.D. (2008) Using bacteriophages to modulate *Salmonella* colonization of the chicken's gastrointestinal tract: lessons learned from *in silico* and *in vivo* modeling. *Avian Diseases* Vol. 52, pp. 599-607.

Klieve, A.V. & Bauchop, T. (1988) Mophological diversity of ruminal bacteriophages from sheep and cattle. *Applied and Environmental Microbiology* Vol. 54, pp. 1637-1641.

Klieve, A.V. & Swain, R.A. (1993) Estimation of ruminal bacteriophage numbers by pulsed-field gel electrophoresis and laser densitometry. *Applied and Environmental Microbiology* Vol. 59, pp. 2299–2303.

Kobland, J.D.; Gale, G.O.; Gustafson, R.H. & Simkins, K.L. (1987) Comparison of therapeutic versus subtherapeutic levels of chlortetracycline in the diet for selection of resistant *Salmonella* in experimentally challenged chickens. *Poultry Science* Vol. 66, pp. 1129.

Kudva, I.T.; Jelacic, S.; Tarr, P.I.; Youderian, P. & Hovde, C.J. (1999) Biocontrol of *Escherichia coli* O157 with O157-specific bacteriophages. *Applied and Environmental Microbiology* Vol. 65, pp. 3767-3773.

Leverentz, B.; Conway, W.S.; Alavidze, Z.; Janisiewicz, W.J.; Fuchs, Y.; Camp, M.J.; Chigladze, E. & Sulakvelidze, A. (2001) Examination of bacteriophage as a biocontrol method for *Salmonella* on fresh fruit: a model study. *Journal of Food Protection* Vol. 64, pp. 1116-1121.

Leverentz, B.; Conway, W.S.; Camp, M.J.; Janisiewicz, W.J.; Abuladze, T.; Yang, M.; Saftner, R. & Sulakvelidze, A. (2003) Biocontrol of *Listeria monocytogenes* on fresh-cut produce by treatment with lytic bacteriophages and a bacterosin. *Applied and Environmental Microbiology* Vol. 69, pp. 4519-4526.

Lowbury, E.J. & Hood, A.M. (1953) The acquired resistance of *Staphylococcus aureus* to bacteriophage. *Journal of General Microbiol* Vol. 9, pp. 524-535.

Mai, V.; Ukhanova, M.; Visone, L.; Abuladze, T. & Sulakvelidze,A. (2010) Bacteriophage administration reduces the concentration of *Lisateria monocytogenes* in the gastrointestinal tract and its translocation to spleen and liver in experimentally infected mice. *International Journal of Microbiology* Article ID 624234 doi:10.1155/2010/624234.

Manning, J.G.; Hargis, B.M.; Hinton, A.; Corrier, D.E.; DeLoach, J.R. & Creger, C.R. (1992) Effect of nitrofurazone or novabiocin on *Salmonella enteritidis* cecal colonization and organ invasion in leghorn hens. *Avian Diseases* Vol. 36, pp. 334.

Manning, J.G.; Hargis, B.M.; Hinton, A.; Corrier, D.E.; DeLoach, J.R. & Creger, C.R. (1994) Effect of selected antibiotics and anticoccidials on *Salmonella enteritidis* cecal colonization and organ invasion in Leghorn chicks. *Avian Diseases* Vol. 38, pp. 256.

Marčuk, L.M.; Nikiforov, V.N.; Ščerbak, Ja.F.; Levitov, T.A.; Kotljanova, R.I.; Nanmŝina, M.S.; Davydov, S.U.; Monsur, K.A.; Rahman, M.A.; Latif, M.A.; Northrup, R.S.; Cash, R.A.; Huq, I.; Dey, C.R. & Phillips, R.A. (1971) Clinical studies in the use of bacteriophage in the treatment of Cholera. *Bulletein of Organization mond Santé Bulletein World Health Organization* Vol. 45, pp. 77-83.

Merril, C.R.; Scholl, D. & Adhya, S.L. (2003) The prospect for bacteriophage therapy in Western medicine. *National Review* Vol. 2, pp. 489-497.

Muirhead, S. (1994) Feed Additive Compendium. Miller Publishing, Minneapolis, MN.

Naylor, S.W.; Low, J.C.; Besser, T.E.; Mahajan, A.; Gunn, G.J.; Pearce, M.C.; McKendrick, I.J.; Smith, D.G.E. & Gally, D.L. (2003) Lymphoid Follicle-Dense Mucosa at the Terminal Rectum Is the Principal Site of Colonization of Enterohemorrhagic

Escherichia coli O157:H7 in the Bovine Host. *Infection and Immunity* Vol. 71, Iss. 3, pp. 1505-1512.

O'Flynn, G.; Ross, R.P.; Fitzgerald, G.F. & Coffey, A. (2004) Evaluation of a cocktail of three bacteriophages for biocontrol of *Escherichia coli* O157:H7. *Applied Environmental Microbiology* Vol. 70, pp. 3417-3424.

Oot, R.A.; Raya, R.R.; Callaway, T.R.; Edrington, T.S.; Kutter, E.M. & Brabban, A.D. (2007) Prevalence of *Escherichia coli* O157:H7-infecting bacteriophages in feedlot cattle feces. *Letters in Applied Microbiology* Vol. 45, pp. 445-453.

Orpin, C.G. & E.A. Munn (1973) The occurrence of bacteriophages in the rumen and their influence on rumen bacterial populations. *Experentia* Vol. 30, pp. 1018-1020.

Persson, U. & Jendteg, S.U. (1992) The economic impact of poultry-borne salmonellosis: how much should be spent on prophylaxis? *International Journal of Food Microbiology* Vol. 15, pp. 207.

Pyle, N.J. (1926) The bacteriophage in relation to *Salmonella Pullora* infection in the domestic fowl. *Journal of Bacteriology* Vol. 12, pp. 245-261.

Rangel, J.M.; Sparling, P.H.; Crowe, C.; Griffin, P.M.; & Swerdlow, D.L. (2005) Epidemiology of *Escherichia coli* O157:H7 outbreaks, United States, 1982-2002. *Emerging Infectious Disease* Vol. 11, pp. 603-609.

Raya, R.R.; Varey, P.; Cot, R.A.; Dyen, M.R.; Callaway, T.R.; Edrington, T.S.; Kutter, E.M.; and Brabban, A.D. (2006) Isolation and characterization of a new T-even bacteriophage, CEV1, and determination of its potential to reduce *Escherichia coli* O157:H7 levels in sheep. *Applied Environmental Microbiology* Vol. 72, No. 9, pp. 6405-6410.

Smith, H.W. & Huggins, M.B. (1982) Successful treatment of experimental *Escherichia coli* infection in mice using phage: its general superiority over antibiotics. *Journal of General Microbiology* Vol. 128, pp. 307-318.

Smith, H.W. & Huggins, M.B. (1983) Effectiveness of phages in treating experimental *Escherichia coli* diarrhea in calves, piglets, and lambs. *Journal of General Microbiology* 129: 2659-2675

Smith, H.W.; Huggins, M.B. & Shaw, K.M. (1987) The control of experimental *Escherichia coli* diarrhea in calves by means of bacteriophages. *Journal of General Microbiology* Vol. 133, pp. 1111-1126.

Sayamov, R.M. (1963) Treatment and prophylaxis of Cholera with bacteriophage. *Bulletin of Organization mond Santé Bulletin Worlld Health Organization* Vol. 28, pp. 361-367.

Sheng, H.; Kneccht, H.J.; Kudva, I.T. & Hovde, Q. (2006) Application of bacteriophages to control intestinal *Escherichia coli* O157:H7 levels in ruminants. *Applied Environmental Microbiology* Vol. 72, pp. 5359-5366.

Slopek, S.; Durlakowa, I.; Weber-Dabrowska, B.; Kucharewicz-Krukowska, A.; Dabrowski, M. & Bisikiewicz, R. (1983a) Results of bacteriophage treatment of supperative bacterial infections. I. General evaluation of the results. *Archivum in Immunologae et Therapiae and Experimentalis* Vol. 31, pp. 267-91.

Slopek, S.; Durlakowa, I.; Weber-Dabrowska, B.; Kucharewicz-Krukowska, A.; Dabrowski, M. & Bisikiewicz, R. (1983b) Results of bacteriophage treatment of supperative bacterial infections. II. Detailed evaluation of the results. *Archivum in Immunologae et Therapiae and Experimentalis Experiments* Vol. 31, pp. 293-327.

Slopek, S.; Durlakowa, I.; Weber-Dabrowska, B.; Kucharewicz-Krukowska, A.; Dabrowski, M.; & Bisikiewicz, R. (1984) Results of bacteriophage treatment of supparative bacterial infections. III. Detailed evaluation of the results obtained in further 150 cases. *Archivum in Immunologae et Therapiae and Experimentalis* Vol. 32, pp. 317-35.

Slopek, S.; Durlakowa, I.; Weber-Dabrowska, B.; Kucharewicz-Krukowska, A.; Dabrowski, M. & Bisikiewicz, R. (1985a) Results of bacteriophage treatment of supparative bacterial infections. IV. Evaluation of the results obtained in 370 cases. *Archivum in Immunologae et Therapiae and Experimentalis* Vol. 33, pp. 219-40.

Slopek, S.; Durlakowa, I.; Weber-Dabrowska, B.; Kucharewicz-Krukowska, A.; Dabrowski, M. & Bisikiewicz, R. (1985b) Results of bacteriophage treatment of supparative bacterial infections. V. Evaluation of the results obtained in children. *Archivum in Immunologae et Therapiae and Experimentalis* Vol. 33, pp. 241-59.

Slopek, S.; Durlakowa, I.; Weber-Dabrowska, B.; Kucharewicz-Krukowska, A.; Dabrowski, M. & Bisikiewicz, R. (1985c) Results of bacteriophage treatment of supparative bacterial infections. VI. Analysis of treatment of suppurative staphylococcal infections. *Archivum in Immunologae et Therapiae and Experimentalis* Vol. 33, pp. 261-73.

Slopek, S.; Weber-Dabrowska, B.; Dabrowski, M. & Kucharewicz-Krukowska, A. (1987) Results of bacteriophage treatment of suppurative bacterial infections in the years 1981 – 1986. *Archivum in Immunologae et Therapiae and Experimentalis* Vol. 35, pp. 569-583.

Soothill, J.S. (1992) Treatment of experimental infections of mice with bacteriophages. *J Med Microbiol* Vol. 37, pp. 258-261.

Soothill, J.S. (1994) Bacteriophage prevents destruction of skin grafts by *Pseudomonas aeruginosa*. *Burns* Vol. 20, pp. 209-211.

Swain, R.A.; Nolan, J.V. Klieve, A.V. (1996) Natural variability and diurnal fluctuations within the bacteriophage population of the rumen. *Applied Environmental Microbiology* 62, 994–997.

Tanji, Y.; Shimada, T.; Fukudomi, H.; Miyanaga, K.; Nakai, Y. & Unno, H. (2005) Therapeutic use of phage cocktail for controlling *Escherichia coli* O157:H7 in gastrointestinal tract of mice. *Journal of Bioscience and Bioengineering* Vol. 100, pp. 280-287.

Toro, H.; Price, S.B.; McKee, S.; Hoerr, F.J.; Krehling, J.; Perdue, M. & Bauermeister, L. (2005) Use of bacteriophages in combination with competitive exclusion to reduce *Salmonella* from infected chickens. *Avian Diseases* Vol. 49, No. 1, pp. 118-124.

Wells, J.G.; Shipman, L.D.; Greene, K.D.; Sowers, E.G.; Green, J.H.; Cameron, D.N.; Downes, F.P.; Martin, M.L.; Griffin, P.M. & Ostroff, S.M. (1991) Isolation of *Escherichia coli* serotype O157:H7 and other Shiga-like-toxinproducing *E. coli* from dairy cattle. *Journal of Clinical Microbiology* Vol. 29, pp. 985–989.

World Health Organization (2006) Subject: Drug resistant *Salmonella*. http://www.who.int/mediacentre/factsheets/fs139/en/

Zhao, T.; Doyle, M.P.; Shere, J. & Garber, L. (1995) Prevalence of enterohemorrhagic *Escherichia coli* O157:H7 in a survey of dairy herds. *Applied Environmental Microbiology* Vol. 61, pp. 1290–1293.

9

Bacteriophages of *Ralstonia solanacearum*: Their Diversity and Utilization as Biocontrol Agents in Agriculture

Takashi Yamada
Department of Molecular Biotechnology, Graduate School of Advanced Sciences of Matter,
Hiroshima University, Higashi-Hiroshima,
Japan

1. Introduction

Bacterial wilt is one of the most important crop diseases, and is caused by the soil-borne Gram-negative bacterium *Ralstonia solanacearum*. *R. solanacearum* was formerly classified as *Pseudomonas solanacearum* or *Bacterium solanacearum* (Smith, 1986; Yabuuchi et al., 1995). This bacterium has an unusually wide host range, infecting more than 200 species belonging to more than 50 botanical families, including economically important crops (Hayward, 1991; Hayward, 2000). *R. solanacearum* strains represent a heterogeneous group, subdivided into five races based on host range, and into five biovars based on physiological and biochemical characteristics (Hayward, 2000). There is no general correlation between races and biovars, and the five races of *R. solanacearum* have different geographical distributions. Race 1 is a poorly defined group with a very wide host range, and is endemic to tropical, subtropical, and warm areas. Strains of race 2 mainly infect bananas, and are found primarily in Southeast Asia and Central America. Race 3 strains are distributed worldwide, and are principally associated with potato. Strains of race 4 infect ginger in areas of Asia and Hawaii, and race 5 strains infect mulberries in China. Recently, a new classification system for *R. solanacearum* strains, based on phylogenetic information, has been proposed, where strains are sub-grouped into four phylotypes roughly corresponding to their geographic origin. Phylotype I includes strains originating primarily from Asia, phylotype II from America, phylotype III from Africa and surrounding islands in the Indian Ocean, and phylotype IV from Indonesia (Fegan & Prior, 2005).

In the field, *R. solanacearum* is easily disseminated via soil, contaminated irrigation water, surface water, farm equipment, and infected material (Janse, 1996). Bacterial cells can survive for many years in association with alternate hosts. Once identified as being infected, plants in cropping fields, gardens, or greenhouses must be destroyed, and soil and water draining systems that could potentially be contaminated with the bacteria must be treated with chemical bacteriocides. Soil fumigation with methyl bromide, vapam, or chloropicrin is of limited efficacy. Methyl bromide depletes the stratospheric ozone layer; therefore, the production and use of this gas was phased out in 2005, under the *Montreal Protocol and the Clean Air Act*. The limited effectiveness of the current integrated management strategies has meant that bacterial wilt continues to be an economically serious problem for field-grown crops in many tropical, subtropical, and warm areas of the world (Hayward, 1991; Hayward, 2000). For example, bacterial wilt of potato has been estimated to affect 3.75 million acres in approximately 80 countries, with global damage estimates currently exceeding $950 million per year (Momol et al., 2006). At present, protection from losses by bacterial wilt is provided mainly by early detection and subsequent eradication by destroying the host. Development of effective disease management strategies and improvement in detection and monitoring tools are required.

Various kinds of bacteriophage with characteristic features have been isolated recently (Yamada et al., 2007), and have paved the way for new methods of biocontrol of bacterial wilt. These phages may be useful as tools for effective detection (diagnosis) of the pathogen in cropping ecosystems and in growing crops. They also have potential uses in eradication of the pathogen from contaminated soil or prevention of bacterial wilt in economically important crops. Like other methods of biological control, one advantage of phage biocontrol is the reduction in the use of chemical agents against pathogens. This prevents the problems of multiple environmental pollution, ecosystem disruption, and residual chemicals on the crops. Phage biocontrol in agricultural settings was extensively explored 40–50 years ago as a means of controlling plant pathogens (Okabe & Goto, 1992). Two major problems arose in those practical trials; (i) extracellular polysaccharides produced by pathogenic bacteria prevented phage adsorption, and (ii) there were various degrees of susceptibility among bacterial strains (Goto, 1992). Nevertheless, over recent decades, the use of phage biocontrol to restrict the growth of plant-based bacterial pathogens has been explored with increasing enthusiasm. Certain bacteriophages of *R. solanacearum* have already been isolated and some of their physical and physiological properties have been characterized. The virulent phage P4282, and its related phages, have a polyhedral head (69 nm in diameter) and a short tail (20 nm in length) (Tanaka et al., 1990), and contain a circular 39.3-kbp dsDNA genome (Ozawa et al., 2001). Another phage, PK-101, was isolated from soil, and has a linear 35-kbp dsDNA genome (Toyoda et al., 1991). Both of these phages show very narrow host ranges and infect only a few strains of *R. solanacearum*. Phage P4282, which infects strain M4S, was used to control bacterial wilt of tobacco plants in laboratory experiments and possible phage-mediated protection was observed (Tanaka et al., 1990). However, for practical use of phages as biocontrol agents against bacterial wilt, multiple phages with wide host-ranges and strong lytic activity are required.

2. Characterization and classification of bacteriophages infecting *R. solanacearum*

Recently, Yamada et al. (2007) isolated and characterized several different kinds of phage that specifically infect *R. solanacearum* strains belonging to different races and/or biovars. Two of the phages, φRSS1 and φRSM1, are filamentous, Ff-like phages (*Inoviridae*). As demonstrated for coliphage M13, filamentous phages are very useful as vectors to display proteins on the virion surface in the bacteriophage-display method (Smith, 1985; Smith, 1991). In the same way, φRSS1 and φRSM1 may be utilized to display a tag protein on the virion surface in diagnosis or monitoring applications (Kawasaki et al., 2007a). Another phage, φRSA1, is a P2-like head-tailed virus (*Myoviridae*), and has the widest host range; all but one of 18 strains tested of races 1, 3, or 4 and biovar 1, N2, 3, or 4 were susceptible to this phage. φRSL1 is another myovirus having a very large genomic DNA of approximately 240 kbp and strongly lysed 17 of 18 different strains. Another lytic phage, φRSB1, shows a T7-like morphology (*Podoviridae*) and also has a wide-host-range (15 of 18 strains of races 1, 3, or 4 were susceptible). φRSB1 vigorously lyses host cells and forms very large clear plaques that extend for 10–15 cm on assay plates. The characteristics and host ranges of these phages are summarized in Table 1. Further surveys for phages infecting different strains of *R. solanacearum* detected many interesting examples, some of which were induced from lysogenic strains. Searching through the genomic databases for the sequences determined for these phages also revealed similar sequences integrated in various bacterial genomes (prophages). Such phages and prophages of *R. solanacearum* and related bacterial species are roughly grouped into six phage-types as follows: φRSS phages, φRSM phages, φRSB phages, φRSA phages, φRSL phages, φRSC phages , and others.

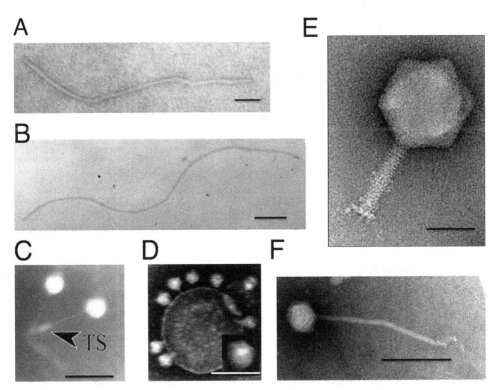

Fig. 1. Electron micrographs showing morphology of bacteriophages infecting *R. solanacearum*.

φRSS1 (A) and φRSM1 (B) are filamentous phages of *Inoviridae*. φRSA1 (C) and φRSL1 (E) are myoviruses (*Myoviridae*) with a contractile tail. φRSB1 is a podovirus (*Podoviridae*) with a head and a short tail (D). Several φRSB1 particles are seen attached to a membrane vesicle. φRSC1 is a λ like phage (*Siphoviridae*) with a non-contractile tail (F). TS, tail sheath. Bar represents 100 nm.

Ralstonia solanacearum		Phages					
Strain	Race, Biovar	φRSS1	φRSM1	φRSA1	φRSB1	φRSL1	φRSC1
C319	1, ND	+	–	+	+	+	+
M4S	1 . 3	–	+	+	+	+	+
Ps29	1 , 3	–	+	+	+	+	+
Ps65	1, ND	–	+	+	+	+	+
Ps72	1, ND	–	+	+	+	+	+
Ps74	1, ND	–	+	+	+	+	+
MAFF106603	1 , 3	+	–	+	+	+	+
MAFF106611	1 , 4	+	–	+	+	+	+
MAFF211270	1, N2	–	+	+	–	+	–
MAFF211271	3, N2	+	–	+	+	+	–
MAFF211272	4 , 4	–	–	+	+	–	–
MAFF301556	1 , 4	+	–	+	+	+	–
MAFF301558	3, N2	+	–	+	+	+	–
MAFF730135	1 , 4	+	–	+	+	+	–
MAFF730138	1 , 3	–	+	+	+	+	+
MAFF730139	1 , 4	+	–	+	+	+	+
RS1002	1 , 4	+	–	+	–	+	–
AA4017	1, ND	+	–	–	–	+	–

Phage susceptibility is shown as sensitive (+) or resistant (-). Some original data are updated (Yamada et al., 2007; Askora et al., 2009). ND, Not determined.

Table 1. Host specificity of bacteriophages infecting R. *solanacearum*

Ralstonia solanacearum		Phages					
Strain	Race, Biovar	φRSS1	φRSS0	φRSS2	φRSS4	φRSM1	φRSM3	φRSM13
C319	1, ND	+	–	–	+	–	+	+
M4S	1 . 3	–	–	+	–	+	–	–
Ps29	1 , 3	–	–	+	–	+	–	–
Ps65	1, ND	–	–	+	–	+	–	–
Ps72	1, ND	–	–	+	–	+	+	+
Ps74	1, ND	–	–	+	–	+	+	+
MAFF106603	1 , 3	+	+	–	+	–	+	+
MAFF106611	1 , 4	+	+	–	+	–	+	+
MAFF211270	1, N2	–	–	+	–	+	+	+
MAFF211271	3, N2	+	–	–	+	–	+	+
MAFF211272	4 , 4	–	–	–	–	–	+	+
MAFF301556	1 , 4	+	–	–	+	–	+	+
MAFF301558	3, N2	+	–	–	+	–	+	+
MAFF730135	1 , 4	+	–	ND	+	–	ND	ND
MAFF730138	1 , 3	–	–	+	–	+	–	–
MAFF730139	1 , 4	+	+	–	+	–	+	+
RS1002	1 , 4	+	+	ND	+	–	ND	ND
AA4017	1, ND	+	–	ND	+	–	ND	ND

Phage susceptibility is shown as sensitive (+) or resistant (-). Some original data are updated (Yamada et al., 2007; Askora et al., 2009). ND, Not determined.

Table 2. Host specificity of different phages of φRSS and φRSM groups

2.1 φRSS phages of the *Inoviridae*

φRSS1 was isolated from a soil sample collected from tomato crop fields. It infected 10 of 18 strains tested and gave relatively small and turbid plaques on assay plates. φRSS1 particles have a flexible filamentous shape of 1100 ± 100 nm in length and 10 ± 0.5 nm in width (Fig. 1), giving a morphology resembling coliphage fd (Buchen-Osmod, 2003; ICTVdB). Infection with φRSS1 phage does not cause host cell lysis, but establishes a persistent association between the host and phage, releasing phage particles from the growing and dividing host cells. The φRSS1 particles contain single-stranded (ss)DNA as the genome. Therefore, φRSS1 belongs to Ff-like phages or inoviruses. In general, the genome of Ff-phages is organized in a module structure, in which functionally related genes are grouped (Hill et al., 1991; Rasched & Oberer, 1986). Three functional modules are always present: the replication module, the structural module, and the assembly and secretion module. The replication module, contains the genes encoding rolling-circle DNA replication and single-strand DNA (ssDNA) binding proteins; gII, gV, and gX (Model & Russel, 1988). The structural module contains genes for the major (gVIII) and minor coat proteins (gIII, gVI, gVII, and gIX), and gene gIII encodes the host recognition or adsorption protein pIII (Armstrong et al., 1981). The assembly and secretion module contains the genes (gI, and gIV) for morphogenesis and extrusion of the phage particles (Marvin,1998). Gene gIV encodes protein pIV, an aqueous channel (secretin) in the outer membrane, through which phage particles exit from the host cells. Some phages encode their own secretins, whereas others use host products (Davis et al., 2000). The genome of φRSS1 is 6,662 nt long (DDBJ accession No. AB259124), with a GC content of 62.6 %, which is comparable to that of R. solanacearum GMI1000 (66.97 %, Saranoubat et al., 2002). There are 11 open reading frames (ORFs), located on the same strand (Fig. 2A). Frequently, the termination codon of the preceding gene overlaps with the initiation codon of the following gene. The coding sequence occupied by these ORFs accounts for 91.6% of the total φRSS1 sequence. A survey of the databases for amino acid sequences of φRSS1 ORFs leads to the gene organization as shown in Fig. 2A. The φRSS1 genes fit well with the general arrangement of Ff-like phages. Homology searches revealed that the dsDNA plasmid pJTPS1, found in a strain R. solanacearum (accession no. AB015669), has significant nucleotide sequence similarity to φRSS1 DNA. The size of pJTPS1 is 6,633 bp; 29 bp smaller than that of φRSS1. Nucleotide sequence identity between the two DNAs was 95%. Compared with pJTPS1, two major different regions in the φRSS1 DNA are evident: One region (φRSS1 positions 2,674-3,014) corresponds to ORF7, putatively encoding pIII, which is a minor coat protein at one end of the phage particle required to recognize and adsorb host cells (Marvin, 1998, Model and Russel, 1988). The other extended change was found at positions 6,632-6,657 corresponding to the IG (intergenic region), which is highly conserved in other Ff-like phages (Model and Russel, 1988). This region may be involved in the rolling circle DNA replication mechanism, producing phage genomic ssDNA molecules. These results suggest that pJTPS1 may have been derived from a φRSS1-like phage, followed by changes in the phage DNA. We tested the phage nature of pJTPS1 in strain M4S, where pJTPS1 was initially identified as a circular dsDNA plasmid (Negishi, et al., 1993). We could not detect dsDNA plasmid in strain M4S, but found a φRSS1-related sequence integrated into the M4S genome. According to the pJTPS1 sequence, the corresponding sequence was amplified by PCR and circularized by self-ligation. When the resulting DNA was introduced in M4S cells, plaques appeared on assay plates. The nucleotide sequence of the phage coincided with pJTPS1 (Kawasaki et al., 2007a). The plaque size was relatively small and the frequency was relatively low, indicating that pJTPS1 is a ccc form of a φRSS1-related phage (designated as φRSS2) and is integrated in strain M4S. There is no immunity in the phage infection. Interestingly, the pJTPS1 phage (φRSS2) infected φRSM1-sensitive strains, including M4S (see below) and did not infect φRSS1-sensitive strains, including strain C319. This may be because of the specific difference in a region of pIII host recognition protein between φRSS1 and φRSS2.

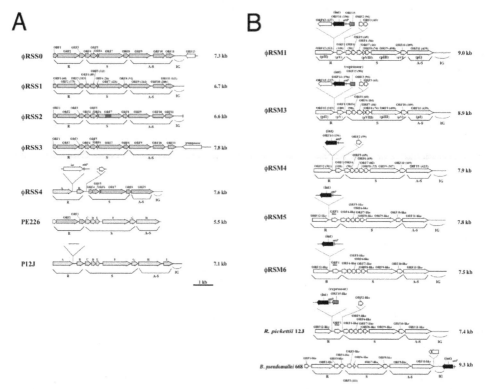

According to the *E. coli* M13-model (Model and Russel, 1988; Marvin, 1998), ORFs identified on the phage genome are grouped in the functional modules for replication (R), structure (S), and assembly-secretion (A-S). IG, large intergenic region. Among φRSS1-related phages (A), φRSS0, φRSS2 and φRSS3 were derived from prophages of strains C319, M4S, and MAFF106611, respectively. P12J and PE226 are phages of *Ralstonia pickettii* (accession no. AY374414) and Korean strains of *R. solanacearum* (Murugaiyan et al., 2010). Of φRSM1-related phages (B), φRSM3, φRSM4, φRSM5, and φRSM6 are prophages of strains MAFF730139, UW551 (race 3, biovar 2; Gabriel et al., 2006), IPO1609 (race 3, biovae 2; Remenant et al., 2010), and CMR15 (phylotype III, Remenant et al., 2010), respectively. A similar prophage found in *Ralstonia pickettii* 12J is also shown.

Fig. 2. Genomic organization of φRSS1-related phages (A) and φRSM1-related phages (B).

Genomic Southern blot hybridization showed frequent examples of φRSS1-related sequences integrated in the genomes of various *R. solanacearum* strains (Yamada et al., 2007). As seen above, φRSS2 is integrated in the genome of strain M4S. φRSS1 was also found to exist as a lysogenic phage in strain C319 (Kawasaki et al., 2007). By analyzing the host genomic sequences flanking the φRSS1 region, its integration site was determined to correspond to φRSS1 position 6,629 in the IG region, 34 nt upstream from ORF1. However, the nucleotide sequence around this φRSS1 showed no significant homology to the core sequences involved in the XerC/D recombination system (Kawasaki et al., 2007a). Tasaka et al. (unpublished data) further characterized the nucleotide sequences in the neighborhood of the φRSS1-integration site in strain C319. It was revealed that φRSS1 is a truncated form of a larger phage (designated as φRSS0) of 7,288 nt in size, 626 nt larger than φRSS1 (Fig. 2A). The 626 nt φRSS0 sequence missing from φRSS1 DNA contains two nucleotide sequence elements that repeat at both *attL* and *attR*, the latter of which has the sequence 5'-TATTT AACAT AAGAT AAAT, corresponding to *dif* of *R. solanacearum* (Carnoy and Roten, 2009). Thus, φRSS0 is integrated at a *dif* site, similarly to CTXφ of *Vibrio cholerae*, which uses the host XerC/D recombination system (Huber and Waldor, 2002). Interestingly, one φRSS0 ORF (ORF13; 156 amino acid residues), located within the region missing in the φRSS1 DNA, shows sequence homology to DNA-binding phage regulator (accession no. B5SCX5, E-value 1e-29). This may function as a phage repressor for immunity, because C319 (φRSS0 lysogen) is resistant to second infection by φRSS0 (Table 2). C319 is susceptible to φRSS1, thus φRSS1 (without ORF13) may be an escaped superinfective phage.

Another form of φRSS1-type prophage was found in strain MAFF106601 (Tasaka et al., unpublished data). This prophage (designated as φRSS3) has an extended size (8,193 nt) caused by an insertion of an IS1405 element (transposase) within the IG (positions 6,250-6,251 of φRSS1). The map of φRSS3 is compared with other related phages in Fig. 2A. This phage was active when its genomic DNA was amplified from the MAFF106601 genome by PCR and introduced into MAFF106603 as a host. It is interesting to note that φRSS3 may exploit the transposase to move or transmit itself among genomes. A chimeric form of a φRSS1-type phage is seen in the structure of φRSS4, a prophage in strain MAFF211271 (Kawasaki et al., unpublished data). This phage (7,618 nt in size) is integrated at the host gene for arginine tRNA (CGG) as *attB* and it contains a 58 nt sequence corresponding to the 3'-portion of arginine tRNA (CGG) as *attP*. An ORF of 363 aa residues is closely associated with *attP* and this unit (*attP*/Int) is almost identical to that of φRSA1 (Fujiwara et al., 2008), a P2-type myovirus as described below. The map of φRSS4 is shown in Fig. 2A. As can be seen, an approximately 2.2 kb region (corresponding to the structure module) containing ORF4 to ORF8 and an ORF of assembly-secretion module (ORF9) are almost identical between φRSS1 and φRSS4, but most of the replication module in φRSS4 is replaced with that of phage P12J infecting *Ralstonia pickettii* (accession no. AY374414). The replication module of φRSS4 is inserted with an ORF (ORF2) and attP; ORF 2 shows amino acid sequence similarity to *Burkholderia pseudomallei* bacteriophage integrase (accession no. Q63PM9, E-value e-100). A similar, but significantly smaller, phage, PE226, infecting *R. solanacearum* strains was recently reported (Murugaiyan et al., 2010). The φRSS4 structure suggests that φRSS phages can exchange each of the Ff-phage modules (Model and Russel, 1988) among related phages and evolve to a new characteristic phage.

2.2 φRSM phages of the *Inoviridae*

φRSM1 was isolated from a soil sample and can form small plaques on assay plates with limited strains (7 of 18 strains tested) as a host. φRSM1 particles show a long fibrous shape of 1400 ± 300 μm in length and 10 ± 0.7 nm in width by electron microscopy (Fig. 1), giving a shape similar to coliphage M13 (Buchen-Osmond, 2003; ICTVdB). The infection cycle of φRSM1 phage resembles that of φRSS1. The genome of φRSM1 is 9,004 nt long (DDBJ accession No. AB259123) with a GC content of 59.9%, which is lower than that of *R. solanacearum* GMI1000 (66.97%). There are 12 putative ORFs located on the same strand and two on the opposite strand. Compared with the conserved gene arrangement of Ff-like phages, the φRSM1 genes can be drawn as Fig. 2B (Kawasaki et al., 2007a). Here, ORF13 and ORF14 (reversely oriented) are inserted between ORF11, corresponding to pII as a replication protein, and ORF1, corresponding to an ssDNA-binding protein like pV, in the putative replication module. ORF13 and ORF14 show amino acid sequence similarity to a proline-rich transmembrane protein and a resolvase/DNA invertase–like recombinase, respectively. There are two additional ORFs (ORF2 and ORF3) between the replication and structural modules. The functions of these ORF-encoded proteins are not known. In genomic Southern blot hybridization, two different types of φRSM1-related prophage sequences were detected in *R. solanacearum* strains. Strains of type A include MAFF211270 and produce φRSM1 itself, and strains of type B (giving different restriction patterns) are resistant to φRSM1 infection, but serve as hosts for φRSS1 with different nature and host range as described above (Kawasaki et al., 2007a). By determining the nucleotide sequences of junction regions of the φRSM1-prophage in the MAFF211270 chromosomal DNA, an *attP/attB* core sequence was identified as 5'-TGGCGGAGAGGGT-3', corresponding to positions 8,544-8,556 of φRSM1 DNA, located between ORF14 and ORF1. Its nucleotide sequence is identical to the 3'-end of the host *R. solanacearum* gene for serine tRNA(UCG) in the reverse orientation. By PCR with appropriate primers containing these *att* sequences, a φRSM1-like prophage (type B) in strain MAFF730139 (designated φRSM3) was obtained (Askora et al., 2009). Compared with the φRSM1 genome, the φRSM3 prophage sequence (8,929 bp) is 75 bp shorter. The sequences show 93% nucleotide identity and major differences are found within two regions; positions 400-600 and positions 2,500-3,000 in the φRSM1 sequence. The former region corresponds to ORF2, which is inserted between the replication module (R) and the structural module (S), and the latter falls into the possible D2 domain of ORF9 (pIII), as described below. All 14 ORFs identified along the φRSM3 sequence show high amino acid sequence homology (more than 90% amino acid identity) with their counterparts on the φRSM1 genomic DNA, except for ORF2 (no similarity) and ORF9 (79% identity). It is interesting that the amino acid sequence of ORF14 (putative DNA invertase/recombinase) is 100% identical in the two phages. The gene arrangements are compared in Fig. 2B. During database searches for homologous sequences, we found that an approximately 8 kbp region of the *R. solanacearum* UW551 genome (accession no. DDBJ/EMBL/GenBank AAKL00000000; RS-UW551-Contig0570-70-86K) at positions 3,039-10,984 shows significant DNA sequence homology with φRSM1 and φRSM3. The prophage of UW551 (designated φRSM4, Askora et al., 2009) contains 7,929 bp flanked by *att* sequences, as in φRSM1, comprising 13-bp of the 3'-end of serine tRNA (UCG). The 7,929-bp φRSM4-prophage sequence shows 72% nucleotide sequence identity with φRSM1 DNA. There are three deletions in φRSM4 compared with the φRSM1 gene arrangement; ORF3, ORF13, and part (positions 5,370-5,800) of the intergenic region (IG) are missing. Other ORFs identified in the φRSM4 sequence show variable amino acid sequence similarity to their counterparts in φRSM1: no similarity in ORF2, moderate similarity in ORF4 (57% amino acid identity), ORF5 (77%), and ORF9 (73%), and high similarity in ORF1 (93%), ORF6 (98%), ORF7 (98%), ORF8 (93%), ORF10 (86%), ORF11 (89%), ORF12 (89%), and ORF14 (81%). The gene arrangement of φRSM4 prophage DNA is also included in the comparison shown in Fig. 2B.

One of the major differences in the predicted genes between φRSM1 and φRSM3 is confined within the middle part of ORF9, corresponding to pIII, the host recognition and adsorption minor coat protein. Interestingly, this internal region of φRSM3 (also 67% identical in φRSM4) shows a significant similarity (79%) to the corresponding region of φRSS1, which shows a different host range from φRSM1 (Yamada et al., 2007). These results predict that φRSM3 and φRSM4 share a common host range, which differs from that of φRSM1, and suggest that the host range determined by the pIII-D2 domain may be exchangeable among phages. This was experimentally confirmed. The circularized φRSM3 PCR product was introduced into MAFF211272 and resulting phage with the expected DNA restriction pattern was subjected to host-range assay. The results are summarized in Table 2. The host range of φRSM3 is incompatible with that of φRSM1; namely φRSM1-susceptible strains are resistant to φRSM3 and vice versa, except for strains Ps74 and MAFF211270, which are susceptible to both phages. φRSM3 has a wider host range, and all 15 strains are susceptible to either φRSM1 or φRSM3. The host range of φRSM3 partially overlaps with that of φRSS1 (Yamada et al., 2007), but φRSS1 recognizes only ten strains (Table 2), suggesting the involvement of some additional factors in φRSS1 host recognition. To confirm the host recognition by the D2 domain of pIII in φRSM1 and φRSM3, this domain of φRSM1 was replaced with the corresponding region of φRSM3. The resulting phage (designated as φRSM13) did not retain the host range of φRSM1 but showed the exactly same host range as that of φRSM3 (Table 2). The entire nucleotide sequence of φRSM13 DNA is completely the same as that of φRSM1, except for a 350-bp region corresponding to the pIII-D2 of φRSM3; therefore, the change of host range must have been determined by the D2 domain of pIII of φRSM3 (Askora et al., 2009). The differences in the pIII-D2 domain among φRSM phages correspond well to the strain-specific pili types. A minor pilin of approximately 30 kDa varies in size, depending on the strain; slightly smaller proteins correspond to φRSM1-susceptible strains and slightly larger proteins correspond to φRSM3-susceptible strains (Askora et al., 2009).

As described above, the genomes of φRSM phages are sometimes found to be integrated in the host genome. Askora et al. (2011) demonstrated that the integration is mediated by the phage-encoded recombinase (ORF14 of φRSM1), which has significant homology to resolvases/DNA invertases (small serine recombinases) at the sites, attP/attB corresponding to the 3′ end of the host serine tRNA (UCG) gene in the reverse orientation. The same unit of integration (φRSM Int/attP) was found in a R. pickettii 12J phage and in B. pseudomallei 668 prophages. Together with these phages, it is not surprising that similar int-containing filamentous phages occur widely in the natural world.

2.3 φRSB phages of the *Podoviridae*

φRSB1 was isolated from a soil sample from a tomato crop field and was selected for its ability to form large clear plaques on plate cultures of R. solanacearum strain M4S. Plaques formed on assay plates were 1.0 to 1.5 cm in diameter. This phage has a wide host range and infected 15 of 18 strains tested, including strains of races 1, 3, and 4, and of biovars 3, 4, and N2. Under laboratory conditions, host cells of R. solanacearum strains lyse after 2.5 to 3 h postinfection (p.i.) (with an eclipse period of 1.5 to 2 h), releasing approximately 30 to 60 pfu of new phage particles per cell (burst size). Electron microscopic observation of negatively stained phage particles revealed short-tailed icosahedral structures resembling those of the family *Podoviridae*. The particles consist of a head of approximately 60 nm in diameter and a stubby tail of 20 nm in length (Fig. 1). The φRSB1 genome is linear double-stranded DNA of 43,079 bp and includes direct terminal repeats of 325 bp (accession no. AB276040). The G + C content of the genome is 61.7%. This value is lower than the G + C values of the large and small replicons of the *R. solanacearum* GMI1000 genome (67.04% and 66.86%, respectively) (Salanoubat et al., 2002). A total of 47 potential ORFs oriented in the same direction were assigned on the genome (Fig. 3A). Patchy or local nucleotide sequence homologies were detected in the genomic sequences of various phages, including *Xanthomonas oryzae* phages Xop411 (accession no. DQ777876) and Xp10 (accession no. AY299121) (Yuzenkova et al., 2003), *Pseudomonas aeruginosa* phages φKMV (accession no. AJ505558) (Lavigne et al., 2003), *Erwinia amylovora* phage Era103 (accession no. EF160123), and *Burkholderia cenocepacia* phage BcepB1A (accession no. AY616033). All of these are members of the family *Podoviridae*. The genome of coliphage T7, the representative of T7-like viruses of the *Podoviridae*, generally consists of three functional gene clusters: one for early functions (class I), one for DNA metabolism (class II), and the other for structural proteins and virion assembly (class III) (Dunn & Studier, 1983). These gene clusters are essentially conserved in the φRSB1 genome. Figure 3A shows the putative ORFs identified on the φRSB1 genome compared with ORFs from other phages: *Xanthomonas* phage Xop411 (giving the highest local similarities), *Pseudomonas* phage φKMV (showing marginal similarity but longest regions of similarity), and coliphage T7. One of the characteristic features found in the φRSB1 gene organization is that the predicted gene for RNA polymerase (RNAP) of φRSB1 (*orf26*) is not located in the early gene region (class I), but at the end of the class II region (Fig. 3A). Another exception is the gene for DNA ligase (DNAL), *orf25*, encoding the φRSB1 DNAL, is in 5' to the RNAP ORF (*orf26*), whereas the gene encoding T7 DNAL is downstream of the gene for RNAP, at the end of the class I cluster (Dunn & Studier, 1983). In *Pseudomonas* phages φKMV, LKD16, and LKA1, the DNAL gene is upstream of the gene for DNA polymerase in the class II gene cluster (Fig. 3A). T7-like phages are generally known as absolute lytic phages, with a few exceptions, such as integrase-coding phages, e.g., prophage 3 of *Pseudomonas putida* (Molineux, 1999) and the cyanophage P-SSP7 (Lindell et al., 2004). Occasionally, nucleotide sequences related to T7-like phages are found in conjunction with other temperate phages, such as λ-like phages that are integrated in various bacterial genomes (Brussow et al., 2004; Canchaya et al., 2003; Casjens, 2003; Hendrix et al., 2003). A sequence highly homologous to φRSB1 was found to be embedded in a large (85-kbp) λ like prophage sequence (φ1026b) integrated in the genome of *Burkholderia pseudomallei* 1710b (accession no. CP000124). The homologous region of the 1710b prophage contains eight ORFs encoding DNA primase, DNA helicase, DNAL, DNA polymerase, exonuclease, and RNAP, etc. These correspond to the class II genes of φRSB1, as shown in Fig. 3A. Both *Ralstonia* and *Burkholderia* belong to the Betaproteobacteria and may share common bacteriophages (Fujiwara et al., 2008).

Searching for core promoter-like sequences conserved in phages T3, T7, or SP6 in the φRSB1 intergenic regions did not reveal any significant homologies. Instead, a set of sequence elements (possible promoter elements) consisting of a GC-rich stretch and TTGT, TCTGG, and CGGGCAC motifs preceding an AG-rich Shine-Dalgarno sequence were found. The activity of transcriptional promoters of such elements on the φRSB1 genome was demonstrated using a green fluorescent protein (GFP)-expressing single-copy plasmid,

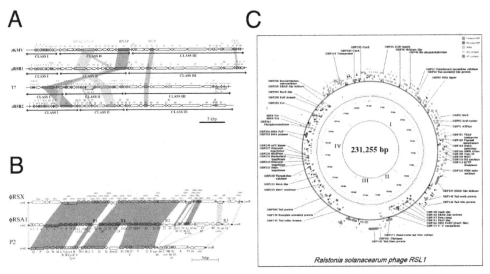

T7-like arrangements of 47 ORFs identified on the φRSB1 genome (43,079 bp, Kawasaki et al., 2009) and 50 ORFs on the φRSB2 genome are compared with those of E. coli T7 (39,937 bp, NC_00164) and Pseudomonas aeruginosa φKMV (42,519 bp, AJ50558)(A). According to the T7-gene-organaization, ORFs are grouped into three classes (Kawasaki et al., 2009). Corresponding major ORFs are connected by shading (DNAL, DNA ligase; DNAP, DNA polymerase; MCP, major capsid protein; LYS lysozyme). P2-like arrangement of 51 ORFs on the φRSA1 genome (38,760 bp, Fujiwara et al., 2008) and 52 ORFs on φRSX (40,713 bp), a prophage found on the genome of strain GMI1000 (Salanoubat et al., 2002). Shading indicates similar regions among the phages (B). In the map of φRSL1 genome (231,255), a total of 343 ORFs are grouped into 4 genomic regions according to their clustering with the same orientation (C).

Fig. 3. Genomic organization of φRSB1-related phages (A), φRSA1-related phages (B), and φRSL1 (C).

pRSS12 (Kawasaki et al., 2007b). A switch from host RNAP to φRSB1 RNAP occurs between 75 min p.i. and 90 min p.i., and the late stages of φRSB1 replication are independent of rifampin. A phage, φRSB3, from our phage collection also showed a similar gene arrangement to φRSB1, but its genome size is significantly larger (tentatively 44,242 bp) than φRSB1 and the nucleotide sequence similarity are entirely marginal between the two phages. The host range of φRSB3 is narrower (5 of 15 strains tested are susceptible). Another phage in our collection, which was obtained from a potato field, gave large clear plaques with 10 of 15 strains tested as hosts. Electron microscopy revealed a typical podoviral morphology of this phage (designated as φRSB2), an icosahedral head of 45 ± 5 nm in diameter, and a short tail of 12.5 ± 2 nm in length. The φRSB2 genome is linear double-stranded DNA of 40,411 bp and includes direct terminal repeats of 214 bp (accession no. AB597179). The G + C content of the genome is 61.7%. A total of 50 ORFs were identified along the genome. Homology searches through the databases revealed a general organization of this phage similar to T7-like phages, consisting of three gene classes; Class I (ORF1-ORF21), Class II (ORF22-ORF34), and Class III (ORF35-ORF50) (Fig. 3A). In contrast to φRSB1, the gene for RNAP associated with dnal (encoding DNAL) is located within Class I. The position of lys (encoding lysozyme) is within Class II in φRSB2, whereas it is near the right terminus in φRSB1.

Searching for core promoter-like sequences conserved in phages T3, T7, or SP6 in the φRSB2 intergenic region identified 14 elements with a consensus sequence 5′ ATTAACCCACACTRYAGGARRRS. The actual activity of the transcriptional promoters of some of these elements was demonstrated using GFP-expressing single-copy plasmid, pRSS12 (Kawasaki et al., unpublished data).

In early studies to control *R. solanacearum*, a few bacteriophages were isolated and their physical and physiological properties were partially characterized. The virulent phage P4282, and related phages, have a polyhedral head (69 nm in diameter) and a short tail (20 nm in length) (Tanaka et al., 1990), and contain a 39.3-kbp dsDNA genome (Ozawa et al., 2001), giving characteristic features of podoviruses. Another phage, PK-101, was isolated from soil, and characterized to have a linear 35-kbp dsDNA genome (Toyoda et al., 1991). The morphological feature of this phage is unknown. Both of these phages show very narrow host ranges and infect only a few strains of *R. solanacearum*. A recently isolated phage, φRSB4 showed a typical podoviral morphology by electron microscopy. Nucleotide sequence of a 40 kbp DNA fragment from the φRSB4 genome perfectly accorded with that of the P4282 gene encoding bacteriolytic protein (accession no. AB048798) (Ozawa et al., 2001). Genomic characterization of this phage is now underway.

2.4 φRSA phages of the *Myoviridae*

φRSA1 spontaneously appeared from strain MAFF211272 (Yamada et al., 2007). It consists of an icosahedral head of 40 ± 5 nm in diameter, a tail of 110 ± 8 nm in length and 3 ± 0.2 nm diameter, and a tail sheath (40 ± 6 nm in length and 17 ± 1.5 nm in diameter) located at the bottom of the tail (Fig. 1). This generates a racket-frame-like structure that resembles the morphology reported for *Burkholderia cepacia* phage KS5 (Seed and Dennis, 2005; Lynch et al., 2010). A tail sheath was often observed attached at the bottom of the tail, but sometimes at intermediate position along the tail, suggesting a movable nature of the sheath along the tail. Sometimes, structures resembling the tail sheath were observed connected in a chain.

φRSA1 has a 38,760-bp dsDNA genome (65.3% G + C) with a 19-bp 5′-extruding cohesive end (cos). The genome contains 51 open reading frames (accession no. AB276040, Fujiwara et al., 2008). Two-thirds of the φRSA1 genomic region on the left side encodes the phage structural modules, which are very similar to those of coliphage P2 and P2-like phages (Fig. 3B). Genes for DNA replication, host lysis, and regulatory functions have been identified on the genome, but there are no apparent pathogenesis-related genes. A late-expression promoter sequence motif (Ogr-binding sequence) was predicted for the φRSA1 genes as 5′ TGTTGT-$(X)_{13}$-ACAACA. It is interesting to compare this sequence with that identified for the P2 family members as TGT-$(N)_{12}$-ACA (Julien & Calendar, 1995). The entire genomic sequence of φRSA1 showed significant similarity to the genomes of *Burkholderia pseudomallei* phage φ52237 (accession no. DQ087285) and *Pseudomonas aeruginosa* phage φCTX (accession no. AB008550, Nakayama et al., 1999). Extended comparison of the φRSA1 sequence with these phage sequences by the matrix plot method revealed characteristic features of the phage gene organization: the sequence homology was broken by small AT-islands (designated R-regions). One such region (R2) on the φRSA1 DNA contained an ORF (ORF34) showing 100% amino acid sequence identity with transposase ISRSO15, which was found in the chromosomal DNA (positions 2,780,153 to 2,781,370; accession no. AL646070) as well as the megaplasmid DNA (positions 111,895 to 113,185; accession no. AL646085) of *R. solanacearum* GMI1000. ORF34 is on an IS of 1,319 bp with a terminal repeat of seven A residues. These regions might have been horizontally transferred and been serving as anchor points for genome rearrangements. φRSA1 uses the lipopolysaccharide core as a receptor site on the cell surface and requires Ca^{2+} ions to bind to the receptor. Its lifecycle takes 60–90 min for one round and the burst size is approximately 200 pfu/cell. Like P2 phages, φRSA1 encodes an ORF for integrase and *att*P, suggesting a lysogenic cycle. In fact, a φRSA1-like phage was found to be integrated into at least three different strains of *R. solanacearum*. In addition, the chromosomal integration site (*att*B) was identified as the 3' portion (45 bases) of the arginine tRNA (CCG) gene (Fujiwara et al., 2008). However, □RSA1 can also infect strains that contain a lysogenized φRSA1-like phage; all 18 strains of *R. solanacearum* tested were susceptible to this phage. Therefore, φRSA1 itself may be a kind of immunity-deficient or super-infective phage. This property of φRSA1 is very important for avoiding the problem of lysogenization by a therapeutic phage, and thus avoids passive import of pathogenicity-related genes (Merril et al., 2003; Brussow 2005). Compared to cells (strain M4S) without φRSA1 sequences, the lysogenic cells (newly established with φRSA1-original phage) showed no obvious changes in growth rate, cell morphology, colony morphology, pigmentation, or extracellular polysaccharide production in culture. No obvious enhanced pathogenicity has been observed so far with φRSA1-related lysogenic cells by plant virulence assays in tobacco plants. It is interesting to note that a prophage previously detected on the genome of *R. solanacearum* strain GMI1000 (Salanoubat et al., 2002) was characterized as a φRSA1-related prophage (φRSX) in the light of the φRSA1 genome sequence. The exact size of φRSX is 40,713 bp, and its ORFs share very high amino acid identity with their φRSA1 counterparts. The φRSX attachment site corresponds to a 15-base 3' portion of the serine tRNA (GGA).

2.5 φRSL phages of the *Myoviridae*

φRSL1 is a large-tailed phage (jumbo phage) belonging to the family *Myoviridae*, and was isolated from crop fields (Yamada et al., 2007; Yamada et al., 2010). Phage particles consist of a 125 nm diameter icosahedral head and a long contractile tail that is 110 nm long and 22.5 nm wide (Fig. 3C). φRSL1 gives clear plaques with various strains of race 1 and 3 and biovar 3 and 4; 17 of 18 strains tested were susceptible to this phage. The infection cycle of φRSL1 with strain M4S as a host has an eclipse phase of 90 min and a latent period of 150 min, followed by a rise period of 90 min (Yamada et al., 2010). The average burst size is 80-90 pfu per infected cell. One characteristic feature of φRSL1 infection is a lasting host killing effect. The large□φRSL1 dsDNA genome of 231,255 bp (G + C = 58.2%) contains 343 ORFs and three tRNA genes (including one pseudogene) (accession no. AB366653). According to the orientation of the ORFs, four major genomic regions are apparent (Fig. 3C): The largest, region I (114.5 kbp, G + C = 58.3%), encodes 193 ORFs mostly located in a clockwise orientation. Region II (15.8 kbp, G + C = 57.4%) encodes 24 ORFs all in a counter clockwise orientation. Region III (27.9 kbp, G + C = 57.4%) encodes 23 ORFs, 19 clockwise and four counterclockwise). The last, and second largest, region is region IV (73.1 kbp, G + C = 58.4%), which encodes 102 ORFs mostly located in a counterclockwise orientation. ORFs are generally tightly organized, with little intergenic space. In many cases, the stop codon of an ORF was found to overlap the start codon of the following ORF. φRSL1 ORFs showed no detectable similarity at the nucleotide level to other viruses or cellular organisms. The proportion of ORFans (i.e. genes lacking detectable homologs in the current databases) was very high; of the 343 ORFs, 251 ORFs (73%) showed no significant sequence similarity in the publicly available databases (E-value < 0.001). Of 83 homologs in UniProt or in the NCBI environmental sequence collection, 53 ORFs (15.5%) showed best sequence similarities in bacteria, 10 ORFs (2.9%) in viruses/plasmids, one ORF (0.3%) in eukaryotes, one ORF (0.3%) in archaea, and 18 ORFs (5.3%) in environmental sequences. It is notable that only a few φRSL1 ORFs were similar to sequences in *R. solanacearum* spp., given that the genomic sequences are available for strain GMI1000 and UW551. Viruses/plasmid best hits include five homologs in myoviruses, three in siphoviruses, one in a eukaryotic virus (Mimivirus), and one in a plasmid. These results suggest that φRSL1 may have access to the gene pools of largely different families of bacteria and viruses. At present, putative functions have been assigned to 47 φRSL1 ORFs by examination of homologous search results, including enzymes for the salvage pathway of NAD^+ and for the biosynthetic pathways of lipid, carbohydrates, and homospermidine in addition to proteins required for phage replication (Yamada et al., 2010). A chitinase-like protein was found to be a potential lysis enzyme. Expression patterns of these φRSL1 genes were characterized using a DNA microarray during the infection cycle. Most genes showing early expression (10-30 min p.i.) and later repression by 90 min p.i. (designated as early-intermediate genes) were confined within region I. In contrast, genes that showed increased expression during 30-90 min p.i. (designated as intermediate-late genes) are mostly concentrated around both extremities of region I, as well as in the entire region IV. The intermediate-late genes also included several genes located in region II and region III. Putative genes for phage structural proteins are intermediate-late genes. Genes involved in DNA metabolism are also intermediate-late, except for ORF065, which encodes NAD^+-dependent DNA ligase. Putative promoter elements for these differential genes' expressions have been identified (Yamada et al., 2010).

Several myoviruses are known to have large genomes over 200 kbp, and are designated as "jumbo phages" (Hendrix, 2009). These include *Pseudomonas aeruginosa* phage φKZ (280 kbp, Mesyanzhinov et al., 2002), EL (211 kbp, Herveldt et al., 2005), and φPA3 (309 kbp, Monson et al., 2011), *Vibrio parahaemolyticus* phage KVP40 (245 kbp, Miller et al., 2003), *Stenotrophomonas maltophila* phage φSMA5 (250 kbp, Chang et al., 2005), and *Yersinia enterocolitica* phage R1-37 (270 kbp, Kiljunen et al., 2005). Jumbo phages were also reported for *Sinorhizobium meliloti* (phage N3, 207 kbp, Martin and Long, 1984) and *Bacillus megaterium* (phage G, 670 kbp, Sun and Serwer, 1997). According to their large genomes, they contain many genes, most of which are unknown and might function in the interaction with their hosts. Jumbo phages have only recently been identified and there is too little information about them to discuss their interrelationships. Currently, only a handful of jumbo phages have been isolated. Jumbo phages might have been missed in ordinary screenings because of their large size: they diffuse too slowly in the top agar gels typically used to plaque phages. Reducing the top agar concentration from 0.7-0.8% to 0.45-0.5% allowed us to plaque φRSL1. A similar observation was reported by Serwer et al. (2007) for *Bacillus* phage G. It is very likely that more jumbo phages will be isolated by adapting these plaquing conditions. φRSL1 is highly virulent, and shows no sign of genomic integration, both of these properties making it suitable for use as a biocontrol agent. Two specific advantages of this phage are its abundant yields (10^{11} pfu/ml) and easy purification using only centrifugation (15,000 × g) from routine cultures.

2.6 φRSC phages of the *Siphoviridae*

φRSC1 spontaneously appeared from a culture of strain MAFF301558. It gave turbid plaques with 15 of 18 strains tested as host (Table 1). Genomic Southern blot hybridization showed that two resistant strains, MAFF211271 and MAFF301558, are lysogenic with φRSC1 and that φRSC1 was easily induced by UV irradiation. These strains were isolated from wilted potato and classified as strains of race 3, biovar N2, and phylotype IV. Electron microscopic observation of negatively stained φRSC1 particles revealed an icosahedral head of 48 ± 3 nm in diameter, a non-contractile tail of 220 ± 15 nm in length, and tail fibers of 30 ± 2 nm, giving a λ-like morphology (Fig. 1). φRSC1 gave a single 40 kbp DNA band by pulsed-field gel electrophoresis, indicating a 40 kbp linear DNA as the genome. Partially determined nucleotide sequences of φRSC1 DNA fragments showed high homology with prophage sequences of *R. solanacesrum* strains; for example, RSc0863 and RSc0875 of strain GMI1000, RSMK00228 of strain Molk2, and RSIPO_02158 of strain IPO1609. Effects of lysogenic integration of φRSC1 on the host strain are largely unknown.

2.7 Other prophages

Recent genomic analyses of several strains of *R. solanacearum* belonging to different races, biovars, and/or phylotypes revealed many strain specific gene clusters (Remenant et al., 2010). Some of these variable regions contained phage-like sequences. There are five phage-related sequences in the chromosome of strain GMI1000 (race 1, biovar 3, and phylotype I), two of which corresponded to φRSX (RSc1896-RSc1948) and φRSC1-like sequences (around RSc0863-RSc0967) (accession no. AL646052), as described above. Another region around RSc1680-RSc1696 resembles λ-like phage HK022 and *Bacillus subtilis* temperate phage φ105. In the chromosome of strain UW551 (race 3, biovar 2, and phylotype II), a cluster of 38 probable prophage genes (RRSL02400-RRSL02437) is remarkable (accession no. NZ_AAKL00000000). This gene cluster was present in all race 3/biovar 2 (R3B2) strains tested, from a wide variety of geographical sources (Gabriel et al., 2006). As described above, φRSM4 is also a filamentous prophage of the φRSM group. There are at least two possible prophage regions in the chromosome of strain IPO1609 (race 3, biovar 2, and phylotype IIB); the region around RSIPO02143-RSIPO002171 (resembling φRSC1) and that around RSIPO04993-RSIPO05020 (accession no. CU694438). Four regions of the chromosome of strain Molk2 (race 2, biovar 1, and phylotype IIB) also contained prophage sequences (accession no. CU644397); regions around RSMK00219-RSMK00259, RSMK01452-RSMK01464 (φ105-like), RSMK01633-RSMK01646 (HK022-like). The recently reported genomes of three other strains, belonging to different phylotypes, contained 2-4 prophage sequences, including strains CMR15 (phylotype III, accession no. FP885895), PSI07 (phylotype IV, accession no. FP885906), and CFBP2957 (phylotype IIA, accession no. FP885897).

It is becoming increasingly clear that phages play important roles in the evolution and virulence of many pathogenic bacteria (Canchaya et al., 2003; Brussow et al., 2004). Phages are important vehicles for horizontal gene exchange between different bacterial species, as well as between different strains of the same species. When a temperate phage integrates into the host genome (lysogenization), it may affect the host cells in several ways: (i) disrupting host genes, (ii) changing the expression levels of host genes, (iii) serving as recombinational hot spots, (iv) protection from lytic infection, (v) lysis of the cells by prophage induction, and (vi) introduction of new fitness factors (lysogenic conversion genes). Such lysogenic conversion genes may change the host phenotype drastically. Such lysogenic conversion genes (cargo genes) are sometimes called "morons". Morons are not required for the phage life cycle. Instead, many morons from prophages in pathogenic bacteria encode proven or suspected virulence factors. Therefore, prophages, especially those found commonly in a certain group of R. solanacearum, are vital for understanding the specific virulence of such a group. Through identification and characterization of lysogenic conversion genes, pathogenesis mechanism of R. solanacearum will be clarified.

3. Phage prophylaxis and treatment of bacterial wilt

For phage biocontrol or therapy, only lytic phages are usually used, thereby avoiding the problem of lysogeny. A phage cocktail has been recommended to prevent the problem of resistance, which contains several phages with different host specificities, different replication mechanisms, and/or different infection cycles (Gill and Abedon, 2003; Jones et al., 2007). Fujiwara et al. (2011) used three lytic phages, φRSA1, φRSB1, and φRSL, for biocontrol of tomato bacterial wilt caused by R. solanacearum. Although φRSA1 and φRSB1 infection resulted in quick lysis of host cells, multi-resistant cells arose approximately 30 h post infection. By contrast, cells infected solely with φRSL1 kept a steady low level of cell density for a long period. Under laboratory culture conditions, when host R. solanacearum cells were quickly lysed by treatment with φRSA1 or φRSB1, resistant cells (presumably pre-existing in the population at a very low frequency) were raised after 30 h post infection (pi). Killing susceptible cells, the majority of the cell population, by phages may allow minor cells to predominate in subsequent generations. The recovering cells were somehow resistant to both φRSA1 and φRSB1: mixed treatment with these phages resulted in the same killing and recovering pattern of bacterial cells as did sole treatment. The resistance mechanisms used by these cells are unknown. A cocktail containing three phages, φRSA1, φRSB1, and φRSL1, also failed to stably prevent bacterial growth. By contrast, cells infected solely with φRSL1 kept a steady low level of cell density for a long period. Pretreatment of tomato seedlings with φRSL1 drastically limited the penetration, growth, and movement of inoculated bacterial cells. Treated plants survived for as long as four months. Either φRSA1 or φRSB1, which kills cells quickly, could not bring about similar plant-protecting effects. Using these observations, Fujiwara et al. (2011) proposed an alternative phage biocontrol method using a unique phage, such as φRSL1, instead of a phage cocktail containing highly lytic phages. With this method, bacterial cells are not killed completely, but a sustainable state of phage-bacteria coexistence (with a low level of bacterial population) is maintained.

Phages are utilized for controlling plant pathogens either in the rhizosphere or phylosphere. Application of phages to the phylosphere, namely directly to aerial tissues of the plant, must involve a serious phage stability problem (Jones et al., 2007). Field and laboratory studies have demonstrated that phages are inactivated rapidly by exposure to sunlight (UV-A and UV-B, 280-400 nm), high temperature, high and low pH, oxidative conditions, and washing-down by water (Ignoffo and Garcia, 1992, Iriarte et al., 2007). However, in the case of bacterial wilt, phages for biocontrol can be applied to the rhizosphere. Sunlight UV, the most destructive environmental factor, and oxidative inactivation, can be relieved in this case. φRSL1 was shown to be relatively stable in soil conditions, especially at higher temperatures (bacterial wilt occurs at higher temperatures). Prolonged disease control may be possible if φRSL1 is applied to plants at the seedling stage.

There is another method of phage biocontrol of bacterial wilt. As described above, filamentous φRSM phages cause loss of virulence in the infected host cells. Infection with φRSM phages does not cause host cell lysis, but establishes a persistent association between the host and phage, releasing phage particles from the growing and dividing host cells. Therefore, these phages are also good candidates for bacterial wilt biocontrol agents. In addition, φRSM-infected bacterial cells protect their pre-treated tomato plants from a second infection of virulent cells (Addy et al., unpublished data). Once plants are treated with φRSM-infected cells, the prevention effect lasts for up to two months. Two months after treatment with φRSM-infected cells, the second infected virulent cells could not cause wilting symptoms. Using a mixture of φRSM1 and φRSM3 seems to be especially effective, because the two phages have different host ranges (complementary to each other) and most strains of different races and/or biovars are expected to be susceptible to either of the two phages (Askora et al., 2009). Furthermore, plants inoculated with φRSM-infected R. solanacearum cells showed stable resistance to virulent bacterial cells inoculated thereafter. This resistance was induced as early as one day post inoculation and lasted for up to two months. Therefore, φRSM phages also give an additional possibility to prevent bacterial wilt, namely utilization as "vaccine against bacterial wilt of many crops".

An additional potential application is the use of phage genes or gene products as therapeutic agents (Loesnner, 2005; Hermoso et al., 2007; Fischetti, 2005). Ozawa et al. (2001) isolated the bacteriolytic gene from a R. solanacearum phage P4282. The 71 kDa phage protein consists of 687 amino acids and showed strong bacteriolytic activities against several field-isolated strains of R. solanacearum. Although the biochemical and enzymatic nature of this protein is not fully characterized, homologous sequences are integrated in several bacterial genomes (database accession no. A1H7Z4 and A4JD43). This phage gene was suggested to be useful for generating transgenic plants that are resistant to bacterial wilt (Ozawa et al., 2001). Phage-encoded endolysin, which disrupts the peptidoglycan matrix of the bacterial cell wall, and phage-encoded holins, which permeabilize the bacterial membrane, could also be effective against bacterial pathogens. Practically, these phage proteins are probably not effective for field use, but could have applications in local plant therapy or disease prevention.

4. Detection of *R. solanacearum* cells in plants and soil and diagnosis of bacterial wilt

Effective bacterial diagnosis is always required for successful biocontrol. A variety of methods have been developed to detect R. solanacearum, including typical bioassays, dilution plating on semi-selective media, fatty-acid analysis, immunofluorescence microscopy, enzyme-linked immunosorbent assay (ELISA), and polymerase chain reaction (PCR) (Janse,1988; Seal et al. 1993; Elphinstone et al., 1996; Van der Wolf et al., 2000; Weller et al., 2000; Priou et al., 2006; Kumar et al., 2002). However, none of these methods can reliably detect the pathogen both in plants and soils, and in soil-related habitats. Recently, interest in the use of phages for direct detection and identification of bacterial pathogens has rapidly increased. A number of phage-based bacterial diagnoses exist: (i) Lysis of bacterial cells by specific phages results in release of intracellular molecules that may be assayed using various methods. For example, ATP release can be easily detected by the use of firefly luciferase/luciferin system (Entis et al., 2001). The presence of bacterial pathogens can also be monitored by measuring specific enzyme activities released by phage-mediated cell lysis. (ii) The phage amplification assay detects the increase in phage particles in target bacterial cells after infection (Mole & Maskell, 2001). (iii) The phage-tagging method is another phage-based approach for bacterial detection. In this method, which is well established with coliphage M13 (Smith 1985; Smith 1991), a tag protein, for example green fluorescent protein (GFP), can be displayed on the phage particle and may then be detected by several methods, including epifluorescence microscopy, flow cytometry, or by a fluorescent plate reader after adsorption into host cells (Goodridge et al., 1999). (iv) Reporter phages, defined here as recombinant phages expressing reporter genes in host cells after infection, are also used to monitor phage infection. A variety of reporter genes are available, such as those encoding GFP and luciferase.

In the case of *R. solanacearum*, the filamentous phages φRSS1 and φRSM1 may be useful tools for phage-based diagnoses of bacterial wilt. In general, the genomes of inoviruses (Ff-like phages) are organized in a three-module structure in which functionally related genes are grouped (Rasched & Oberer, 1986; Model & Russel, 1988). The replication module contains the genes encoding rolling-circle DNA replication and the ssDNA binding proteins, gpII, gpV, and gpX. The structural module contains genes for the major (gpVIII) and minor (gpIII, gpVI, gpVII, and gpIX) coat proteins. Among these, gpIII is the host recognition or adsorption protein (Armstrong et al., 1981). The assembly and secretion module contains genes gI and gIV for morphogenesis and extrusion of the phage particles. The genome of φRSS1 is 6,662 nt and encodes 11 ORFs arranged in a generally conserved module structure (Kawasaki et al., 2007a). The genomic DNA of φRSM1 is a little longer (9,004 nt), and encodes 14 ORFs, 12 of which are located on the same strand in a similar manner to φRSS1, and two of which are in the opposite orientation in the replication module (Kawasaki et al., 2007a). All strains tested were susceptible to either φRSM1 or φRSM3. Similar compensating host ranges are also detected between φRSS1 and φRSS2. Selective recognition of the host with different types of pili is mediated by the minor coat protein pIII of these phages. When the gene for GFP was inserted in the intergenic region (IG) of both φRSS1 and φRSM1 genomic DNAs, the resulting phages exhibited strong green fluorescence in phage-infected host cells, and each phage caused large plaques to appear on the host bacterial lawn (Fig. 4). The efficiency of infection and host specificity of the phages were unchanged (unpublished results). These phages can be used as reporter phages to quantify bacterial cell number in the natural environment, because they propagate in the host cells, but do not cause death. A

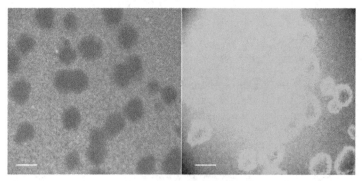

Fig. 4. Green fluorescence emission from plaques of φRSS1-GFP-infected *R. solanacearum* cells (right). ND, Not determined.

tag protein on φRSS1 and φRSM1 particles, such as GFP or luciferase, would lead to rapid and direct detection of the pathogen. For practical use of this method, the following three requirements should be met: (i) a set of phages with different host specificities covering all phylotypes of *R. solanacearum* strains should be prepared; (ii) the intensity of tag-signals should be increased; and (iii) an appropriate, simple device for sensitive detection of the tag-signals should be developed.

These phages would also be very useful in both basic and applied research for monitoring infection of bacterial cells in plant tissues, and for observing bacterial behavior in ecological systems. Kawasaki et al. (2007b) obtained mini-replicons from φRSS1 and φRSM1. pRSS11 is a 2.2 kbp-region of φRSS1 containing the entire replication module (ORF1-ORF3) and IG connected to the Km-cassette (1.5 kbp). pRSM12 is a 3.4 kbp-φRSM1 fragment containing ORF1, ORF12, and IG connected to the Km cassette (1.5 kbp). Both plasmids are very stably maintained in *R. solanacearum* cells of different races and biovars, even without selective pressure. Almost 100% of transformed cells retained the plasmids after cultivation for 100 generations (12 d) in CPG medium without Km. This stability makes pRSS11 and pRSM12 valuable vectors for studies on *R. solanacearum* in natural ecosystems, where selective pressure cannot be applied. To demonstrate the usefulness of these plasmids, a GFP-expressing plasmid (pRSS12, 4.7 kbp) was derived from pRSS11 by connecting the GFP gene, and was introduced into various strains of *R. solanacearum* (Kawasaki et al., 2007b). As expected, pRSS12 was stably maintained in all the transformants and expressed strong green fluorescence. To monitor cell behavior, pRSS12-transformed cells were infected into tomato plants and tobacco BY-2 cells, and were also introduced into soil samples. The strong green fluorescence emitted from pRSS12-transformed cells was easily observed in tomato stems, petioles, and roots (Kawasaki et al., 2007b; Fujie et al., 2010). Bacterial cells adhered to BY-2 cell surfaces preferentially by one pole, possibly via pili on the cell surface. These phage-derived plasmids can serve as an easy-to-use GFP-tagging tool for any given strain of *R. solanacearum* in cytological or field studies. Although there have been several reports on the expression of GFP-fused proteins in *R. solanacearum* cells, all the vectors used were selective-pressure-dependent (Huynh, 1989; Aldon et al., 2000). Transposons have also been used for the constitutive expression of GFP in *R. solanacearum* cells. Random chromosomal insertion of the pAG408 mini-transposon (Suarez et al., 1997) was used to label the wild-type strain GMI1000 (Aldon et al., 2000). To monitor the movement of individual cells, and their chemotactic behaviors, Tn5-GFP-tagged *R. solanacearum* strains were examined (Liu et al., 2001; Yao & Allen, 2006). However, there are intrinsic problems in using transposition techniques. Transposon insertion may affect the genetic background of the host cells. The GFP expression itself may be affected by the genetic environment around the insertion site (position effects). Moreover, under natural environmental conditions with various physical and biological stresses, some transposons are unstable, and are easily moved or lost. pRSS12 is easily introduced by electroporation and is stably maintained in *R. solanacearum* cells of different races and biovars; therefore, it serves as an easy-to-use GFP-tagging tool for any given *R. solanacearum* strain in the wild-type background. By monitoring pRSS12-transformed cells, the following may be studied in detail: (i) differences in the virulence traits among strains; (ii) differences in the resistance level (responses) of plant hosts against a given bacterial strain; (iii) effects of environmental factors during establishment of infection; and (iv) evaluation of therapeutic effects in the development of new agricultural chemicals for bacterial wilt disease.

5. Phage biocontrol in other phytopathogen systems

To date, phage-mediated biocontrol of plant pathogens has been successfully attempted in several other pathogen–plant systems. Historical applications in this area are described in the reviews by Gill and Abedon (2003) and Jones et al., (2007). Effective phage applications have been observed in systems using *Streptomyces* phage to disinfest *Streptomyces scabies*-infected potato seed-tubers (McKenna et al., 2001), in *Xanthomonas pruni*-associated bacterial spot of peaches (Sacchardi et al., 1993; Civerolo & Kiel, 1969; Randhawa & Civerolo, 1986), in *Xanthomonas* leaf bright of onion (Lang et al., 2007), to control soft rot caused by *Erwinia* spp. (Eayre et al., 1990), in fire blight of pear and apple associated with *Erwinia amylovora* (Gill et al., 2003; Schnabel et al., 1998; Schnabel, 2001), and using phage Xav to manage bacterial leaf spot of mungbean (Borah et al., 2000). Phage biocontrol has also been successfully extended using host-range phage mutants (h-mutants) of *Xanthomonas* to bacterial blight of geraniums caused by *Xanthomonas hortorum* pv. *pelargonii* (Flaherty et al., 2000) and bacterial spot of tomatoes caused by *X. perforans* (Balogh et al., 2003). Moreover, phages have been used against bacterial blotch of mushrooms caused by *Pseudomonas tolassii* (Munisch & Olivier, 1995).

6. Future prospects

To meet increasing food demands, there is a need to reconstruct agricultural systems that are much more efficient, economical, sustainable, and free from the problems arising from agrichemical use. Phage biocontrol has many advantages: the application is relatively easy; it is relatively low cost; it does not disturb larger ecological systems; and it is safe for humans, animals, and plants. However, it should be acknowledged that phage biocontrol is not a panacea against plant pathogens. Extrapolation of practices from one pathogen-plant system, even if fully successful, to other systems may not always be applicable. Several factors can influence the success of phage biocontrol: (i) the niche where the target pathogen population resides; (ii) stability, decay, and diffusion of phage particles in the applied ecosystems; (iii) timing of phage application during the crop-growing cycle; (iv) phage host-range and emergence of phage-resistant host derivatives; and (v) the density of target bacteria and applied phages. The optimal conditions for the most effective phage biocontrol should be established for each pathogen–host system. Furthermore, detailed understanding of the properties and behavior of each specific phage-bacterium system will help to optimize phage biocontrol. Like an arms race, both phages and their host bacteria have evolved a variety of mechanisms to resist each other during their long history of interaction. Such mechanisms can be deduced from the genomes of both phages and host bacteria. Genomic information of major pathogenic bacterial species is accumulating rapidly, and newly isolated phages are also subjects for immediate genomic analysis. Molecular mechanisms deduced from genomic information on phage–host interactions, no matter how general or specific, could be useful for establishing long-lasting phage biocontrol systems. These include, for example, general interactions (Comeau & Krisch, 2007), restriction/antirestriction systems (Tock & Dryden, 2005; Hoskisson & Smith, 2007), and phage receptor/host adsorption interactions (Goldberg et al., 1994; Tetart et al., 1998; Thomassen et al., 2003).

7. References

Aldon, D., Brito, B., Boucher, C., & Genin, S. (2000). A bacterial sensor of plant cell contact controls the transcriptional induction of *Ralstonia solanacearum* pathogenicity genes. *EMBO J.*, 19, pp. 2304-2314.

Armstrong, J., Perharm, R. N., & Walker, J. E. (1981). Domain structure of bacteriophage fd adsorption protein. *FEBS Lett.*, 135, pp. 167-172.

Askora, A., Kawasaki, T., Usami, S., Fujie, M., & Yamada, T. (2009). Host recognition and integration of filamentous phage φRSM in the phytopathogen, *Ralstonia solanacearum*. *Virology*, 384, pp. 69-76.

Askora, A., Kawasaki, T., Fujie, M., & Yamada, T. (2011). Resolvase-like serine recombinase mediates integration/excision in the bacteriophage φRSM. *J. Biosci. Bioeng.*, 111, pp. 109-116.

Borah, P. K., Jindal, J. K., & Verma, J. P. (2000). Integrated management of bacterial leaf spot of mungbean with bacteriophages of Xav and chemicals. *J. Mycol. Plant Pathol.*, 30, pp. 19-21.

Balogh, B., Jones, J. B., Momol, M. T., Olson, S. M., Obradovic, A., King, P., & Kackson, L. E. (2003). Improved efficacy of newly formulated bacteriophages for management of bacterial spot of tomato. *Plant Dis.*, 87, pp. 949-954.

Brussow, H. (2005). Phage therapy: the *Escherichia coli* experience. *Microbiology*, 151, pp. 2133-2140.

Brussow, H., Canchaya, C., & Hardt, W.-D. (2004). Phages and the evolution of bacterial pathogens: from genomic rearrangements to lysogenic conversion. *Microbiol. Mol. Biol. Rev.*, 68, pp. 560-602.

Buchen-Osmond, C. (2003). Inoviridae. In: *ICTVdB-The Universal Virus Database, version 3*. A. Z. Oracle, (Ed.), ICTVdB Management, The Earth Institute, Biosphere 2 Center, Columbia University.

Buchen-Osmond, C. (2003). Myoviridae. In: *ICTVdB-The Universal Virus Database, version 3*. A. Z. Oracle, (Ed.), ICTVdB Management, The Earth Institute, Biosphere 2 Center, Columbia University.

Canchaya, C., Proux, C., Fournous, G., Bruttin, A., & Brussow, H. (2003). Prophage genomics. *Microbiol. Mol. Biol. Rev.*, 67, pp. 238-276.

Carnoy, C. & Roten, C.-A. (2009). The dif/Xer recombination systems in Proteobacteria. *PLoS ONE*, 4. Issue 9, e6531.

Casjens, S. (2003). Prophages and bacterial genomics: what have we learned so far? *Mol. Microbiol.*, 49, pp. 277-300.

Chang, H.-C., Chen, C.-R., Lin, J.-W., Shen, G.-H., Chang, K.-M., Tseng, Y.-H., & Weng, S.-F., (2005). Isolation and characterization of novel giant *Stenotrophomonas maltophila* phage φSMA5. *Appl. Environ. Microbiol.*, 71, pp. 1387-1393.

Civerolo, E. L., & Kiel, H. L. (1969). Inhibition of bacterial spot of peach foliage by *Xanthomonas pruni* bacteriophage. *Phytopathology*, 59, pp. 1966-1967.

Comeau, A. M., & Krisch, H. M. (2007). War is peace- dispatches from the bacterial and phage killing fields. *Curr. Opin. Microbiol.*, 8, pp. 488-494.

Davis, B. M., Lawson, E. H., Sandkvist, M., Ali, A., Sozhamannan, S., & Waldor, M. K. (2000). Convergence of the secretory pathways for cholera toxin and the filamentous phage, CTXφ. *Science*, 288, pp. 333-335.

Dunn, J. J., & Studier, F. W. (1983). Complete nucleotide sequence of bacteriophage T7 DNA and the locations of T7 genetic elements. *J. Mol. Biol.*, 166, pp. 477-535.

Eayre, C. G., Concelmo, D. E., & Bartz, J. A. (1990). Control of soft rot *Erwinia* with bacteriophages. *Phytopathology*, 80, pp. 994-994.

Elphinstone, J. G., Hennessy, J., Willson, J. K., & Stead, D. E. (1996). Sensitivity of different methods for the detection of *Ralstonia solanacearum* in potato tuber extracts. *Bull. OEPP/EPPO*, 26, pp. 663-678.

Entis, P., Fung, D. Y. C., Griffiths, M. W., McIntyre, L., Russel, S., Sharpe, A. N., & Tortello, M. L. (2001). Rapid methods for detection, identification, and enumeration. In: *Compendium of methods for the microbiological examinations of foods*, F. P. Downes & K. Ito, (Eds.), 89-126, American Public Health Association, Washington, D. C., USA.

Fegan, M., & Prior, P. (2005). How complex is the *Ralstonia solanacearum* species complex? In: *Bacterial wilt: the disease and the Ralstonia solanacearum species complex.*, C. Allen, P. Prior & A. C. Hayward, (Eds.) 449-461. American Phytopathology Society, St. Paul, USA.

Fischetti, V. A. (2005). Bacteriophage lytic enzymes: novel anti-infectives. *Trends Microbiol.*, 13, pp. 491-496.

Flaherty, J. E., Jones, J. B., Harbaugh, B. K., Somodi, G. C., & Jackson, L. E. (2000). Control of bacterial spot on tomato in the greenhouse and field with h-mutant bacteriophages. *HortScience*, 35, pp. 882-884.

Fujie, M., Takamoto, H., Kawasaki, T., Fujiwara, A. & Yamada, T. (2010). Monitoring growth and movement of *Ralstonia solanacearum* cells harboring plasmid pRSS12 derived from bacteriophage φRSS1. *J. Biosci. Bioeng.*, 109, pp. 153-158.

Fujiwara, A., Kawasaki, T., Usami, S., Fujie, M., & Yamada, T. (2008). Genomic characterization of *Ralstonia solanacearum* phage φRSA1 and its related prophage (φRSX) in strain GMI1000. *J. Bacteriol.*, 190, pp. 143-156.

Fujiwara, A., Fujisawa, M., Hamasaki, R., Kawasaki, T., Fujie, M. & Yamada, T. (2011). Biocontrol of *Ralstonia solanacearum* by treatment with lytic bacteriophages. *Appl. Environ. Microbil.*, 77, 4155-4162.

Gabriel, D., Allen, C., Schell, M., Denny, T., Greenberg, J., Duan, Y., Flores-Cruz, Z., Huang, Q., Clifford, J., Presting, G., González, E., Reddy, J., Elphinstone, J., Swanson, J., Yao, J., Mulholland, V., Liu, L., Farmerie, W., Patnaikuni, M., Balogh, B., Norman, D., Alvarez, A., Castillo, J., Jones, J., Saddler, G., Walunas, T., Zhukov, A., & Mikhailova, N. (2006). Identification of open reading frames unique to a select agent: *Ralstonia solanacearum* race 3 biovar 2. *Mol. Plant- Microbe Interact.*, 19, pp. 69-79.

Gill, J., & Abedon, S. T. (2003). Bacteriophage ecology and plants. APSnet (http://apsnet.org/online/feature/phages/)

Gill, J. J., Svircev, A. M., Smith, R., & Castle, A. J. (2003). Bacteriophages of *Erwinia amylovora*. *Appl. Environ. Microbiol.*, 69, pp. 2133-2138.

Goodridge, L., Chen, J., & Griffiths, M. (1999). Development and characterization of a fluorescent-bacteriophage assay for detection of *Escherichia coli* O157:H7. *Appl. Environ. Microbiol.*, 65, pp. 1397-1404.

Goldberg, E., Grinius, L., & Letellier, L. (1994). Recognition, attachment, and injection. In: *Molecular Biology of Bacteriophage T4*, J. D. Karam, (Ed.), 347-356, American Society for Microbiology, New York, USA.

Goto, N. (1992). *Fundamentals of bacterial plant physiology*. Academic Press, New York, USA.

Hayward, A. C. (1991). Biology and epidemiology of bacterial wilt caused by *Pseudomonas solanacearum*. *Annu. Rev. Phytopathol.*, 29, pp. 65-87.

Hayward, A. C. (2000). *Ralstonia solanacearum*. In: *Encyclopedia of Microbiology*, vol. 4, J. Lederberg (Ed.), 32-42, Academic Press. San Diego, CA, USA.

Hendrix, R. W. (2009). Jumbo bacteriophages. *Curr. Top. Microbiol. Immunol.*, 328, pp. 229-240.

Hendrix, R. W., Hatfull, G. F., & Smith. M. C. M. (2003). Bacteriophages with tails: chasing their origins and evolution. *Res. Microbiol.*, 154, pp. 253-257.

Hermoso, J. A., Garcia, J. L., & Garcia, P. (2007). Taking aim on bacterial pathogens: from phage therapy to enzybiotics. *Curr. Opin. Microbiol.*, 10, pp. 461-472.

Hertveldt, K., Lavigne, R., Pleteneva, E., Sernova, N., Kurochkina, L., Korchevskii, R., Robben, J., Mesyanxhinov, V., Krylov, V. N., & Volckaert, G. (2005). Genome comparison of *Pseudomonas aeruginosa* large phages. *J. Mol. Biol.*, 354, pp. 536-545.

Hill, D. F., Short, J., Perharm, N. R., & Petersen, G. B. (1991). DNA sequence of the filamentous bacteriophage Pf1. *J. Mo.l Biol.*, 218, pp. 349-364.

Hoskisson, P. A., & Smith, M. C. M. (2007). Hypervariation and phase variation in the bacteriophage 'resistome'. *Curr. Opin. Microbiol.*, 10, 396-400.

Huber, K. E., & Waldor, M. K. (2002). Filamentous phage integration requires the host recombinases XerC and XerD. *Nature*, 417, pp. 656-659.

Huynh, T. V., Dahlbeck, D.,& Staskawicz, B. J.(1989). Bacterial bright of soybean; regulation of a pathogen gene determining host cultivar specificity. *Science*, 245, pp. 1374-1377.

Ignoffo, C. M., & Garcia, C. (1992). Combination of environmental factors and simulated sunlight affecting activity of inclusion bodies of the heliothis (lepidoptera: Noctuidae) nucleopolyhedrosis virus. *Environ. Entomol.*, 21, pp. 210-213.

Iriarte, F. B., Balogh, B., Momol, M. T., Smith, L. M., Wilson, M. & Jones, J. B. (2007). Factors affectin survival of bacteriophage on tomato leaf surfaces. *Appl. Environ. Microbiol.*, 73, pp. 1704-1711.

Janse, J. (1996). Potato brown rot in western Europe-history, present occurrence and some remarks on possible origin, epidemiology and control strategies. *Bull. OEPP/EPPO*, 26, pp. 679-695.

Janse, J. D. (1988). A detection method for *Pseudomonas solanacearum* in symptomless potato tubers and some data on its sensitivity and specificity. *Bull. OEPP/EPPO*, 18, 343-351.

Jones, J. B., L. E. Jackson, B. Balogh, A. Obradovic, F. B. Iriarte, and M. T. Momol. (2007). Bacteriophages for plant disease control. *Annu Rev Phytopathol.*, 45: 245-262.

Julien, B., and Calendar, R. (1995). Purification and characterization of the bacteriophage P4 d protein. *J. Bacteriol.*, 177, 3743-3751.

Kawasaki, T., Nagata, S., Fujiwara, A., Satsuma, H., Fujie, M., Usami, S., & Yamada, T. (2007a). Genomic characterization of the filamentous integrative bacteriophages φRSS1 and φRSM1, which infect *Ralstonia solanacearum. J. Bacteriol.*, 189, 5792-5802.

Kawasaki, T., Nagata, S., Fujiwara, A., Satsuma, H., Fujie, M., Usami, S., & Yamada, T. (2007b). Genomic characterization of the filamentous integrative bacteriophage φRSS1 and φRSM1, which infect *Ralstonia solanacearum. J. Bacteriol.*, 189, pp. 5792-5802.

Kiljunen, S., Hakala, K., Pinta, E., Huttunen, S., Pluta, P., Gador, A., Lonnberg, & Skurnik, M. (2005). Yersinophage φR1-37 is a tailed bacteriophage having a 270 kb DNA genome with thymidine replaced by deoxyuridine. *Microbiology*, 151, pp. 4093-4102.

Kumar. A., Sarma, Y. R., & Priou, S. (2002). Detection of *Ralstonia solanacearum* in ginger rhizome using post enrichment ELISA. *J. Spec. Arm. Crop.*, 11, pp. 35-40.

Lang, J. M., Gent, D. H., & Schwartz, H. F. (2007). Management of *Xanthomonas* leaf blight of onion with bacteriophage and a plant activator. *Plant Dis.*, 91, pp. 871-878.

Lavigne, R., Burkal'tseva, M. V., Robben, J., Sykilinda, N. N., Kurochkina, L. P., Grymonprez, B., Jonckx, B., Krylov, V. N., Mesyanzhinov, V. V., & Volckaert, G. (2003). The genome of bacteriophage φKMV, a T7-like virus infecting *Pseudomonas aeruginosa.Virology*, 312, pp. 49-59.

Lindell, D., Sullivan, M. B., Johnson, Z. I., Tolonen, A. C., Rohwer, F., & Chisholm, S. W. (2004). Transfer of photosynthesis genes to and from *Prochlorococcus* viruses. *Proc. Natl. Acad. Sci. USA*, 101, pp. 11013-11018.

Liu, H., Kang,Y., Genin, S., Schell, M. A., & Denny, T. P. (2001). Twitching motility of *Ralstonia solanacearum* requires a type IV pilus system. *Microbiology*, 147, pp. 3215-3229.

Loesnner, M. J. (2005). Bacteriophage endolysins-current state of research and applications. *Curr. Opin. Microbiol.*, 8, pp. 480-487.

Lynch, K. H., Stothard, P., & Dennis, J. J. (2010). genomic analysis and related ness of P2-like phages of the *Burkholderia cepacia* complex. *BMC Genomics*, 11, 599.

McKenna, F., El-Tarabily, K. A., Hardy, G. E. S. T., & Dell, B. (2001). Novel in vivo use of a polyvalent *Streptomyces* phage to disinfest *Streptomyces scabies*-infected seed potatoes. *Plant Pathol.*, 50, pp. 666-675.

Martin, M. O., & Long, S. R. (1984). Generalized transduction of *Rhizobium meliloti*. *J. Bacteriol.*, 159, pp. 125-129.

Marvin, D. A. (1998). Filamentous phage structure, infection and assembly. *Curr. Opin. Struct. Biol.*, 8, pp. 150-158.

Merril, C. R., Scholl, D., & Adhya, S. L. (2003). The prospect for bacteriophage therapy in western medicine. *Nature Rev.*, 2, pp. 489-497.

Mesyanzhinov, V. V., Robben, J., Grymonprez, B., Kostyuchenko, V. A., Bourkaltseva, M. V., Sykilinda, N. N., Krylov, V. N., & Volckaert, G. (2002). The genome of bacteriophage φKZ of *Pseudomonas aeruginosa*. *J. Mol. Biol.*, 317, pp. 1-19.

Miller, E. S., Heidelberg, J. F., Eisen, J. A., Nelson, W. C., Durkin, A. S., Ciecko, A., Feldblyum, T. V., White, O., Paulsen, I. T., Nierman, W. C., Lee, J., Szczypinski, B., & Fraser, C. M. (2003). Complete genome sequence of the broad-host-rang vibriophage KVP40: Comparative genomics of a T4-related bacteriophage. *J. Bacteriol.*, 185, pp. 5220-5233.

Model, P., & Russel., M. (1988). Filamentous bacteriophages. In: *The Bacteriophages*, vol. 2, R. Calendar (Ed.), 375-456, Plenum Press, New York, USA.

Mole, R. J., & Maskell, T. W. O. C. (2001). Phage as a diagnostic- the use of phage in TB diagnosis. *J. Chem. Technol. Biotechnol.*, 76, pp. 683-688.

Molineux I. J. (1999). T7-like phages (Podoviridae), p.1722-1729. In: *Encyclopedia of Virology*, A. Granoff & R. Webster (Eds.), Academic Press, Ltd., London, UK.

Momol, T. M., Pingsheng, J., Jones, J. B., Allen, C., Bell, B., Floyd, J. P., Kaplan, D., Bulluck, R., Smith, K., & Cardwell, K. (2006). *Ralstonia solanacearum* race 3 biovar. 2 causing brown rot of potato, bacterial wilt of tomato, and southern wilt of geranium Recovery Plan, *National Plant Disease Recovery System (NPDRS), HSPD-9*, pp. 1-17.

Monson, R., Foulds, I., Foweraker, J., Welch, M., & Salmond, G. P. C. (2011). The *Pseudomonas aeruginosa* generalized transducing phage φPA3 is a new member of the φKZ-like group of 'jumbo' phages, and infects model laboratory strains and clinical isolates from cystic fibrosis patients. *Microbiology*, 157, pp. 859-867.

Munisch, P., & Olivier, J. M. (1995). Biocontrol of bacterial blotch of the cultivated mushroom with lytic phages: some practical considerations. In: *Science and Cultivation of Edible Fungi*, T. J. Elliott, (Ed.), 595-602, A. A. Balkema, Rotterdam, The Netherland.

Murugaiyan, S., Bae, J. Y., Wu, J., Lee, S. D., Um, H. Y., Choi, H. K., Chung, E., Lee, J. H., & Lee, S.-W. (2010). Characterization of filamentous bacteriophage PE226 infecting *Ralstonia solanacearum* strains. *J. Appl. Microbiol.*, 110, pp. 296-303.

Nakayama, K., Kanaya, S., Ohnishi, M., Terawaki, Y., & Hayashi, T. (1999). The complete nucleotide sequence of ϕCTX, a cytotoxin-converting phage of *Pseudomonas aeruginosa*: implications for phage evolution and horizontal gene transfer via bacteriophages. *Mol. Microbiol.*, 31, pp. 399-419.

Negishi, H., Yamada, T, Shiraishi,T., Oku, H., & Tanaka, H. (1993). *Pseudomonas solanacearum*: plasmid pJTPS1 mediates a shift from the pathogenic to nonpathogenic phenotype. *Mol. Plant Microbe Interact.*, 6, pp. 203-209.

Okabe, N., & Goto, M. (1963). Bacteriophages of plant pathogens. *Annu, Rev. Phytopathol.*, 1, pp. 397-418.

Ozawa, H., Tanaka, H., Ichinose, Y., Shiraishi, T., & Yamada, T. (2001). Bacteriophage P4282, a parasite of *Ralstonia solanacearum*, encodes a bacteriolytic protein important for lytic infection of its host. Mol. Genet. Genom.., 265, pp. 95-101.

Priou, S., Gutarra, L., & Aley, P. (2006). An improved enrichment broth for the sensitive detection of *Ralstonia solanacearum* (biovar 1, and 2A) in soil using DAS-ELISA. *Plant Pathol.*, 55, pp. 36-45.

Randhawa, P. S., & Civerolo, E. L. (1986). Interaction of *Xanthomonas campestris* pv *pruni* with pruniphages and epiphytic bacteria on detached peach trees. *Phytopathology*, 76, 549-553.

Rasched, I. & Oberer, E. (1986). Ff colifages: structural and functional relationships. *Microbiol. Rev.*, 50, pp. 401-427.

Remenant, B., Coupat-Goutaland, B., Guidot, A., Cellier, G., Wicker, E., Allen, C., Fegan, M., Pruvost, O., Elbaz, M., Calteau, A., Salvignol, G., Mornico, D., Mengenot, S., Barbe, V., Medigue, C. & Proir, P. (2010). Genomes of three tomato pathogens within the *Ralstonia solanacearum* species complex reveal significant evolutionary divergence. *BMC Genomics*, 11, 319.

Sacchardi, A. E., Gambin, M., Zaccardelli, G., Barone, G., & Mazzucchi, U. (1993). *Xanthomonas campestris* p.v. *pruni* control trials with phage treatments on peaches in the orchard. *Phytopathol. Mediterr.*,32, pp. 216-210.

Salanoubat, M., Genin, S., Artiguenave, F., Gouzy, J., Mangenot, S., Ariat, M., Billault, A., Brottier, P., Camus, J. C., Cattolico, L., Chandler, M., Choisene, N., Claudel-Renard, S., Cunnac, N., Gaspin, C., Lavie, M., Molsan, A., Robert, C., Saurin, W., Schlex, T., Siguier, P., Thebault, P., Whalen, M., Wincker, P., Levy, M., Weissenbach, J., & Boucher, C. A. (2002). Genome sequence of the plant pathogen *Ralstonia solanacearum*. *Nature*, 415, pp. 497-502.

Seal, S. E., Jackson, L. A., Yong, J. P., & Daniels, M. J. (1993). Differentiation of *Pseudomonas solanacearum*, *Pseudomonas syzygii*, *Pseudomonas picketii* and the blood disease bacterium by partial 16S rRNA sequencing: construction of oligonucleotide primers for sensitive detection by polymerase chain reaction. *J. Gen. Microbiol.*, 139, pp. 1587-1594.

Seed, K. D. & Dennis, J. J. (2005). Isolation and characterization of bacteriophages of the *Burkholderia cepacia* complex. *FEMS Microbiol. Lett.*, 251, pp. 273-280.

Smith, E. F. (1986). A bacterial disease of tomato, pepper, eggplant, and Irish potato (*Bacillus solanacearum* nov. sp.). *U. S. Dept. Agric. Div. Vegetable Physiol. Pathol. Bull.*, 12, pp. 1-28.

Schnabel, E. L., & Jones, A. L. (2001). Isolation and characterization of five *Erwinia amylovora* bacteriophages and assessment of phage resistance in strains of *Erwinia amylovora*. *Appl. Environ. Microbiol.*, 67, pp. 59-64.

Smith, G. P. (1985). Filamentous fusion phage: novel expression vectors that display cloned antigens on the virion surface. *Science*, 228, pp. 1315-1317.

Smith, G. P. (1991). Surface presentation of protein epitopes using bacteriophage expression systems. *Curr. Opin. Biotechnol.*, 2, pp. 668-673.

Schnabel, E. L., Fernando, G. D., Jackson, L. L., Meyer, M. P., & Jones, A. L. (1998). Bacteriophages of *Erwinia amylovora* and their potential for biocontrol. *Acta Hortic.*, 489, pp. 649-654.

Suarez, A., Gutter, A., Stratz, M., Staendner, L. H., Timmis, K. N., & Guzman, C. A. (1997). Green fluorescent protein-based reporter systems for genetic analysis of bacteria including monocopy applications. *Gene*, 196, pp. 69-74.

Sun, M., & Serwer, P. (1997). The conformation of DNA packaged in bacteriophage G. *Biophys. J.*, 72, pp. 958-963.

Tanaka, H., Negishi, H., & Maeda, H. (1990). Control of tobacco bacterial wilt by an avirulent strain of *Pseudomonas solanacearum* M4S and its bacteriophage. *Ann. Phytopathol. Soc. Japan*, 56, pp. 243-246.

Tetart, E., Desplats, C., & Krisch, H. M. (1998). Genome plasticity in the distal tail fiber locus of the T-even bacteriophage: recombination between conserved motifs swaps adhesin specificity. *J. Mol. Biol.*, 282, pp. 543-556.

Thomassen, E., Gielen, G., Schultz, M., Schoehn, G., Abrahams, J. P., Miller, S., & van Raaij, M. J. (2003). The structure of the receptor-binding domain of the bacteriophage T4 short tail fibre reveals a knitted trimer metal-binding fold. *J. Mol. Biol.*, 331, 361-373.

Tock, M. R., & Dryden, D. T. F. (2005). The biology of restriction and anti-restriction. *Curr. Opin. Microbiol.*, 8, pp. 466-472.

Toyoda, H., Kakutani, K., Ikeda, S., Goto, S., Tanaka, H., & Ouchi, S. (1991). Characterization of deoxyribonucleic acid of virulent bacteriophage and its infectivity to host bacterium, *Pseudomonas solanacearum*. *J. Phytopathol.*, 131, pp. 11-21.

Van der Wolf, J. M., Vriend, S. G., Kastelein, O., Nijhuis, E. H., Van Bekkum, P. J., & Van Vuurde, J. W. L. (2000). Immunofluorescence colony-staining (IFC) for detection and quantification of *Ralstonia* (*Pseudomonas*) *solanacearum* biovar 2 (race 3) in soil and verification of positive results by PCR and dilution plating. *Eur. J. Plant Pathol.*, 106, pp. 123-133.

Weller, S. A., Elphinstone, J. G., Smith, N. C., Boonham, N., & Stead, D. E. (2000). Detection of *Ralstonia solanacearum* strains with a quantitative, multiplex, real-time, fluorogenic PCR (TaqMan) assay. *Appl. Environ. Microbiol.*, 66, pp. 2853-2858.

Yabuuchi, E., Kosako, Y., Yano, I., Hotta, H., & Nishiuchi, Y. (1995). Transfer of two *Burkholderia* and an *Alcaligenes* species to *Ralstonia* gen. nov.: proposal of *Ralstonia picketii* (Ralston, Palleroni, and Doudorff, 1973) comb. nov. *Ralstonia solanacearum* (Smith 1986) comb. nov. and *Ralstonia eutropha* (Davis 1969) comb. nov. *Microbiol. Immunol.*, 39, pp. 897-904.

Yamada, T., Kawasaki, T., Nagata, S., Fujiwara, A., Usami, S., & Fujie, M. (2007). New bacteriophages that infect the phytopathogen *Ralstonia solanacearum*. *Microbiology*, 153, pp. 2630-2639.

Yamada, T., Satoh, S., Ishikawa, H., Fujiwara, A., Kawasaki, T., Fujie, M., & Ogata, H. (2010). A jumbo phage infecting the phytopathogen *Ralstonia solanacearum* defines. *Virology*, 398, pp. 135-147.

Yao, J., & Allen, C.(2006). Chemotaxis is required for virulence and competitive fitness of the bacterial wilt pathogen *Ralstonia solanacearum*. *J. Bacteriol.*, 188, pp. 3697-3708.

Yuzenkova, J., Nechaev, S., Berlin, J., Rogulja, D., Kuznedelov, K., Inman, R. Mushegian, A., & Severinov, K. (2003). Genome of *Xanthomonas oryzae* bacteriophage XP10: an odd T-odd phage. *J. Mol. Biol.*, 330, pp. 735-748.

10

Bacteriophages of *Clostridium perfringens*

Bruce S. Seal et al.*
Poultry Microbiological Safety Research Unit, Agricultural Research Service, USDA,
USA

1. Introduction

Bacterial viruses were first reported in 1915 by Fredrick William Twort when he described a transmissible "glassy transformation" of micrococcus cultures that resulted in dissolution of the bacteria (Twort, 1915). Subsequently, Felix Hubert d'Hérelle reported a microscopic organism that was capable of lysing *Shigella* cultures on plates that resulted in clear spaces in the bacterial lawn that he termed "plaques" (d'Hérelle, 1917). The term "bacteriophage" was introduced by d'Hérelle (1917) as he attributed the replicate nature of this phenomenon to bacterial viruses. During 1919 d'Hérelle utilized phages isolated from poultry feces as a therapy to treat chicken typhus and further utilized this approach to successfully treat dysentery among humans (Summers, 2001). Prior to the discovery and widespread use of antibiotics, bacterial infections were treated by administering bacteriophages and were marketed by L'Oreal in France (Bruynoghe & Maisin, 1921). Although Eli Lilly Co. sold phage products for human use up until the 1940's, early clinical studies with bacteriophages were not extensively undertaken in the United States and Western Europe after the 1930's and '40's. Bacteriophages were and continue to be sold in the Russian Federation and Eastern Europe as treatments for bacterial infections (Sulakvelidze *et al.*, 2001).

Bacteriophages have been identified in a variety of forms and may contain RNA or DNA genomes of varying sizes that can be single or double-stranded nucleic acid (Ackermann, 1974; 2003; 2006; 2007). Of all the bacteriophages examined by the electron microscope, 95% of those reported are tailed with only 3.7% being filamentous, polyhedral or pleomorphic (Ackermann, 2007). The tailed bacteriophages contain a linear, double-stranded DNA genome that can vary from 11 to 500 kb in the order *Caudovirales* which is further divided into three families based on tail morphology (Ackermann, 2003; 2006). These bacterial viruses have icosohedral heads while those phages with contractile tails are placed in the *Myoviridae*, those phages with a long non-contractile tail are placed in the *Siphoviridae* and phages with short tail structures are members of the *Podoviridae*. Although bacteriophages of the *Caudovirales* (tailed-phages) may be physically similar it has been difficult to classify them by use of DNA or protein sequences due to the tremendous diversity because of

* Nikolay V. Volozhantsev[2], Brian B. Oakley[1], Cesar A. Morales[1], Johnna K. Garrish[1], Mustafa Simmons[1], Edward A. Svetoch[2] and Gregory R. Siragusa[3]
[1]*Poultry Microbiological Safety Research Unit, Agricultural Research Service, USDA, USA*
[2]*State Research Center for Applied Microbiology & Biotechnology, Russian Federation*
[3]*Danisco USA Inc., USA*

horizontal gene transfer (Casjens, 2005). Also, Unlike the case for *Bacteria* and *Archaea*, both of which can be classified using the 16S rRNA gene (Woese and Fox, 1977), due to the mosaic nature of bacteriophage genomes (Hendrix *et al.*, 1999), there appears not to be one candidate conserved gene that can be utilized to categorize all phages for a suitable classification scheme (Nelson, 2004). One approach has been to construct a "phage proteomic tree" based on predicted protein sequences of a bacterial virus (Rohwer and Edwards, 2002) while another approach is to divide bacteriophages based on genome type (ssRNA or DNA) with a further demarcation by physical characteristics such as tailed or filamentous types (Lawrence *et al.*, 2002). Proux *et al.* (2002) proposed a phage taxonomy scheme based on comparative genomics of a single structural gene module (head or tail genes). This partially phylogeny-based taxonomical system purportedly parallels many aspects of the current *International Committee on Taxonomy in Virology* (ICTV) classification system.

2. Antibiotics, antibiotic resistance and the future role of bacteriophages

There is worldwide concern over the present state of antimicrobial resistance (AMR) issues with zoonotic bacteria potentially circulating among food-producing animals, including poultry (McDermott *et al.*, 2002; Gyles, 2008). This has resulted in the general public's perception that antibiotic use by humans and in food animals selects for the development of AMR among food-borne bacteria that could complicate public health therapies (DuPont, 2007). A major issue is that antibiotic resistance may not only occur among disease-causing organisms but also become an issue for other resident organisms in the host which may accumulate in the environment (Yan & Gilbert, 2004). Sub-therapeutic use of antibiotics as growth promoters has been discontinued in the European Union (Regulation EC No. 1831/2003 of the European parliament and the council of 22 September 2003 on additives for use in animal nutrition; Castanon, 2007). This concern is justified due to the increase in antibiotic resistance among bacterial pathogens (NAS, 2006; Gyles, 2008), including bacteria from healthy broiler chickens (Persoons *et al.*, 2010). Consequently, there is a need for developing novel intervention methods including narrow-spectrum antimicrobials and probiotics that selectively target pathogenic organisms while avoiding killing of beneficial organisms (NAS, 2006).

There has been a resurgent interest in bacteriophage biology and their use or use of phage gene products as antibacterial agents (Merril *et al.*, 1996; Wagner and Walder, 2002; Liu *et al.*, 2004; Fischetti, 2010). The potential use of lytic bacteriophage and/or their lytic enzymes is of considerable interest for medicine, veterinary and bioindustry worldwide due to antibiotic resistance issues. Recently, the U.S. Food and Drug Administration approved a mixture of anti-*Listeria* viruses as a food additive to be used in processing plants for spraying onto ready-to-eat meat and poultry products to protect consumers from *Listeria monocytogenes* (Bren, 2007). In veterinary practice, experimental alimentary *E. coli* infections in mice and cattle were controlled by bacteriophage therapy (Smith & Huggins, 1982; 1987). Similarly Barrow *et al.* (1998) reported the use of lytic bacteriophages to protect against *E. coli* septicemia and meningitis in chickens and young cattle. Huff *et al.* (2002a,b; 2003) reported the use of a lytic bacteriophage to reduce effects of *E. coli* respiratory illness in chickens and bacteriophages have been proposed as a strategy for control of food-borne pathogens (Hudson *et al.*, 2005). Joerger (2002) reviewed the literature for application of lytic bacteriophage to control specific bacteria in poultry and concluded that evidence from several trials indicated that phage therapy may be effective under certain circumstances. However, obstacles for the use of phage as antimicrobials remain due to reasons such as limited host-range for many bacteriophages (Labrie et al., 2010).

In the European Union (EU) antimicrobial growth promoters have been banned from animal feeds because of concerns over the spread of antibiotic resistances among bacteria (Bedford, 2000; Moore et al.,2006) and the EU-wide ban on the use of antibiotics as growth promoters in animal feed entered into effect on January 1, 2006 (Regulation 1831/2003/EC). Removal of these antimicrobials will induce changes within the chicken intestinal microbial flora, dictating the need to further understand the microbial ecology of this system (Knarreborg et al., 2002; Wise and Siragusa, 2007), so that appropriate antibiotic alternatives may be developed based on this knowledge (Cotter et al., 2005; Ricke et al., 2005). There has been a limited number of new antibiotic drugs marketed recently with only two, linezolid which targets bacterial protein synthesis and daptomycin wherein the mechanism of action is unknown, appearing since 2000. This is disconcerting considering that this is happening at a time when there is an increasing emergence of antibiotic resistant bacteria with a meager number of new drugs being developed active against such agents (Projan et al., 2004). The view that there is no compelling reason to pursue development of novel therapeutic agents is unwise (Projan & Youngman, 2002), especially considering emergence of "pan-resistant" or multiple-antibiotic resistant strains of Gram-positive bacteria (French, 2010). Consequently, bacteriophage or perhaps more importantly their gene products may provide us with new antimicrobials to combat antibiotic resistant bacteria or that could be used synergistically with traditional antibiotics.

3. Biology of *Clostridium perfringens*, human and veterinary medical issues

Clostridium perfringens is a Gram-positive, spore forming, anaerobic bacterium that is commonly present in the intestines of people and animals. *C. perfringens* is classified into one of five types (A, B, C, D, or E) based on toxin production (Smedley et al., 2004; Sawires & Songer, 2006). Spores of the pathogen can persist in soil, feces or the environment and the bacterium causes many severe infections of animals and humans. The bacterium can cause food poisoning, gas gangrene (clostridial myonecrosis), enteritis necroticans and non-foodborne gastrointestinal infections in humans and is a veterinary pathogen causing enteric diseases in both domestic and wild animals (Smedley et al., 2004; Sawires & Songer, 2006). Spores of the pathogen can persist in soil, feces, and in the environment causing many severe infections in humans and animals. Clinical symptoms and pathogenesis of the infection is determined by enterotoxins produced by *C. perfringens* strains of type A (CPE strains). If a sufficient number of pre-formed *C. perfringens* cells are ingested from contaminated food, these cells are capable of passage from the stomach to the intestinal tract where upon sporulation (spore formation) CPE is released causing the disease state of *C. perfringens* food poisoning (Smedley et al., 2004; Sawires & Songer, 2006). Many heat processes are incapable of inactivating the *C. perfringens* endospores. Survival of spores in these products allows the subsequent outgrowth where spores can germinate and commence growth at temperatures of 43 to 47°C. In foods such as meats with gravy, heating reduces the oxygen tension (lowered redox) to cause sufficient anaerobiosis in which greater numbers of *C. perfringens* will rapidly divide. Importantly, *C. perfringens* has been documented to have very rapid doubling times, in some cases as low as 7 to 9 minutes in beef broth (Smedley *et al.*, 2004; Sawires & Songer, 2006).

Clostridium perfringens plays a significant role in food-borne human disease and is among the most common food-borne illnesses in industrialized countries (Brynestad & Granum, 2002; Lindström *et al.*, 2011). It can be the second or third most frequent cause of bacterial foodborne illness in the United States and is responsible for approximately one million domestic cases annually (Mead *et al.*, 1999; Scallan *et al.*, 2011). Outbreaks are frequently associated with temperature-abused meat or poultry dishes and typically involve a large number of victims (Lindström *et al.*, 2010). If a sufficient number of *C. perfringens* cells are ingested from contaminated food, these cells are capable of passage from the stomach to the intestinal tract where, upon sporulation, CPE is released causing the disease state of *C. perfringens* food poisoning (Wen & McClane, 2004). In addition to food poisonings, CPE-positive *C. perfringens* type A has been implicated in other diseases such as antibiotic-associated and sporadic diarrhea in humans that also may be food-related or non-food sources (Lindström *et al.*, 2010). The Centers for Disease Control and Prevention (CDC) collects data on food-borne disease outbreaks (FBDOs) from all states and territories through the Food-borne Disease Outbreak Surveillance System (FBDSS). The 12 June 2009 issue of Morbidity and Mortality Weekly Report (http://www.cdc.gov/mmwr/preview/mmwrhtml/mm5822a1.htm) states that one of the pathogen-commodity pairs responsible for the most outbreak-related cases was *C. perfringens* in poultry (902 cases). Although *C. perfringens* is considered in the "medium" risk category, it can become a high risk pathogen/product combination with temperature abused poultry-meat products during extended shelf life or when cross-contaminated by *Listeria monocytogenes* (Mataragas *et al.*, 2008). It was reported that improper retail and consumer refrigeration accounted for approximately 90% of the *C. perfringens* illnesses (Crouch *et al.*, 2009) and poultry meat can be a frequently implicated food vehicle during outbreaks (Gormley *et al.*, 2010; Nowell *et al.*, 2010).

Necrotic enteritis is a peracute disease syndrome and the subclinical form of *C. perfringens* infection in poultry are caused by *C. perfringens* type A producing the alpha toxin, and some strains of *C. perfringens* type A produce an enterotoxin at the moment of sporulation that are responsible for food-borne disease in humans. The mechanisms for colonization of the avian small intestinal tract and the factors involved in toxin production are largely unknown. Unfortunately, few tools and strategies are available for prevention and control of *C. perfringens* in poultry. Vaccination against the pathogen and the use of probiotic or prebiotic products has been suggested, but are not available for practical use in the field at the present time (Van Immerseel *et al.*, 2004). Since most poultry harbor intestinal *C. perfringens* commensally as a component of the gut microflora, these issues lend credence to the hypothesis that as subtherapeutic usage of antibiotics is discontinued during poultry production, food-borne illness associated with *C. perfringens* will likely increase. This could potentially become a greater problem for the U.S. poultry industry as antibiotics are withdrawn from animal feeds as has been done in the European Union (Casewell *et al.*, 2003; Van Immerseel *et al.*, 2004). Control of clostridia in commercial poultry has traditionally been accomplished by feeding sub-therapeutic amounts of antibiotics in feed (Jones & Ricke, 2003; Collier *et al*, 2003). Antibiotics have been utilized for over thirty years (Maxey and Page, 1977; George *et al.*, 1982; Engberg *et al.*, 2000; Brennan *et al.*, 2003) and resistance of *C. perfringens* to growth-enhancing antibiotics has been detected among isolates from poultry

(Diarra *et al.*, 2007). Consequently, there is a need for developing on-farm interventions to reduce populations of this bacterial pathogen that lead to peracute flock disease and possibly greater numbers of CPE+ isolates of *C. perfringens* entering the human food chain.

4. Early literature reporting bacteriophages of *Clostridium perfringens*

There is a paucity of genomics data for *C. perfringens* bacteriophages, but it has been known that both temperate and lytic phages are associated with the pathogen, while the Russian literature compiled by Spencer (1953) reported the use of clostridial bacteriophages to treat gas gangrene. Investigators at the Institute Pasteur reported bacteriophages that could be induced from lysogeny among isolates of *C. perfringens* that were long-tailed viruses of the *Siphoviridae* (Kreguer *et al.*, 1947; Guelin & Kreguer, 1950; Guelin, 1953; Elford *et al.*, 1953; Hirano & Yonekura, 1967). Subsequently, a member of the *Podoviridae* designated bacteriophage 80 was isolated with a distinct tail structure that was considered morphologically different from previously reported viruses of anaerobic bacteria (Vieu *et al.*, 1965). Intracellular replication of this virus was examined by Bradley & Hoeniger (1971) who reported that the bacteriophage had a head of approximately 40 nm in size with a 30 nm tail. Intact viruses could be detected within the bacterial cell by 75 minutes post-infection (p.i.) with cell lysis beginning at 105 to 115 min p.i.

Gaspar & Tolnai (1959) published isolation of a virulent *C. perfringens* phage, while Ionesco *et al.* (1974) reported isolation of lysogenic bacteriophages. Lysogenic cultures could be induced by UV irradiation, nitrogen mustard [mechlorethamine; 2-chloro-N-(2-chloroethyl)-N-ethyl-ethanamine, a nonspecific DNA alkylating agent] and to a lesser extent by mercaptoacetic acid. Twelve bacteriophages were induced from type A *C. perfringens* strains, ten from type B and 26 from type C strains of the bacterium, many of the phages were highly host specific with a high proportion of the *C. perfringens* strains resistant to infection by the viruses (Smith, 1959). Smith (1959) also reported that several viruses were apparently unable to enter into lysogeny and hence those were classified as 'virulent' bacteriophages. Following UV irradiation one lysogenic strain of *C. perfringens* resulted in isolation of a long-tailed, DNA-containing bacteriophage with a non-contractile tail, designated CPT1 that produced turbid plaques. This phage had an eclipse phase of approximately 45 min with a maximum rise in titer 45 min following initial release of progeny virus (Mahony and Kalz, 1968). A second bacteriophage designated CPT4 with similar characteristics, but with a shorter tail as compared with CPT1, was also isolated by these investigators (Mahony & Easterbrook, 1970). However, U.V. irradiation did not result in release of viruses from the indicator strain and it was reported that spontaneous release of the virus occurred with all resultant plaques that were clear.

Lysogenic bacteriophages were isolated specifically from *C. perfringens* type C that were induced using mitomycin C treatment on specific isolates of the bacterium (Grant & Riemann, 1976). All the viruses had a similar morphology with polyhedral heads of 55 nm and long flexible tails of 130 to 190 nm. Paquette & Fredette (1977) reported four lysogenic phages from *C. perfringens* type A that were induced with UV irradiation for 5 sec and had 0.5 mm plaques with outer lysis rings. One phage was a podovirus, while the others were siphoviruses (Paquette & Fredette, 1977). Stewart & Johnson (1977) reported that lysogenic phages can have a positive effect on *C. perfringens* sporulation and Canard & Cole (1990) demonstrated that two different lysogenic phages had separate attachment sites that did not share sequence similarity. Also, Shimizu *et al.* (2002) reported at least 20 phage-related sequence elements in the complete *C. perfringens* Strain 13, a gas gangrene isolate.

A bacteriophage isolated from a *C. perfringens* fecal strain was adapted to a number of host strains from clinical swab and fecal isolates to develop a typing scheme using nine host modified phages (Yan, 1989). Of 109 strains, the phage types of 57 (52.3%) were identified, while nine (8.2%) other strains were sensitive to the phages at varying degrees. The remaining 43 (39.4%) strains were resistant and eleven of the 57 typable strains yielded cell-surface mutants which belonged to different phage types from their parent strains (Yan, 1989). Another phage-typing method for the bacterium was developed, but little or no information was available from the report (Satija & Narayan, 1980).

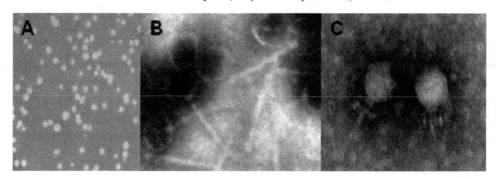

Fig. 1. Plaques and bacteriophages of *Clostridium perfringens* isolated from a joint Russian Federation-USA collaborative research project. (A) Clear plaques produced by bacteriophages from a series of isolates reported in the text and references. (B) Long-tailed phages of the *Siphoviridae*. (C) Short-tailed phages of the *Podoviridae*.

Initial screening for bacteriophages lytic for *C. perfringens* at the Poultry Microbiological Safety Research Unit, of the ARS, USDA and at the State Research Center for Applied Microbiology and Biotechnology in the Russian Federation was performed using filtered samples obtained from poultry (intestinal material), soil and processing drainage water (Figure 1). Bacterial viruses capable of lysing strains of *C. perfringens* type A and producing clear plaques were identified by spot-testing and titration of strains susceptible to the isolated phages (Fig. 1A). Lytic phage preparations were initially characterized morphologically utilizing plaque purified (3X) phage by electron microscopy using the modified method of Horne (1973) where both siphoviruses (Fig. 1B) and podoviruses (Fig. 1C) have been discovered that are virulent for *C. perfringens* (Seal *et al.*, 2011; Volozhantsev *et al.*, 2011).

5. Characteristics of *Clostridium perfringens* bacteriophage and prophage genomes

Zimmer *et al.* (2002a) isolated two temperate bacteriophages by UV irradiation (phi3626 and phi8533) from lysogenic *C. perfringens*. The linear, double-stranded DNA genome of phi3626 was reported to be 33.5 kb with nine nucleotide 3′ protruding cohesive ends and a G+C content of 28.4% (Zimmer *et al.*, 2002a) which is essentially equivalent to its host DNA of 28.6% (Shimizu *et al.*, 2002; Myers *et al.*, 2006). The phage phi3626 had a 55 nm diameter isometric capsid with a 170 nm flexible, non-contractile tail (Zimmer *et al.*, 2002a) that conformed to the *Siphoviridae* phage family in the order *Caudovirales* (Ackermann, 2006). Phage phi3626 was reportedly easier to propagate, so no genomics data were provided for phi8533 (Zimmer *et al.*, 2002a). Physical characteristics of other *C. perfringens* bacteriophages are similar to phi3626 in that they were reported to have polyhedral heads of 55 nm in diameter with long flexible tails (Grant & Riemann, 1976; Paquette & Fredette, 1977) that also presumably had double-stranded DNA genomes. Only nineteen gene products could be assigned to the phage phi3626 genome based on bioinformatics analyses. Those were identified as encoding DNA-packaging proteins, structural components, a dual lysis system, a putative lysogeny switch, and proteins involved with replication, recombination, and modification of phage DNA. Several of the genes potentially influence cell spore-formation due to availability of the phage genes in the bacterial genome. Also, the phi3626 attachment site, *attP*, lies in a non-coding region immediately downstream of *int* encoding the integrase protein. Integration of the viral genome occurred into the bacterial attachment site *attB*, which is located within the 3' end of a *C. perfringens guaA* gene homologue. Subsequently, a phage-specific enzyme, a murein hydrolase, was expressed which had lytic activity against forty-eight test cultures of *C. perfringens*, but was not active against other clostridial species or bacteria belonging to other genera (Zimmer *et al.*, 2002a).

Bacteriophage genomes from viruses isolated from broiler chicken offal washes (O) and poultry feces (F), designated phiCP39O and phiCP26F, respectively, produced clear plaques on host strains (Seal *et al.*, 2011). Both bacteriophages had isometric heads of 57 nm in diameter with 100-nm non-contractile tails characteristic of members of the family *Siphoviridae* in the order *Caudovirales*. The double-strand DNA genome of bacteriophage phiCP39O was 38,753 base pairs (bp), while the phiCP26F genome was 39,188 bp, with an average GC content of 30.3%. Both viral genomes contained 62 potential open reading frames (ORFs) predicted to be encoded on one strand of the DNA (Table 1). Among the ORFs, 29 predicted proteins had no known similarity to other reported proteins while others encoded putative bacteriophage capsid components such as a pre-neck/appendage, tail, tape measure and portal proteins. Other genes encoded a predicted DNA primase, single-strand DNA-binding protein, terminase, thymidylate synthase and a potential transcription factor. Lytic proteins such as a fibronectin-binding autolysin, an amidase/hydrolase and a holin were encoded in the viral genomes. Several ORFs encoded proteins that gave BLASTP matches with proteins from *Clostridium* spp. and other Gram-positive bacterial or bacteriophage genomes as well as unknown putative *Collinsella aerofaciens* proteins that were detected in the virion. Proteomics analysis of the purified viruses resulted in the identification of the putative pre-neck/appendage protein and a minor structural protein encoded by large open reading frames. Variants due to potential phosphorylation of the portal protein were identified in the virion, and several mycobacteriophage gp6-like protein variants were detected in large amounts relative to other virion proteins. The predicted amino acid sequences of the pre-neck/appendage proteins had major differences in the central portion of the protein between the two phage gene products indicating that it may be the potential anti-receptor for the virus. Based on phylogenetic analysis of the large terminase protein, these phages were predicted to be *pac*-type phages, using a head-full DNA packaging strategy. Table 1 summarizes the gene products common to currently known *C. perfringens* siphoviral bacteriophages.

Function ID	Function Name
COG0629	Single-stranded DNA-binding protein
COG0860	N-acetylmuramoyl-L-alanine amidase
COG4722	Phage-related protein
COG5412	Phage-related protein
COG1351	Predicted alternative thymidylate synthase
COG2333	Predicted hydrolase (metallo-beta-lactamase superfamily)
COG3561	Phage anti-repressor protein
COG3645	Uncharacterized phage-encoded protein
COG5545	Predicted P-loop ATPase and inactivated derivatives
COG5546	Small integral membrane protein
COG3617	Prophage antirepressor
COG3747	Phage terminase, small subunit
COG4626	Phage terminase-like protein, large subunit
COG4695	Phage-related protein
COG0175	3'- PAPS reductase/FAD synthetase and related enzymes
COG2369	Uncharacterized protein, homolog of phage Mu protein gp30
COG2755	Lysophospholipase L1 and related esterases
COG4926	Phage-related protein
COG4974	Site-specific recombinase XerD
COG0338	Site-specific DNA methylase
COG0740	Protease subunit of ATP-dependent Clp proteases
COG1476	Predicted transcriptional regulators
COG3757	Lyzozyme M1 (1,4-beta-N-acetylmuramidase)
COG5283	Phage-related tail protein
COG5614	Bacteriophage head-tail adaptor
COG0305	Replicative DNA helicase
COG1191	DNA-directed RNA polymerase specialized sigma subunit
COG1783	Phage terminase large subunit
COG3064	Membrane protein involved in colicin uptake
COG3740	Phage head maturation protease
COG4824	Phage-related holin (Lysis protein)

Note: Domains are observed in all *Clostridium perfringens* siphoviral genomes.

Table 1. The *Siphoviridae* pan-genome encoded proteins representative of *Clostridium perfringens* bacteriophages. The table shows the union of all COGs present in the genomes of phages SM101, 3626, 9O, 13O, 26F, 34O, and 39O.

Function ID	Function Name
COG0417	DNA polymerase elongation subunit (family B)
COG0739	Membrane proteins related to metalloendopeptidases
COG0860	N-acetylmuramoyl-L-alanine amidase
COG1196	Chromosome segregation ATPases
COG2088	Uncharacterized protein, involved in the regulation of septum location
COG3023	N-acetyl-anhydromuramyl-L-alanine amidase
COG3772	Phage-related lysozyme (muraminidase)
COG5434	Endopolygalacturonase
pfam00246	Peptidase_M14
pfam01391	Collagen
pfam05352	Phage Connector
pfam05894	Podovirus_Gp16 (DNA encapsidation)
pfam12841	YvrJ protein family
PHA00144	major head protein
PHA00148	lower collar protein
PHA00380	tail protein

Note: Domains are observed in all *Clostridium perfringens* podoviral genomes.

Table 2. *Podoviridae* pan-genome protein products representative of *Clostridium perfringens* bacteriophages. The table shows the union of all conserved domains present in the genomes of phages CPV1, CPV4, ZP2, CP7R, and CP24R.

Other bacteriophages lytic for *C. perfringens* were isolated from sewage, feces and broiler intestinal contents and phiCPV1, a virulent bacteriophage, was classified in the family *Podoviridae* (Volozhantsev *et al.*, 2011). The purified virus had an icosahedral head and collar of approximately 42nm and 23nm in diameter, respectively, with a structurally complex tail of 37nm lengthwise and a basal plate of 30nm. The phiCPV1 double-stranded DNA genome was 16,747 base pairs with a GC composition of 30.5%, similar to its host. Twenty-two open reading frames (ORFs) coding for putative peptides containing 30 or more amino acid residues were identified in the genome. Amino acid sequences of the predicted proteins from the phiCPV1 genome ORFs were compared with those from the NCBI database and potential functions of 12 proteins were predicted by sequence homology. Three putative proteins were similar to hypothetical proteins with unknown functions, whereas seven proteins did not have similarity with any known bacteriophage or bacterial proteins. Identified ORFs formed at least four genomic clusters that accounted for predicted proteins involved with replication of the viral DNA, its folding, production of structural components and lytic properties. One bacteriophage genome encoded lysin was predicted to share homology with N-acetylmuramoyl-l-alanine amidases and a second structural lysin was predicted to be a lysozyme-endopeptidase. These enzymes probably digest peptidoglycan of the bacterial cell wall and could be considered potential therapeutics to control *C. perfringens*. Table 2 summarizes the gene products common to currently known *C. perfringens* podoviral bacteriophage genomes.

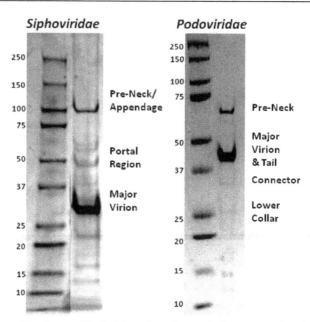

Fig. 2. Protein gel profiles for purified virions from bacteriophages virulent for *Clostridium perfringens* representing the *Siphoviridae* and *Podoviridae* from a joint Russian Federation-USA collaborative research project.

Three structural proteins were identified in the lysogenic phi3626 by N-terminal sequencing of proteins following SDS-PAGE of the purified virions (Zimmer *et al.*, 2002a). The major capsid component was estimated to be 43.3% of total phage protein and was determined to be post-translationally processed resulting in a decrease in size from 47.7 to 34.3 kDa. The major tail protein represented approximately 12.7% of the total protein, with an apparent size of 27 kDa while a minor structural protein composing 2.1% of the virion protein was reported with a predicted size of 55.1kD. More recently the proteins of virulent bacteriophages infecting *C. perfringens* have been described in detail (Seal *et al.*, 2011; Volozhantsev *et al.*, 2011). From the siphoviruses (Seal *et al.*, 2011), four principle virion protein regions were identified (Fig. 2) that included a portal protein, mycobacteriophage gp6-like protein which was the major virion protein, a pre-neck appendage protein and several lower molecular weight minor structural proteins with no known function. The portal protein was identified as a protein that was also highly variable with respect to isoelectric point and size at approximately 50kDa. This was attributed to potential differences in phosporylation and myristilation of the portal protein due to the large number of post-translational modification sites on the molecule. The podoviruses identified to date have virion proteins essentially indicative of those types of bacteriophages (Volozhantsev *et al.*, 2011). These viruses encode for a collar protein with a predicted size of approximately 27kDa and a connector protein with a predicted size of approximately 35.9kDa. The major head or major capsid protein was predicted to have a size of 43.3kDa and was found in the greatest abundance in the purified virus. A large pre-neck protein of 75 kDa and a tail protein of a size similar to the major capsid protein were also identified in *C. perfringens* phages of the *Podoviridae* (Fig. 2).

C. perfringens is an important agricultural as well as human pathogen and because biotechnological uses of bacteriophage gene products as alternatives to conventional antibiotics will require a thorough understanding of their genomic context, we sequenced and analyzed the genomes of four more closely related viruses isolated from the bacterium, then compared the known phage genomes (Oakley *et al.*, 2011). Phage whole-genome tetra-nucleotide signatures and proteomic tree topologies correlated closely with host phylogeny. Comparisons of our phage genomes to 26 others revealed three shared COGs of which one of particular interest within this core genome was an endolysin (PF01520, an N-acetylmuramoyl-L-alanine amidase) and a holin (PF04531). Comparative analyses of the evolutionary history and genomic context of these common phage proteins revealed two important results. One was a strongly significant, host-specific sequence variation within the endolysin and secondly is the protein domain architecture apparently unique to our phage genomes in which the endolysin is located upstream of its associated holin among certain members of the *Siphoviridae* (Oakley *et al.*, 2011). Endolysin sequences from our viruses were one of two very distinct genotypes distinguished by variability within the putative enzymatically-active domain. The shared or core genome was comprised of genes with multiple sequence types belonging to five pfam families, and genes belonging to 12 pfam families, including the holin genes, which were nearly identical.

6. Potential use of bacteriophages or their gene products to control *Clostridium perfringens*

Bacteriophages have been utilized experimentally in an attempt to control a variety of pathogens and there has been increased interest to control disease among poultry (Joerger, 2002). *In vivo* studies were conducted to determine if a cocktail of *C. perfringens* bacteriophages (INT-401) would be capable of controlling necrotic enteritis (NE) caused by *C. perfringens* (Miller *et al.*, 2010). The first study investigated the efficacy of INT-401 and a toxoid-type vaccine in controlling NE among *C. perfringens*-challenged broiler chickens reared until 28 days old. Compared with the mortality observed with the bacterium-challenged, but untreated chickens, oral administration of INT-401 significantly reduced mortality of the *C. perfringens*-challenged birds by 92%. Overall, INT-401 was more effective than the toxoid vaccine in controlling active *C. perfringens* infections of chickens. When the phage cocktail was administered via oral gavages, feed, or drinking water it significantly reduced mortality due to the bacterium and weight gain as well as feed conversion ratios were significantly better in the *C. perfringens*-challenged chickens treated with bacteriophages than in the *C. perfringens*-challenged, phage-untreated control birds (Miller *et al.*, 2010).

In order to repeat a similar study by Miller *et al.*, (2010) and to determine optimal schemes for application of bacteriophage formulations to cure or prevent disease from *C. perfringens* infection in poultry, investigators at the *State Research Center for Applied Microbiology and Biotechnology* (Obolensk, Moscow Region, Russian Federation) completed a series of experiments to monitor the persistence of *C. perfringens* lytic bacteriophage phiCPV1 in broiler gastrointestinal tracts (GIT). The phage suspension was administered *per os* once to 14-17 days old chicks (6×10^8 pfu/bird). To determine concentrations of the phage, materials from each section of the gastrointestinal tract (the crop, glandular stomach, the upper department of the small intestine, ileum, cecum, and the large intestine) were suspended in

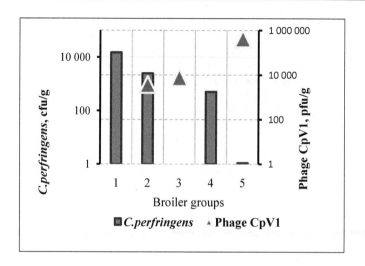

Note: Fourteen-day old broilers were inoculated with a suspension of two phiCPV1-sensitive *C. perfringens* Rif[R]-strains in the volume of 0.2ml (3×10[7] - 1×10[8] CFU for broilers) given *per os* to each broiler at day 19 (Groups 1 and 2) and at days 19, 20, 21 and 22 (Groups 4 and 5). The phiCPV1 in the volume of 0.2ml was administered *per os* twice a day to birds of Group 2 (10[8] pfu/bird) when they reached the age of 19 days, and to broilers of Groups 3 and 5 (10[9] PFU/bird) at days 19, 20, 21 and 22.

Fig. 3. Titres of *Clostridium perfringens* and phiCPV1 in lower sections of the gastrointestinal tract (ileum/cecum) of broiler chickens experimentally infected with the bacterium.

phage buffer followed by agar layer titration on a lawn produced by a *C. perfringens* phiCPV1- susceptible strain. Two independent experiments revealed that the highest concentration of the phage titer at 7×10[7] pfu/g was in the crop one hour after the administration. In the glandular stomach its concentration varied between 2×10[3] and 3×10[5] pfu/g. In the interval between 3 and 12 hours after treatment, phage concentration reached 10[7]pfu/g both in cecum and ileum of all birds. Such high concentrations of the phage in the GIT are extremely important from the standpoint of the phage therapy for *C. perfringens*-associated infection. Ileum and cecum are known to be main sites for the bacterium to colonize and proliferate. In the ileum and cecum, as well as in the large intestine, the maximal phage concentration (>10[6] pfu/g) was detected 6 hours after the administration of viruses and retained at a rather high level (>10[5] pfu/g) at least for the next 6 hours. The following day after administration of the phage in the GIT, the concentration decreased markedly. However, the phage was not fully eliminated even from the crop and was detected at the concentration of 500 pfu/g 48 hours later. The assessment of therapeutic and prophylactic effects of bacteriophage formulations in broilers during model experiments has demonstrated that phiCPV1 reduced intestinal colonization of the phage-sensitive *C. perfringens* in broiler chickens, with the phage titer being increased (Fig.3). At the same time experiments on phage therapy of broilers carrying natural *C. perfringens* infection by means of a phage cocktail were not successful and this was associated with the narrow lytic spectra of the phages. Consequently, natural *C. perfringens* isolated from the broiler chickens were resistant to the bacteriophages used during the experimentation, demonstrating the need for libraries of bacteriophage isolates to therapeutically eliminate the bacterium in animals.

Zimmer *et al.* (2002b) investigated the cell wall lysis system of *C. perfringens* bacteriophage phi3626, whose dual lysis gene cassette consisted of a holin gene and an endolysin gene. The Hol3626 had two predicted membrane-spanning domains (MSDs) and was designated a group II holin. A positively charged beta turn between the two MSDs indicated that both the amino-terminus and the carboxy-terminus of Hol3626 protein might be located outside the cell membrane which is a very unusual holin topology (Young, 2002). The holin function was experimentally demonstrated by using the ability of the peptide to complement a deletion of the heterologous phage lambda S holin in lambda delta-Sthf. The endolysin gene *ply3626* was cloned into an *E. coli* expression system. However, protein synthesis occurred only when the *E. coli* were supplemented with rare tRNA(Arg) and tRNA(Ile) genes required for proper codon usage of Gram+ genes in a Gram- system (Kane, 1995). Amino-terminal modification by a six-histidine tag did not affect enzyme activity and enabled purification by Ni-chelate affinity chromatography. The Ply3626 had an N-terminal amidase domain and a unique C-terminal portion that was hypothesized to be responsible for the specific lytic range of the enzyme. A total of 48 *C. perfringens* strains were sensitive to the murein hydrolase, whereas other clostridia and bacteria belonging to other genera were generally not affected by the lysin (Zimmer *et al.*, 2002b).

Two putative phage lysin genes (*ply*) from the clostridial phages phiCP39O and phiCP26F were cloned, expressed in *E. coli* and the resultant proteins were purified to near homogeneity (Simmons *et al.*, 2010). Gene and protein sequencing revealed that the predicted and chemically determined amino acid sequences of the two recombinant proteins were homologous to N-acetylmuramoyl-L-alanine amidases. The proteins were identical in the C-terminus cell-wall binding domain, but only 55 per cent identical to each other in the N-terminal catalytic domain. Both recombinant lytic enzymes were capable of lysing both parental phage host strains of *C. perfringens* as well as other type-strains of the bacterium in spot and turbidity reduction assays. The observed reduction in turbidity was correlated with up to a 3 log cfu/ml reduction in viable *C. perfringens* on brain heart infusion agar plates. However other member species of the clostridia were resistant to the enzymes by both assay methods. Interestingly, diversity exists even among closely-related bacteriophages, holins and endolysins represent conserved functions across divergent phage genomes and endolysins can have significant variability with host-specificity even among closely-related genomes. Endolysins of phage genomes in the presented study may be subject to different selective pressures than the rest of the genome and these findings may have important implications for potential biotechnological applications of phage gene products (Oakley *et al.*, 2011). Interestingly, a variety of encoded potential gene products have been detected in the genomes of *C. perfringens* bacteriophages that could potentially be utilized as antimicrobials to control the bacterium (Fig. 4).

The number of known genes encoding these peptidoglycan hydrolases has increased markedly in recent years, due in large part to advances in DNA sequencing technology. As the genomes of more bacterial species/strains are sequenced, lysin-encoding open reading frames (ORFs) can be readily identified in lysogenized prophage regions such as in the genomes of *C. perfringens* (Shimizu, *et al.*, 2002; Myers *et al.*, 2006). The genomes of nine *C. perfringens* strains were computationally mined for prophage lysins and lysin-like ORFs, revealing several dozen

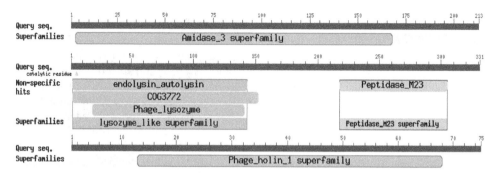

Fig. 4. Lytic proteins discovered in *Clostridium perfringens* bacteriophages from a joint Russian Federation-USA collaborative research project.

proteins of various enzymatic classes (Schmitz *et al.*, 2011). Of these lysins, a muramidase from strain ATCC 13124 (termed PlyCM) was chosen for recombinant analysis based on its dissimilarity to previously characterized *C. perfringens* lysins. Following expression and purification, various biochemical properties of PlyCM were determined *in vitro*, including pH/salt-dependence and temperature stability. The enzyme exhibited activity at low µg/ml concentrations, a typical value for phage lysins. It was active against 23 of 24 strains of *C. perfringens* assayed, with virtually no activity against other clostridial or non-clostridial species (Schmitz *et al.*, 2011). Also, an endolysin predicted to encode an N-acetylmuramidase was identified as encoded by the episomal phage phiSM101 of *C. perfringens* (Nariya *et al.*, 2011). Homologous genes were identified in the genomes of all five *C. perfringens* toxin types and the phiSM101 muramidase gene (psm) was cloned, then expressed in *E. coli* as a protein histidine-tagged at the N-terminus (Psm-his). Similar to other *C. perfringens* phage lysins the purified protein lysed cells of all *C. perfringens* toxin types, but not other clostridial species tested as demonstrated by a turbidity reduction assay (Nariya *et al.*, 2011). Consequently, more potential antimicrobials remain to be discovered utilizing genomics approaches.

Immobilization and separation of bacterial cells by replacing antibodies with cell wall-binding domains (CBDs) of bacteriophage-encoded peptidoglycan hydrolases (endolysins) has been accomplished for use as a potential diagnostic (Kretzer *et al.*, 2007). Paramagnetic beads coated with recombinant phage endolysin-derived CBD molecules and bacterial cells could be immobilized and recovered from diluted suspensions within 20 to 40 min. The CBD-based magnetic separation (CBD-MS) procedure was evaluated for capture and detection of *Listeria monocytogenes* from contaminated food samples and this approach was demonstrated by using specific phage-encoded CBDs specifically recognizing both *Bacillus cereus* and *C. perfringens* cells (Kretzer *et al.*, 2007). Consequently, the use of bacteriophage lysin cell-wall binding domains could be utilized for other applications as well as for improving diagnostic detection of Gram+ bacteria.

7. Conclusions

Bacteriophages have been utilized as potential interventions to treat bacterial infections. However, the development of bacterial resistances to their viruses occurs that include evolution of phage receptors, super-infection exclusion, restriction-modification systems and abortive infection systems such as genomic CRISPR sequences (Labrie *et al.*, 2010). These phenomena substantiate the inevitable need to constantly search for new bacteriophage isolates to use therapeutically. Also, it should be noted that although bacteriophage therapy has been utilized and examined as a treatment, it was pointed out early on by Smith (1959) that a large proportion of *C. perfringens* strains remained insusceptible to many of the bacteriophages isolated during those studies. This has routinely been observed during our investigations wherein most bacteriophages virulent for *C. perfringens* have a restricted host range (Fig. 5). Host specificity has routinely been observed relative to the bacteriophages isolated from various *C. perfringens* isolates that is most likely due to evolution of the receptor and anti-receptor molecules (Seal *et al.*, 2011; Volozhantsev *et al.*, 2011; Oakley *et al.*, 2011). Therefore, selection of appropriate 'bacteriophage cocktails' may not necessarily be effective against many of the various bacterial isolates that exist in the environment and cause disease.

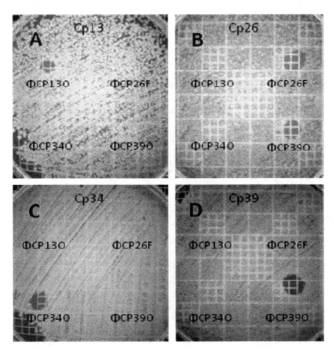

Fig. 5. Spot-assay with *Clostridium perfringens* bacteriophages on their respective hosts. Note that most all phages are restricted in their respective host-ranges.

Many enzymes are added to monogastric animal feeds to increase the digestibility of nutrients, leading to greater feed efficiency during the production of meat animals and eggs. Enzymes are added to monogastric animal feed for digesting carbohydrates and for metabolizing phytate to produce free phosphorus (Cowieson *et al.*, 2006; Olukosi *et al.*, 2010). There are a wide variety of enzymes marketed commercially for poultry feed additives, many of which are produced as a recombinant proteins in yeast and sold as a lysate which argues for the economic feasibility of developing enzyme additives (see DSM:

http://www.dsm.com/en_US/html/dnp/anh_enzymes.htm; Bio-Cat: http://www.bio-cat.com/applicationDetails.php?application_id=8; Ultra BioLogics: http://www.ublcorp.com/; Danisco: http://www.danisco.com/; Novozymes: http://www.novozymes.com/en/solutions/agriculture/animal-nutrition/). Consequently, production of enzymes by *Pichia pastoris* can serve as a potential source for structural or animal feed studies (Johnson *et al.*, 2010) and lysozyme can be encapsulated (Zhong & Jin, 2009) which has been utilized as a feed additive in the diet of chickens to significantly reduce the concentration of *C. perfringens* in the ileum and reduce intestinal lesions due to the organism (Liu *et al.*, 2010). Therefore, it is conceivable that bacteriophage proteins capable of lysing *C. perfringens* could be expressed in yeast and added as lysates to animal feed for reducing the bacterium to improve health and food safety for monogastric animals.

Clostridium perfringens (formerly known as *C. welchii*) is a ubiquitous Gram+ anaerobic, spore-forming bacterium that causes debilitating diseases in both humans and a wide variety of animals resulting in both personal tragedy and economic losses. Although the bacterium can cause severe diseases in most animals including domestic chickens, non-virulent forms of the bacillus are commonly found in the intestinal tracts of warm-blooded species as well as the environment. Several antibiotics can be utilized to treat clostridial diseases and sub-therapeutic amounts of antibiotics have been used in animal feeds as antibiotic growth promoters (AGP). Removal of AGP's from animal feed has resulted in the need for increased use of antibiotics therapeutically to treat diseases among food-producing animals, in particular necrotic enteritis in poultry. Consequently, this situation along with a concern as how to treat antibiotic resistant bacteria has provided the impetus to develop alternative antimicrobials or new antimicrobials that can be used synergistically with antibiotics. Prior to the discovery and widespread use of antibiotics, bacterial infections were often treated with bacteriophages, which were marketed and sold commercially for human use up until the 1940's. Following discovery of antibiotics, the use of phages to treat bacterial diseases was discontinued in Western Europe and the United States. Bacteriophages continue to be sold in the Russian Federation and Eastern Europe as treatments for bacterial infections and there is renewed interest in utilizing bacterial viruses to prevent or treat bacterial infections. Bacteriophages which infect *C. perfringens* that are both lysogenic and virulent have been discovered that have long tails, members of the *Siphoviridae*, and those with short tails, members of the *Podoviridae*. If these bacteriophages or their gene products are to be used as antimicrobials, it is essential to have a blueprint of the genomic machinery underlying phage-mediated bacterial lysis. As genome sequencing costs are reduced in price, genomics-enabled approaches to utilizing bacteriophages, or perhaps more importantly their gene products, as naturally occurring antimicrobials will become increasingly more common.

8. Acknowledgements

Support for the research was provided by the U.S. Department of Agriculture, Agricultural Research Service (CRIS project no. 6612-32000-046), the Russian Federation State Research Center for Applied Microbiology and Biotechnology, and the U.S. Department of State via International Science and Technology Center (ISTC) grants administered by the ARS, USDA Office of International Research Programs (OIRP).

9. References

Ackermann, H.W. (1974) Classification of Bacillus and Clostridium bacteriophages. *Pathologie Biologie* (Paris). Dec; Vol. 22, No. 10, pp.909-917.

Ackermann, H-W. (2003) Bacteriophage observations and evolution. *Research in Microbiology* 154(4):245-251.

Ackermann, H-W. (2006) Classification of Bacteriophages. p. 8-16. *In* R. Calendar and S.T. Abedon (ed.), *The Bacteriophages*, Oxford University Press, Oxford.

Ackermann, H-W. (2007) 5500 Phages examined in the electron microscope. *Archives of Virology* 152(2):227-243.

Barrow, P., Lovell M., Berchieri Jr A. (1998) Use of lytic bacteriophage for control of experimental *Escherichia coli* septicemia and meningitis in chickens and calves. *Clinical Diagnostic Laboratory Immunology* 5(3):294-298.

Bedford, M. (2000) Removal of antibiotic growth promoters from poultry diets: implications and strategies to minimize subsequent problems. *World Poultry Science Journal* 56:347-365.

Bradley D.E., Hoeniger J.F.M. (1971) Structural changes in cells of *Clostridium perfringens* infected with a short-tailed bacteriophage. *Canadian Journal of Microbiology* 17(3):397-402.

Bren, L. (2007) Bacteria-eating virus approved as food additive. *FDA Consumer* 41:20-22.

Brennan J., Skinner J., Barnum D.A., Wilson J. (2003) The efficacy of bacitracin methylene disalicylate when fed in combination with narasin in the management of necrotic enteritis in broiler chickens. *Poultry Science* 82(3):360-363.

Bruynoghe R., Maisin J. (1921) Essais de thérapeutique au moyen du bacteriophage. *Comptes Rendus des Seances de la Societe de Biologie et de ses Filiales* 85:1120-1121.

Brynestad S., Granum P.E. (2002) *Clostridium perfringens* and foodborne infections. *International Journal of Food Microbiology* 74(3):195-202.

Canard B., Cole S.T. (1990) Lysogenic phages of *Clostridium perfringens*: mapping of the chromosomal attachment sites. *FEMS Microbiology Letters* 54(1-3):323-326.

Casewell M., Friis C., Marco E., McMullin P., Phillips I. (2003) The European ban on growth-promoting antibiotics and emerging consequences for human and animal health. *Journal of Antimicrobial Chemotherapy* 52(2):159-161.

Casjens S.R. (2005) Comparative genomics and evolution of the tailed-bacteriophages. *Current Opinion in Microbiology* 8(4):451-458.

Castanon J.I. (2007) History of the use of antibiotic as growth promoters in European poultry feeds. *Poultry Science* 86(11):2466-2471.

Collier C.T., van der Klis J.D., Deplancke B., Anderson D.B., Gaskins H.R. (2003) Effects of tylosin on bacterial mucolysis, *Clostridium perfringens* colonization, and intestinal barrier function in a chick model of necrotic enteritis. *Antimicrobial Agents and Chemotherapy* 47(10):3311-3317.

Cotter P.D., Hill C., Ross R.P. (2005) Bacteriocins: developing innate immunity for food. *Nature Reviews Microbiology* 3(10):777-788.

Cowieson A.J., Hruby M., Pierson E.E. (2006) Evolving enzyme technology: impact on commercial poultry nutrition. *Nutrition Research Reviews* 19(1):90-103.

Crouch E.A., Labarre D., Golden N.J., Kause J.R., Dearfield K.L. (2009) Application of quantitative microbial risk assessments for estimation of risk management metrics: *Clostridium perfringens* in ready-to-eat and partially cooked meat and poultry products as an example. *Journal of Food Protection* 72(10):2151-61.

Diarra M.S., Silversides F.G., Diarrassouba F., Pritchard J., Masson L., Brousseau R., Bonnet C., Delaquis P., Bach S., Skura B.J., Topp E. (2007) Impact of feed supplementation with antimicrobial agents on growth performance of broiler chickens, *Clostridium perfringens* and enterococcus counts, and antibiotic resistance phenotypes and distribution of antimicrobial resistance determinants in *Escherichia coli* isolates. *Applied and Environmental Microbiology* 73(20):6566-6576.

DuPont H.L. (2007) The growing threat of foodborne bacterial enteropathogens of animal origin. *Clinical Infectious Diseases* 45(10):1353-1361.

Elford W.J., Guelin A.M., Hotchin J.E., Challice C.E. (1953) The phenomenon of bacteriophagy in the anaerobes; *Clostridium perfringens*. *Annals Institute Pasteur* (Paris). 84(2):319-327.

Engberg R.M., Hedemann M.S., Leser T.D., Jensen B.B. (2000) Effect of zinc bacitracin and salinomycin on intestinal microflora and performance of broilers. *Poultry Science* 79(9):1311-1319.

Fischetti VA. (2010) Bacteriophage endolysins: a novel anti-infective to control Gram-positive pathogens. *International Journal of Medical Microbiology* 300(6):357-362.

French G.L. (2010) The continuing crisis in antibiotic resistance. *International Journal of Antimicrobial Agents* 36 Suppl 3:S3-7.

Gaspar G, Tolnai G. (1959) Studies on a *Clostridium perfringens* phage. *Acta Microbiologica Academy of Sciences Hungary* 6:275-281.

George B.A, Quarles C.L., Fagerberg D.J. (1982) Virginiamycin effects on controlling necrotic enteritis infection in chickens. *Poultry Science* 61(3):447-50.

Gormley F.J., Little C.L., Rawal N., Gillespie I.A., Lebaigue S., Adak G.K. (2011) A 17-year review of foodborne outbreaks: describing the continuing decline in England and Wales (1992-2008). *Epidemiology and Infection* 139(5):688-699.

Grant R.B., Riemann H.P. (1976) Temperate phages of *Clostridium perfringens* type C1. *Canadian Journal of Microbiology* 22(5):603-610.

Guelin A., Kreguer A. (1950) Action of bacteriophage on the toxicity of young cultures of *Cl. perfringens*, type A. *Annals Institute Pasteur* (Paris). 78(4):532-537.

Guelin A. (1953) The appearance of new forms in bacteriophagic filtrates of *Clostridium perfringens* (Welchia perfringens). *Annals Institute Pasteur* (Paris). 84(3):562-76.

Gyles C.L. (2008) Antimicrobial resistance in selected bacteria from poultry. *Animal Health Research Reviews* 9(2):149-158.

Hendrix R.W., Smith M.C., Burns R.N., Ford M.E., Hatfull G.F. (1999) Evolutionary relationships among diverse bacteriophages and prophages: all the world's a phage. *Proceedings of the National Academy of Sciences USA.* 96(5):2192-2197.

d'Hérelle, F. (1917) Sur un microbe invisible antagoniste des bac. dysentÈriques. *Comptes Rendus des Seances de la Societe de Biologie et de ses Filiales* Paris 165:373-375.

Hirano S., Yonekura Y. (1967) The structure of *Clostridium perfringens* bacteriophages. *Acta Medical University of Kagoshima* 9:41–56.

Horne R.W. (1973) Contrast and resolution from biological objects examined in the electron microscope with particular reference to negatively stained specimens. *Journal of Microscopy* 98(3):286-298.

Hudson J.A., Billington C., Carey-Smith G., Greening G. (2005) Bacteriophages as biocontrol agents in food. *Journal of Food Protection* 68(2):426-437.

Huff W.E., Huff G.R., Rath N.C., Balog J.M., Donoghue A.M. (2003) Bacteriophage treatment of a severe *Escherichia coli* respiratory infection in broiler chickens. *Avian Diseases* 47(4):1399-1405.

Huff W.E., Huff G.R., Rath N.C., Balog J.M., Donoghue A.M. (2002a) Prevention of *Escherichia coli* infection in broiler chickens with a bacteriophage aerosol spray. *Poultry Science* 81(10):1486-1491.

Huff W.E., Huff G.R., Rath N.C., Balog J.M., Xie H., Moore Jr P.A., Donoghue A.M. (2002b) Prevention of *Escherichia coli* respiratory infection in broiler chickens with bacteriophage. *Poultry Science* 81(4):437-441.

Ionesco H, Wolff A, Sebald M. (1974) The induced production of bacteriocin and bacteriophage by the BP6K-N-5 strain of *Clostridium perfringens* (author's transl). *Annals Microbiology* (Paris). 125B(3):335-346.

Joerger R.D. (2002) Alternatives to antibiotics: bacteriocins, antimicrobial peptides and bacteriophages. *Poultry Science* 82:640-647.

Johnson S.C., Yang M., Murthy P.P. (2010) Heterologous expression and functional characterization of a plant alkaline phytase in *Pichia pastoris*. *Protein Expression and Purification* 74(2):196-203.

Kane J.F. (1995) Effects of rare codon clusters on high-level expression of heterologous proteins in *Escherichia coli*. *Current Opinion in Biotechnology*. 6(5):494-500.

Knarreborg A., Simon M.A., Engberg R.M., Jensen B.B, Tannock G.W. (2002) Effects of dietary fat source and sub-therapeutic levels of antibiotic on the bacterial community in the ileum of broiler chickens at various ages. *Applied and Environmental Microbiology* 68(12):5918-5924.

Kreguer A., Guelin A., Le Bris J. (1947) Isolation of a bacteriophage active on *Clostridium perfringens* type A. *Annals Institute Pasteur* (Paris). 73(10):1038.

Kretzer J.W., Lehmann R., Schmelcher M., Banz M., Kim K.P., Korn C., Loessner M.J. (2007) Use of high-affinity cell wall-binding domains of bacteriophage endolysins for immobilization and separation of bacterial cells. *Applied and Environmental Microbiology* 73(6):1992-2000.

Labrie S.J., Samson J.E., Moineau S. (2010) Bacteriophage resistance mechanisms. *Nature Reviews Microbiology* 8(5):317-327.

Lawrence J.G., Hatfull G.F., Hendrix R.W. (2002) Imbroglios of viral taxonomy: genetic exchange and failings of phenetic approaches. *Journal of Bacteriology* 184(17):4891-4905.

Lindström M., Heikinheimo A., Lahti P., Korkeala H. (2011) Novel insights into the epidemiology of *Clostridium perfringens* type A food poisoning. *Food Microbiology* 28(2):192-198.

Liu D., Guo Y., Wang Z., Yuan J. (2010) Exogenous lysozyme influences *Clostridium perfringens* colonization and intestinal barrier function in broiler chickens. *Avian Pathology* 39(1):17-24.

Liu J., Dehbi M., Moeck G., Arhin F., Bauda P., Bergeron D., Callejo M., Ferretti V., Ha N., Kwan T., McCarty J., Srikumar R., Williams D., Wu J.J., Gros P., Pelletier J., DuBow M. (2004) Antimicrobial drug discovery through bacteriophage genomics. *Nature Biotechnology* 22(2):185-191.

Mahony D.E., Easterbrook K.B. (1970) Intracellular development of a bacteriophage of *Clostridium perfringens*. *Canadian Journal of Microbiology* 16(10):983-988.

Mahony D.E., Kalz G.G. (1968) A temperate bacteriophage of *Clostridium perfringens*. *Canadian Journal of Microbiology* 14(10):1085-1093.

Mataragas M., Skandamis P.N., Drosinos E.H. (2008) Risk profiles of pork and poultry meat and risk ratings of various pathogen/product combinations. *International Journal of Food Microbiology* 126(1-2):1-12.

Maxey B.W., Page R.K. (1977) Efficacy of lincomycin feed medication for the control of necrotic enteritis in broiler-type chickens. *Poultry Science* 56(6):1909-1913.

McDermott P.F., Zhao S., Wagner D.D., Simjee S., Walker R.D., White D.G. (2002) The food safety perspective of antibiotic resistance. *Animal Biotechnology* 13(1):71-84.

Mead P.S., Slutsker L., Dietz V., McCaig L.F., Bresee J.S., Shapiro C., Griffin P.M., Tauxe R.V. (1999) Food related illness and death in the United States. *Emerging Infectious Diseases* 5(5):607-625.

Merril C.R., Biswas B., Carlton R., Jensen N.C., Creed G.J., Zullo S., Adhya S. (1996) Long-circulating bacteriophage as antibacterial agents. *Proceedings of the National Academy of Sciences USA.* 93(8):3188-3192.

Miller R.W., Skinner E.J., Sulakvelidze A., Mathis G.F., Hofacre C.L. (2010) Bacteriophage therapy for control of necrotic enteritis of broiler chickens experimentally infected with *Clostridium perfringens*. *Avian Diseases* 54(1):33-40.

Moore J.E., Barton M.D., Blair I.S., Corcoran D., Dooley J.S., Fanning S., Kempf I., Lastovica A.J., Lowery C.J., Matsuda M., McDowell D.A., McMahon A., Millar B.C., Rao J.R., Rooney P.J., Seal B.S., Snelling W.J., Tolba O. 2006. The epidemiology of antibiotic resistance in *Campylobacter*. *Microbes and Infection* 8(7):1955-1966.

Myers G.S., Rasko D.A., Cheung J.K., Ravel J., Seshadri R., DeBoy R.T., Ren Q., Varga J., Awad M.M., Brinkac L.M., Daugherty S.C., Haft D.H., Dodson R.J., Madupu R., Nelson W.C., Rosovitz M.J., Sullivan S.A., Khouri H., Dimitrov G.I., Watkins K.L., Mulligan S., Benton J., Radune D., Fisher D.J., Atkins H.S., Hiscox T., Jost B.H., Billington S.J., Songer J.G., McClane B.A., Titball R.W., Rood J.I., Melville S.B., Paulsen I.T. (2006) Skewed genomic variability in strains of the toxigenic bacterial pathogen, *Clostridium perfringens*. *Genome Research* 16(8):1031-1040.

National Academy of Sciences. Treating Infectious Diseases in a Microbial World: Report of Two Workshops on Novel Antimicrobial Therapeutics. ISBN: 0-309-65490-4, 2006 (http://www.nap.edu/catalog.php?record_id=11471).

Nariya H., Miyata S., Tamai E., Sekiya H., Maki J., Okabe A. (2011) Identification and characterization of a putative endolysin encoded by episomal phage phiSM101 of *Clostridium perfringens*. *Applied Microbiology and Biotechnology* 90(6):1973-1979.

Nelson D. (2004) Phage taxonomy: we agree to disagree. *Journal of Bacteriology* 186(21):7029-7031.

Nowell V.J., Poppe C., Parreira V.R., Jiang Y.F., Reid-Smith R., Prescott J.F. (2010) *Clostridium perfringens* in retail chicken. *Anaerobe* 16(3):314-315.

Oakley B.B., Talundzic E., Morales C.A., Hiett K.L., Siragusa G.R., Volozhantsev N.V., Seal B.S. (2011) Comparative genomics of four closely related *Clostridium perfringens* bacteriophages reveals variable evolution among core genes with therapeutic potential. *BMC Genomics* 12(1):282.

Olukosi O.A., Cowieson A.J., Adeola O. (2010) Broiler responses to supplementation of phytase and admixture of carbohydrases and protease in maize-soyabean meal diets with or without maize Distillers' Dried Grain with Solubles. *British Poultry Science* 51(3):434-43.

Paquette G., Fredette V. (1977) Properties of four temperate bacteriophages active on *Clostridium perfringens* type A. *Reviews of Canadian Biology* 36(3):205-215.

Persoons D., Dewulf J., Smet A., Herman L., Heyndrickx M., Martel A., Catry B., Butaye P., Haesebrouck F. (2010) Prevalence and persistence of antimicrobial resistance in broiler indicator bacteria. *Microbial Drug Resistance* 16(1):67-74.

Projan S.J., Gill D., Lu Z., Herrmann S.H. (2004) Small molecules for small minds? The case for biologic pharmaceuticals. *Expert Opinion on Biological Therapy* 4(8):1345-1350.

Projan S.J., Youngman PJ. (2002) Antimicrobials: new solutions badly needed. *Current Opinion in Microbiology* 5(5):463-465.

Proux C., van Sinderen D., Suarez J., Garcia P., Ladero V., Fitzgerald G.F., Desiere F., Brussow H. (2002) The dilemma of phage taxonomy illustrated by comparative genomics of Sfi21-like Siphoviridae in lactic acid bacteria. *Journal of Bacteriology* 184(21):6026-6036.

Ricke S.C., Kundinger M.M., Miller D.R., Keeton J.T. (2005) Alternatives to antibiotics: chemical and physical antimicrobial interventions and foodborne pathogen response. *Poultry Science* 84(4):667-675.

Rohwer F., Edwards R. (2002) The Phage Proteomic Tree: a genome-based taxonomy for phage. *Journal of Bacteriology* 184(16):4529-4535.

Satija K.C., Narayan K.G. (1980) Bacteriophage typing of food poisoning strains of C. perfringens type A. *Indian Journal of Pathology* and Microbiology 23(4):261-5-A.

Sawires Y.S., Songer J.G. (2006) *Clostridium perfringens*: insight into virulence evolution and population structure. *Anaerobe* 12:23-43.

Shimizu T., Ohtani K., Hirakawa H., Ohshima K., Yamashita A., Shiba T., Ogasawara N., Hattori M., Kuhara S., Hayashi H. (2002) Complete genome sequence of Clostridium perfringens, an anaerobic flesh-eater. *Proceedings of the National Academy of Sciences* USA 99(2):996-1001.

Scallan E., Griffin P.M., Angulo F.J., Tauxe R.V., Hoekstra R.M. (2011) Foodborne illness acquired in the United States--unspecified agents. *Emerging Infectious Diseases* 17(1):16-22.

Schmitz J.E., Ossiprandi M.C., Rumah K.R., Fischetti V.A. (2011) Lytic enzyme discovery through multigenomic sequence analysis in *Clostridium perfringens*. *Applied Microbiology and Biotechnology* 89(6):1783-1795.

Seal B.S., Fouts D.E., Simmons M., Garrish J.K., Kuntz R.L., Woolsey R., Schegg K.M., Kropinski A.M., Ackermann H-W., Siragusa G.R. (2011) *Clostridium perfringens* bacteriophages phiCP39O and phiCP26F: genomic organization and proteomic analysis of the virions. *Archives of Virology* 156(1):25-35.

Simmons M., Donovan D.M., Siragusa G.R., Seal B.S. (2010) Recombinant expression of two bacteriophage proteins that lyse *Clostridium perfringens* and share identical sequences in the C-terminal cell wall binding domain of the molecules but are dissimilar in their N-terminal active domains. *Journal of Agricultural and Food Chemistry* 58(19):10330-10337.

Smedley 3rd J.G., Fisher D.J., Sayeed S., Chakrabarti G., McClane B.A. (2004) The enteric toxins of *Clostridium perfringens*. *Reviews of Physiology, Biochemistry and Pharmacology* 152:183-204.

Smith HW. (1959) The bacteriophages of *Clostridium perfringens*. *Journal of General Microbiology* 21:622-630.

Smith H.W., Huggins M.B. (1982) Successful treatment of experimental *Escherichia coli* infections in mice using phages: its general superiority over antibiotics. *Journal of General Microbiology* 128(2):307-318.

Smith H.W., Huggins M.B. (1987) The control of experimental *E. coli* diarrhea in calves by means of bacteriophage. *Journal of General Microbiology* 133(5):1111-1126.

Spencer M.C. (1953) Gas gangrene and gas gangrene organisms 1940-1952. An annotated bibliography of the Russian literature, 1940-1952, and the non-Russian literature for 1952. Armed Forces Medical Library Reference Division, Washington, D.C. Titles: 38, 39, 40, 41, 42, 158, 162, 210, 265.

Stewart A.W., Johnson M.G. (1977) Increased numbers of heat-resistant spores produced by two strains of Clostridium perfringens bearing temperate phage s9. Journal of General Microbiology 103(1):45-50.

Sulakvelidze A., Alavidze Z., Morris Jr. J.G. (2001) Bacteriophage therapy. Antimicrobial Agents and Chemotherapy 45(3):649-659.

Summers W.C. (2001) Bacteriophage Therapy. Annual Reviews of Microbiology 55:437-451.

Twort F.W. (1915) An investigation on the nature of ultra-microscopic viruses. Lancet II:1241-1243.

Van Immerseel F., De Buck J., Pasmans F., Huyghebaert G., Haesebrouck F., Ducatelle R. (2004) Clostridium perfringens in poultry: an emerging threat for animal and public health. Avian Pathology 33(6):537-549.

Vieu J.F., Guélin A., Dauguet C. (1965) Morphology of the Welchia perfringens bacteriophage 80. Annals Institute Pasteur (Paris) 109(1):157-60.

Volozhantsev N.V., Verevkin V.V., Bannov V.A., Krasilnikova V.M., Myakinina V.P., Zhilenkov E.L., Svetoch E.A., Stern N.J., Oakley B.B., Seal B.S. (2011) The genome sequence and proteome of bacteriophage phiCPV1 virulent for Clostridium perfringens. Virus Research 155(2):433-439.

Wagner P.L., Waldor M.K. (2002) Bacteriophage control of bacterial virulence. Infection and Immunity 70 (8):3985-3993.

Wen Q., McClane B.A. (2004) Detection of enterotoxigenic Clostridium perfringens type A isolates in American retail foods. Applied and Environmental Microbiology 70(5):2685-2691.

Wise M.G., Siragusa G.R. (2007) Quantitative analysis of the intestinal bacterial community in one- to three-week-old commercially reared broiler chickens fed conventional or antibiotic-free vegetable-based diets. Journal of Applied Microbiology 102(4):1138-1149.

Woese C.R., Fox G.E. (1977) Phylogenetic structure of the prokaryotic domain: the primary kingdoms. Proceedings of the National Academy of Sciences USA 74(11):5088-5090.

Yan S.S., Gilbert J.M. (2004) Antimicrobial drug delivery in food animals and microbial food safety concerns: an overview of in vitro and in vivo factors potentially affecting the animal gut microflora. Advanced Drug Delivery Reviews 56(10):1497-521.

Yan WK. (1989) Use of host modified bacteriophages in development of a phage typing scheme for Clostridium perfringens. Medical Laboratory Sciences 46(3):186-93.

Young R. (2002) Bacteriophage holins: deadly diversity. Journal of Molecular Microbiology and Biotechnology 4(1):21-36

Zhong Q., Jin M. (2009) Nanoscalar structures of spray-dried zein microcapsules and in vitro release kinetics of the encapsulated lysozyme as affected by formulations. Journal of Agricultural and Food Chemistry 57(9):3886-3894.

Zimmer M., Scherer S., Loessner M.J. (2002a) Genomic analysis of Clostridium perfringens bacteriophage phi3626, which integrates into guaA and possibly affects sporulation. Journal of Bacteriology 184(16):4359-4368.

Zimmer M., Vukov N., Scherer S., Loessner M.J. (2002b) The murein hydrolase of the bacteriophage phi3626 dual lysis system is active against all tested Clostridium perfringens strains. Applied and Environmental Microbiology 68(11):5311-5317.

11

Phages as Therapeutic Tools to Control Major Foodborne Pathogens: *Campylobacter* and *Salmonella*

Carla M. Carvalho[1]*, Sílvio B. Santos[1]*, Andrew M. Kropinski[2],
Eugénio C. Ferreira[1] and Joana Azeredo[1]
[1]*IBB - Institute for Biotechnology and Bioengineering,
Centre of Biological Engineering, Universidade do Minho,*
[2]*Laboratory for Foodborne Zoonoses, Public Health Agency of Canada,*
[1]*Portugal*
[2]*Canada*

1. Introduction

Foodborne diseases are a growing public health problem worldwide with *Campylobacter* and *Salmonella* being the most common and widely distributed causative agents. These Gram-negative bacteria are common inhabitant of the gut of warm-blooded animals, especially livestock, being transmitted to humans primarily through the consumption of contaminated food of animal origin. Poultry meat and derivatives are regarded as the most common source of human salmonellosis and campylobacteriosis.

In addition to the high prevalence of such pathogens and the consequent health problems caused, control of these pathogens has become increasingly difficult due to the emergence of antibiotic-resistant strains. This emergence is a result of the misuse of antimicrobials in food animals, compromising the action of once effective antibiotics in the treatment of foodborne diseases in humans.

Recent legislation restricting the use of antibiotics as growth promoters in animal production, together with the risk of antibiotic-resistant bacteria entering the human food chain have been the driving force for the development of alternative methods for pathogen control. (Bacterio)phages are naturally occurring predators of bacteria, ubiquitous in the environment, with high host specificity and capacity to evolve to overcome bacterial resistance which makes them an appealing option for the control of pathogens. Several studies have been carried to assess the potential use of phages in the control of *Campylobacter* and *Salmonella* in animals and food material in order to prevent transmission of these pathogens to humans. Overall, although eradication of the target bacteria is an extremely unlikely event, the proof of principle, that phages are able to reduce the number of these pathogens has been established. Even so, some considerations should be taken into account for an efficient application of phages.

---

* These authors contributed equally to this chapter

This chapter aims at giving an overview of the two major foodborne pathogens (*Campylobacter* and *Salmonella*), discussing the problems and concerns related to their prevalence and control, focusing mainly on the potential use of phages as an alternative to other control measures. Consequently, the successes and drawbacks of different studies on the use of phages to control *Campylobacter* and *Salmonella* will be explored. Moreover several aspects of phage biocontrol will be addressed. These include considerations on phage characterization, phage dose and route of administration, and ways of overcoming the emergence of phage resistant-bacteria. Finally the requisites for an acceptable phage product and the issues related to its public acceptance will be discussed.

2. Foodborne diseases

Foodborne diseases are of major concern worldwide. The Centres for Disease Control and Prevention (CDC) estimates that 76 million cases of foodborne diseases occur every year in the United States causing roughly 5000 deaths (Nyachuba, 2010). In Australia the number of cases (5.4 million) has been estimated to have an associated cost of 1.2 billion dollars per year (OzFoodNet Working Group, 2009). The European Food Safety Authority (EFSA) reported a total of 5,550 foodborne outbreaks, causing 48,964 human cases, 4,356 hospitalizations and 46 deaths in 2009 (European Food Safety Authority, 2011). While significant attention is usually given to major foodborne outbreaks, studies indicate that outbreaks only account for a small fraction of *Campylobacter* and *Salmonella* infections in humans (European Food Safety Authority, 2009). While a steady decline in the number of cases attributed to *Salmonella* has been observed since 2004, the number of *Campylobacter* infections has remained constant (Figure 1). Campylobacteriosis caused 198,252 confirmed human cases in 2009 with a fatality rate of 0.02 %, continuing to be the most commonly reported zoonosis in the European Union. A total of 108,614 confirmed human cases were attributed to *Salmonella* with a fatality rate of 0.08 % in the same year. Moreover there is a considerable underreporting, and the true number of cases of illness caused by these two pathogens is likely to be 10-100 times higher than the reported number (European Food Safety Authority, 2011).

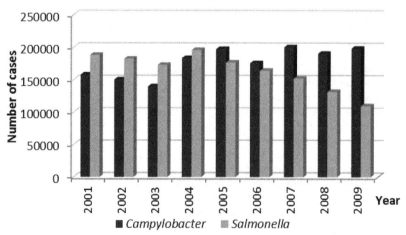

Fig. 1. Number of reported campylobacteriosis and salmonellosis cases in humans 2001-2009 (European Food Safety Authority, 2006; European Food Safety Authority, 2011)

2.1 The pathogens

2.1.1 *Campylobacter*

Campylobacter was first described in 1880 by Theodore Escherich and belongs to the Epsilonproteobacteria, in the order *Campylobacterales* which includes *Helicobacter* and *Wolinella* (Friedman *et al.*, 2000; Keener *et al.*, 2004). The name *Campylobacter* is derived from the Greek "kampulos" = curved and "bacter" = rod. In fact, bacteria belonging to the genus *Campylobacter* are non-spore forming, oxidase-positive, Gram-negative, curved or spiral (occasionally straight) rods with 0.2 μm to 0.8 μm wide and 0.5 μm to 5 μm long. When they form short or long chains they can appear as S-shaped, V-shaped or more rarely comma shaped. *Campylobacter* ssp. usually displays a long unsheathed polar flagellum at one (polar) or both (bipolar) ends of the cell which confer to this microorganism a rapid, darting and reciprocating motility (Keener *et al.*, 2004). *Campylobacter* spp. cells tend to form coccoid and elongated forms on prolonged culture or upon exposure to oxygen (Moran & Upton, 1987). These cells may be associated to a viable but not culturable state (VBNC). However the association between cell culturability and cell morphology remains controversy (Keener *et al.*, 2004).

The genus *Campylobacter* now comprises 17 member species. The most commonly isolated are *C. jejuni ssp. jejuni*, *C. coli* and *C. lari* which are referred as thermophilic species. They can grow at 37°C to 42°C with a pH in the range of 4.9 to 9.0, but their optimum growth conditions include a temperature of 42°C and a pH of 6.5 to 7.5. It is known that they cannot multiply below 30°C and that they require a microaerobic atmosphere (approximately 5% oxygen and 10% carbon dioxide) (Butzler, 2004).

Contrary to most bacteria, *Campylobacter* species do not obtain their energy from the metabolism of carbohydrates but instead from amino acids or tricarboxylic acid cycles intermediates. *C. jejuni* hydrolyzes hippurate and indoxyl metabolizes acetate and reduces nitrate (Butzler, 2004).

2.1.2 *Salmonella*

Investigations on the etiologic agent of the "swine plague" led Theobald Smith, in 1885, to the isolation of a Gram-negative bacillus named *Bacterium suipestifer*. The bacterium was further characterized by D. E. Salmon from whom the name *Salmonella* is derived. The non-spore forming cells possess a straight rod-shaped morphology with sizes varying from 0.7 μm to 1.5 μm in diameter and 2 μm to 5 μm in length. These cells are usually motile presenting peritrichous flagella. *Salmonella* spp. belong to the *Enterobacteriaceae* family and are chemoorganotrophs (organisms which use organic compounds as their energy source), facultative anaerobes and hydrogen sulphide producers (Bell & Kyriakides, 2002).

The outer membrane (OM) of *Salmonella*, as with almost all Gram-negative bacteria, is composed of OM proteins (OMPs) and lipopolysaccharides (LPS). LPS plays an essential role in maintaining the cell structural integrity and protection from chemicals. In the host organisms they act as endotoxins and as a pyrogen displaying a strong immune response. Structurally they are composed by three distinct components: lipid A, core oligosaccharide and O-polysaccharide (Raetz & Whitfield, 2002). The O-polysaccharide (also O-antigen or O-side-chain) together with the H-antigen (from flagella) and Vi (capsular antigens) are the basis for the Kauffman-White classification scheme, enabling the different *Salmonella* to be grouped in serotypes according to their agglutination pattern when reacted with specific commercial antisera (Bell & Kyriakides, 2002; Brenner et al., 2000). This classification led to the recognition of more than 2500 serotypes (Bell & Kyriakides, 2002), a number that increases every year. A revision of the nomenclature has established two species (S. enterica and S. bongori) with the majority of the serotypes grouped into one of the six Salmonella subspecies of S. enterica (Bell & Kyriakides, 2002; Brenner et al., 2000; Velge et al., 2005).

2.2 The diseases

Campylobacter and Salmonella are common inhabitant of the gut of warm-blooded animals mainly livestock (such as cattle, sheep, pigs and chickens), domestic pets and wild animals, where they asymptomatically colonize and multiply (Antunes et al., 2003; Bell & Kyriakides, 2002; Bryan & Doyle, 1995; Doyle & Erickson, 2006; Newell & Fearnley, 2003). As zoonotic agents, Campylobacter and Salmonella can be transferred between humans and other animals. The common route of these pathogens is the consumption of contaminated food of animal origin, particularly meat from pigs, cattle and poultry (and derivatives) and milk. Poultry and derivatives are repeatedly pointed out as the most common sources of infection since the pathogens are present at a high level in fresh poultry meat. Campylobacter and Salmonella strains may also reach humans via routes other than food, directly by the contact with contaminated animals, carcasses or the environment, for example, through drinking water (European Food Safety Authority & European Centre for Disease Prevention and Control, 2011). Therefore, horizontal transmission appears to have a major role in the transmission of these foodborne pathogens. In contrast to Salmonella, vertical transmission of Campylobacter is generally considered a relatively unimportant route of flock colonization with the consequent general absence of Campylobacter in eggs, one of the most common routes of contamination by Salmonella (European Food Safety Authority, 2011; Newell & Fearnley, 2003).

These microorganisms have the ability to survive for considerable periods, especially in conditions that are moist, cool and out of direct sunlight. As a result, they can readily contaminate other hosts, as for example, humans where Campylobacter infection is usually associated with illness and for which doses as low as 500 organisms have been reported to cause gastrointestinal disorders (Friedman et al., 2000; Newell & Fearnley, 2003; Robinson, 1981). As a consequence, bird-to-bird transmission within flocks is very rapid and it was demonstrated that once a broiler flock becomes infected with Campylobacter, close to 100% of birds are reported to become colonized in a very short time (Allen et al., 2007; Newell & Fearnley, 2003). Moreover it is known that, after in vivo-passage, organisms can exhibit an enhancement of colonization potential of at least 1,000-fold in most strains and up to 10,000-fold in some strains (Berndtson et al., 1992; Keener et al., 2004). The most important Campylobacter species associated with human infections are C. jejuni, C. coli, C. lari and C. upsaliensis (European Food Safety Authority, 2011; Friedman et al., 2000; Robinson, 1981). Campylobacter has become the most recognized antecedent cause of Guillain-Barré syndrome (GBS), an acute post-infectious immune-mediated disorder affecting the peripheral nervous system that can be permanent, fatal or last several weeks and usually requires intensive care (Butzler, 2004; Nachamkin, 2002).

The factors contributing for the high prevalence of these pathogens in poultry meat are bad slaughter conditions, cross-contamination, inadequate heat treatment, raw meat and inappropriate food storage (European Food Safety Authority, 2011; Gorman *et al.*, 2002; Hansson *et al.*, 2005; Jacobs-Reitsma, 2000; Johannessen *et al.*, 2007; Luber *et al.*, 2006; van de Giessen *et al.*, 2006). A European Union-wide baseline survey on *Campylobacter* demonstrated that in EU 71.2% of broilers are colonized by *Campylobacter* at the slaughterhouse (European Food Safety Authority, 2010). Therefore, controlling *Campylobacter* and *Salmonella* infections has become an important goal particularly for the poultry industry.

Human infection by these pathogens results in a gastrointestinal infection, which is usually characterized by an inflammatory reaction, watery (sometimes bloody) diarrhoea, fever, vomiting, abdominal cramps and dehydration which can become severe and life-threatening as a result of tissue invasion and toxin production (Bell & Kyriakides, 2002; Butzler, 2004; Friedman *et al.*, 2000; Nachamkin, 2002; Uzzau *et al.*, 2000). *Salmonella* infections are influenced by the bacterium's host range or degree of host adaptation, enabling the division of the bacteria in two groups: host adapted and ubiquitous. The higher the adaptation of *Salmonella* to a host, the higher the pathogenicity, with a consequent severity of the disease, usually leading to septicaemia (Bell & Kyriakides, 2002; Uzzau *et al.*, 2000; Velge *et al.*, 2005). The most prevalent and important *Salmonella enterica* serotypes reported worldwide are Enteritidis and Typhimurium. These are responsible for 99% of salmonellosis in humans and warm-blooded animals (Bell & Kyriakides, 2002; Brenner *et al.*, 2000).

2.3 Antibiotic resistance

Antibiotics were introduced in the 1940s and have been widely used in the United States (US) and Europe (EU) in livestock and poultry since the 1950s. In the US at least 17 antimicrobials were approved to be used in food animals. In Europe, all classes of antibiotics licensed for human medicine were allowed for use in animals. As a consequence, antibiotics were used in food animals therapeutically, prophylactically and as food supplements to promote faster growth by improving feed efficiency. The discovery of antibiotics growth-enhancing effect became an important element of intense animal husbandry leading to their increased use, often in sub-therapeutic doses in healthy animals and without veterinary prescription (Castanon, 2007; Mathew *et al.*, 2007; Sapkota *et al.*, 2007; World Health Organization, 2002).

The amount of antibiotics used in the absence of disease for non-therapeutic purposes in livestock far exceeds the amount of antimicrobials used in human medicine. It was estimated that 60 to 80% of the antibiotics produced in the US is used for this purpose. The use of antibiotics in livestock has become a major source of concern because of the possibility that they contribute to the declining efficacy of antibiotics used to treat bacterial infections in humans (Smith *et al.*, 2002). This may happen because antimicrobial agents used for food-producing animals are frequently the same or belong to the same classes as those used in human medicine. The later includes tetracyclines, macrolides and fluoroquinolones (Aarestrup *et al.*, 2008; Mellon *et al.*, 2001; Sapkota *et al.*, 2007).

Two EU agencies, the European Food Safety Authority (EFSA) and the European Centre for Disease Prevention and Control (ECDC), reported recently on the high incidence of antibiotic resistance in *Salmonella* and *Campylobacter*, and stated their concern. In fact, the high percentage of *Salmonella* and *Campylobacter* isolates displaying resistance to ciprofloxacin is alarming since it represents one of the drugs of choice in human treatment.

High resistance of *Salmonella* to tetracyclines, ampicillin and sulphonamides was also reported, as was *Campylobacter* resistance to high levels of tetracyclines. The EFSA report concluded that the animal antimicrobial usage might be an important factor accounting for the high proportion of resistant isolates (European Food Safety Authority, 2011; European Food Safety Authority & European Centre for Disease Prevention and Control, 2011; Gyles, 2008; Rabsch *et al.*, 2001).

The resistance problem has led the World Health Organization (WHO) in 2002 to advise and encourage all countries to reduce the use of antibiotics outside human medicine and has already established some measures in the surveillance of foodborne diseases in order to reduce the emergence of resistant bacteria with special concern for *Salmonella* and *Campylobacter* (Smith *et al.*, 2005; World Health Organization, 2002). This concern was already present in the EU where several countries have banned the use of antimicrobials that are used in human medicine as growth promoters. The consequent reduction of the selective pressure, has already resulted in a reduction of antimicrobial resistance in a national population of food animals (Aarestrup *et al.*, 2001; Castanon, 2007; Emborg *et al.*, 2003; Smith *et al.*, 2005; Swann, 1969; Tacconelli *et al.*, 2008; Wierup, 2001; World Health Organization, 2002).

Antibiotics are usually the last resource in pathogens control leaving no hope in the treatment of multiresistant bacteria for which no effective antimicrobial exists. Consequently, it can be concluded that an efficient alternative to antibiotics is critical and urgent. In order to control foodborne pathogens, the poultry industry decontaminates carcasses using both chemical and physical treatments. Chemical treatments include washing of carcasses in electrolyzed or chlorinated water, dipping carcasses in a solution containing acidified sodium chlorite before chilling, immersion in acetic or lactic acid or in sodium triphosphate solutions. Physical treatments include freezing of contaminated carcasses, heat-treatment of fresh broiler carcasses, dipping of fresh carcasses in hot water immediately before chilling, radiation, exposure to dry heat, and ultrasonic energy in combination with heat (Corry & Atabay, 2001; Keener *et al.*, 2004). In spite of the effort that has been done to control these pathogens, they are still a major cause of foodborne diseases (European Food Safety Authority, 2011; Nyachuba, 2010)(Figure 1).

Other possible control measures to eliminate or reduce the contamination of birds are still being developed and their cost-effectiveness and applicability to large-scale production remain to be determined. It includes: vaccination, the use of competitive exclusion, improving genetic resistance of birds and the use of probiotics, bacteriocins and bacteriophages (Chen & Stern, 2001; García *et al.*, 2008; Joerger, 2001).

3. Bacteriophages: Novel therapeutic agents

(Bacterio)phages are viruses that are able to infect Bacteria. Phages are able to infect more than 150 bacterial genera, including aerobes and anaerobes, exospore and endospore formers, cyanobacteria, spirochetes, mycoplasmas, and chlamydias (Ackermann, 2001; Ackermann, 2009).

Structurally they consist of a nucleic acid genome enclosed within a protein or lipoprotein coat and like all viruses are absolute parasites, inert particles outside their hosts, deprived of their own metabolism. Inside their hosts, phages are able to replicate using the host cell as a factory to produce new phages particles identical to its ascendant, leading to cell lysis and consequent death of the host (Guttman *et al.*, 2005). As a result of their bacterial parasitism, phages can be found wherever bacteria exist and have already colonized every conceivable habitat. Phages are an extremely diversified group and it has been estimated that ten phage particles exist for each bacterial cell. This fact accounts for an estimated size of the global phage population to be approximately 10^{31} particles making phages the most abundant living entities on earth (Rohwer, 2003).

Their presence in the biosphere is especially predominant in the oceans presenting an excess of 10^7 to 10^8 phage particles per millilitre in coastal sea and in non-polluted water and comparably high numbers in other sources like sewage and faeces, soil, sediments, deep thermal vents and in natural bodies of water (Rohwer, 2003). In the absence of available hosts to infect, and as long as they are not damaged by external agents, phages can usually maintain their infective ability for decades (Guttman *et al.*, 2005; Sharp *et al.*, 1986).

3.1 Phage therapy versus chemotherapy

Phage therapy presents many potential advantages over the use of antibiotics which are intrinsic to the nature of phages. Phages are highly specific and very effective in lysing the target pathogen, preventing dysbiosis, that is, without disturbing the normal flora and thus reducing the likelihood of super-infection and other complications of normal-flora reduction that can often result following treatment with chemical antibacterials. This high specificity means that diagnosis of the bacteria involved in the infection is required before therapy is employed (Guttman *et al.*, 2005; Marks & Sharp, 2000; Matsuzaki *et al.*, 2009). The specificity of phages also enables their use in the control of pathogenic bacteria in foods since they will not harm useful bacteria, like starter cultures. Moreover, phages do not affect eukaryotic cells, or cause adverse side effects as revealed through their extensive clinical use in the former Soviet Union. Furthermore, phages are equally effective against multidrug-resistant pathogenic bacteria.

It was also found that phages can rapidly distribute throughout the body reaching most organs including the prostate gland, bones and brain, that are usually not readily accessible to drugs and then multiply in the presence of their hosts (Dabrowska *et al.*, 2005). The self-replicating nature of phages reduces the need for multiple doses to treat infection diseases since they will replicate in their pathogenic host increasing their concentration over the course of treatment leading to a higher efficacy. This also implies that phages will be present and persist at a higher concentration where their hosts are present, which is where they are more needed, in the place of infection. Reciprocally, where and if the target organism is not present the phages will not replicate and will be removed from the system showing the other side of the self-replicating nature of phages, their self-limiting feature (Goodridge & Abedon, 2003; Petty *et al.*, 2007).

As it happens with antibiotics, bacteria also develop resistance to phages. The latter usually occurs through loss or modification of cell surface molecules (capsules, OMPs, LPS, pili, flagella) that the phage uses as receptors. Since some of these also function as virulence determinants their loss may in consequence dramatically decrease the virulence of the bacterium or reduce its competitiveness (Levin & Bull, 2004). A good example is that of Smith (1987) that used phages against the K1 capsule antigen of *Escherichia coli* and verified that resistant K1 bacteria were far less virulent (Smith *et al.*, 1987b). Furthermore, different phages binding to the same bacteria may recognize different receptors and resistance to a specific phage does not result in resistance to all phages. Phages are able to rapidly change in response to the appearance of phage-resistant mutants making them efficient in combating the emergence of newly arising bacterial threats (Matsuzaki *et al.*, 2009; Sulakvelidze *et al.*, 2001). In addition, the isolation of a new phage able to infect the resistant bacteria can be easily accomplished. It is much cheaper, faster and easier to develop a new phage system than a new antibiotic which is a long and expensive process (Petty *et al.*, 2007).

4. Phage potential in food safety

Phages can be used to combat pathogens in food at all stages of production in the classic 'farm-to-fork' continuum in the human food chain (García *et al.*, 2008). Accordingly, in order to prevent transmission to humans, phages can be used:

i.   in livestock to prevent diseases or reduce colonization;
ii. in food material (such as carcasses and other raw products) or in equipment and contact surfaces to reduce bacterial loads;
iii. in foods as natural preservatives to extend their shelf life.

Several studies have been carried to assess the potential use of phages in the control of *Campylobacter* and *Salmonella* in animals and foodstuff. Although very different results have been obtained it seems that the proof of principle has been established: phage therapy has potential in the control of foodborne pathogens (Johnson *et al.*, 2008). The large scale, high throughput and mechanization of poultry production and industry, made poultry and products the most commonly used models for phage biocontrol (Atterbury, 2009). This is reflected in the studies that will be addressed below.

4.1 *Campylobacter* and *Salmonella* phages

4.1.1 *Campylobacter*

There are relatively few reports on *Campylobacter* phages probably due to the fastidious growth conditions of their host and to unique features that their phages exhibit. This has hindered the use of conventional methods of phage isolation, propagation and characterization (Bigwood & Hudson, 2009; Tsuei *et al.*, 2007). Recently Carvalho *et al.* (2010) described an improved method for *Campylobacter* phage isolation in which a pre-enrichment of the phages with potential host strains supplemented with divalent cations was used to promote phage adherence to the host (Carvalho *et al.*, 2010b). *Campylobacter*-specific phages have been isolated from excreta of both broiler and layer chickens, retail poultry, and other sources including pig, cattle and sheep manure, abattoir effluents, and sewage (Connerton *et al.*, 2011). Some of these phages have been characterized and form the basis of the United Kingdom phage typing scheme (Frost *et al.*, 1999; Sails *et al.*, 1998).

The most frequently encountered *Campylobacter* phages belong to *Caudovirales* order, *Myoviridae* family with a double-stranded DNA genome enclosed in icosahedral heads (Connerton *et al.*, 2008). *Campylobacter* phages have been characterized into three groups according to their genome size and head diameter (Sails *et al.*, 1998): Group I - head diameters of 140 nm – 143 nm and large genome sizes of 320 kb; Group II - average head diameters of 99 nm and average genome sizes of 184 kb; Group III - average head diameters of 100 nm and average genome sizes of 138 kb. Hansen *et al.* (2007) characterized *Campylobacter* phages according to their genome size and susceptibility of digestion by HhaI (Hansen *et al.*, 2007).

The DNA of most *Campylobacter* phages is difficult to extract, clone and sequence and is refractory to restriction enzyme digestion, which is probably due to tightly adherent and proteinase K resistant proteins (Carvalho *et al.*, 2011; Hammerl *et al.*, 2011; Kropinski *et al.*, 2011; Timms *et al.*, 2010). As a consequence, the genome sequence of only five *Campylobacter* phages have been reported so far (Carvalho *et al.*, 2011; Hammerl *et al.*, 2011; Kropinski *et al.*, 2011; Timms *et al.*, 2010). Interestingly, the phage genomes known are all related and also part of the T4 superfamily of phages (Petrov *et al.*, 2010).

There is little information available regarding the prevalence and influence of phages on *Campylobacter*-colonized poultry flocks. In fact, the prevalence of *Campylobacter* phages in poultry has essentially only been described in the UK. It was reported that *C. jejuni* phages were isolated from 20% of the caeca of chickens sampled in which there was a correlation between the presence of natural environmental phages and a reduction in the numbers of colonizing *Campylobacter*. Interestingly, birds that harbored phages had a significant difference (P < 0.001) in *Campylobacter* colony forming units (CFU) per gram in relation to those that did not have phages (Atterbury *et al.*, 2005). *Campylobacter* phages were prevalent in the caecal contents of organic birds with 51% of *Campylobacter*-positive sampled birds also carrying phages. The higher value of phage positive samples in organic flocks can be explained by the fact that these birds are more exposed to the environment and therefore to a greater range of *Campylobacter* types and phages (El-Shibiny *et al.*, 2005).

It was also showed that, like *Campylobacter*, their phages are also transferred between flocks (Connerton *et al.*, 2004). Moreover, phages were also recovered from chilled retail poultry, meaning that these phages can survive on retail chicken under commercial storage condition (Atterbury *et al.*, 2003b).

4.1.2 *Salmonella*

Numerous phages infecting *Salmonella* have been isolated. The first report of a *Salmonella* phage dates back to 1918 and was described by Félix d'Hérelle. Since then, *Salmonella* phages have been isolated from different sources: wastewater plants, sewage, manure, faeces and caecal contents from different animals (e.g. poultry, turkey, pig, humans), zoo ponds, nests from poultry farms and many others (Andreatti Filho *et al.*, 2007; Santos *et al.*, 2010; Sillankorva *et al.*, 2010). The search for different *Salmonella* phages from different sources may be attributed to the interest prompted by the medical and veterinary significance of their pathogenic host.

The great number and different specificity of *Salmonella* phages has enabled *Salmonella* classification through phage typing, a useful typing tool for subcategorizing the more common *Salmonella enterica* serotypes (i.e. *S.* Typhimurium, *S.* Enteritidis, *S.* Heidelberg) recommended by the WHO.

Probably the best known *Salmonella* phages are the lytic phage Felix 01 and the temperate virus P22. Felix 01 is characterized by its broad lytic spectrum among *Salmonella* and has been used as a diagnostic tool in the identification of *Salmonella*. Recently, a phage with a broader host range than Felix 01 has been described which presents great potential not only as a therapeutic agent but also as a diagnostic tool (Santos *et al.*, 2010; Santos *et al.*, 2011; Sillankorva *et al.*, 2010) (Figure 2).

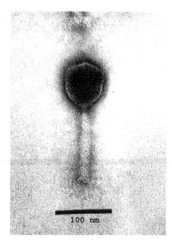

Fig. 2. TEM observation of the broad host range myovirus *Salmonella* phage PVP-SE1 (Santos *et al.*, 2010)

A survey conducted by Ackermann identified 177 *Salmonella* phages of which 91% belong to the *Caudovirales* order. Their distribution by families is roughly equal with 24% *Myoviridae*, 31% *Siphoviridae* and 33% *Podoviridae*. The 9% left are distributed in the *Inoviridae*, *Leviviridae*, *Microviridae*, and *Tectiviridae* families (Ackermann, 2007). A phylogenetic analysis relying on a proteomic comparison resulted in the recognition of at least five groups: P27-like, P2-like, lambdoid, P22-like, and T7-like. Nevertheless, three *Salmonella* phages (epsilon15, KS7, and Felix O1) are outliers since they could not be attributed to any of the previous groups (Kropinski *et al.*, 2007).

4.2 Phage biocontrol in livestock

As already outlined above, contamination of meat products with *Campylobacter* and *Salmonella*, as well as other foodborne pathogens, often results from cross-contamination between carcasses and feces from infected animals during slaughter and processing and also during transportation, leading to an increase of bacterial loads. Phage biocontrol in livestock presents two major purposes: i) treatment of bacterial pathogens in animals to minimize its impact on animal health and production and, ii) control of foodborne pathogens contamination to humans through foodstuff or other vectors. Therefore, the use of phages to control pathogens in livestock seems to be a feasible and efficient approach.

In this chapter the term phage therapy will refer only to the use of phages to control bacterial infections in living animals, whereas, the term biocontrol will be used where phages control pathogens in animals and foodstuff (independent of infection) (Hagens & Loessner, 2010).

4.2.1 Campylobacter

There are only a few reports on phage biocontrol against *Campylobacter*-infected livestock, with all the studies being conducted on poultry. Wagenaar *et al.* (2005) assessed both preventive and therapeutic phage treatment (Wagenaar *et al.*, 2005). In their study, multiple doses of a group III phage were administered to chicks before or after being orally challenged with a *C. jejuni* strain. In order to access the effect of phage administration on broilers, the *Campylobacter* colony forming units (CFU) and the phage plaque forming units (PFU) from caecal contents were enumerated throughout the experimental period. These values were obtained from the group receiving phages and from the control group (in which birds did not received phages). In both treatments, birds were orally challenged with a dose of 1×10^5 CFU *C. jejuni* ten days after hatching. Preventive treatment consisted in the oral administration of phage 71 (4×10^9 to 2×10^{10} PFU) for ten consecutive days, starting seven days after hatching. The phage treatment did not prevent the colonization of the caecum, but may had delayed it. In fact, initially the numbers of *Campylobacter* were reduced by 2 $\log_{10}$ CFU/g but after one week the numbers leveled out at approximately 1 $\log_{10}$ below that of the controls. In the therapeutic treatment the phage was orally administrated five days after birds being challenged with *C. jejuni* and consecutively during the next six days. The numbers of *Campylobacter* had decreased 3 $\log_{10}$ CFU/g at 48h, but after five days stabilized to approximately 1 $\log_{10}$ CFU/g below the control group. In order to mimic the "farm condition" in which birds are normally slaughter at 42 days, birds were orally challenged with a dose of 1×10^5 CFU *C. jejuni* at 32 days after hatching. Seven days later phages 71 and 69 were orally administered to these birds and for four consecutive days. As occurred with the previously described treated group, the values of *Campylobacter* counts dropped initially 1.5 $\log_{10}$ CFU/g but then stabilized at 1 $\log_{10}$ unit lower than in the controls.

In the study by Loc Carrillo *et al.* (2005), broiler chicks at 20 to 22-day-old were challenged with *C. jejuni* strains isolates HPC5 and GIIC8 from United Kingdom broiler flocks that have proved to be good colonizers (Loc Carrillo *et al.*, 2005). The chicks received four different doses of HPC5 (2.7 $\log_{10}$ CFU, 3.8 $\log_{10}$ CFU, 5.8 $\log_{10}$ CFU and 7.9 $\log_{10}$ CFU) and after 48h the *Campylobacter* numbers in the caeca, upper and lower intestine were enumerated. The highest dose led to more reproducible colonization levels of all intestinal sites examined, with a mean value of 6.3 $\log_{10}$ CFU/g, 6.7 $\log_{10}$ CFU/g and 7.4 $\log_{10}$ CFU/g in the upper intestine, lower intestine and caeca respectively. Moreover these colonization levels, which are similar to those of naturally colonized birds (Rosenquist *et al.*, 2006), were maintained over nine days. The phage treatment occurred five days after the *C. jejuni* challenge. Birds were treated orally with two phages (CP8 or CP34) independently at a dose of 10^5, 10^7 or 10^9 PFU. The phages were administered in an antacid suspension ($CaCO_3$) since it proved to protect phages from exposure to low pH during passage through the gastrointestinal tract. The administration of 10^7 PFU was the dose which led to the highest reduction (3.9 $\log_{10}$ CFU/g) in the upper and lower intestine and caecal counts of *Campylobacter* at 24 h. The highest dose was less effective in the treatment probably due to the aggregation and nonspecific association of phages with digesta, non-host bacteria or bacterial cell debris resulting from "lysis from without" (Rabinovitch *et al.*, 2003). Different host-phage combinations were tested *in vivo*, leading to different results. In fact, and contrarily to the results obtained *in vitro*, phage CP34 was more effective in the reduction of *C. jejuni* HPC5 at all intestinal sites compared to CP8. Conversely phage CP8 was efficient in the reduction of *Campylobacter* counts by more than 5 $\log_{10}$ CFU/g in caecal *Campylobacter* counts. Despite this initial reduction, the *C. jejuni* numbers started to increase 72 h after phage administration with the lower intestinal counts exhibiting significant differences of 2.1 $\log_{10}$ and 1.8 $\log_{10}$ CFU/g but with the upper intestinal counts showing no significant difference with the *Campylobacter* levels recorded for the control group.

El-Shibiny *et al.* (2009) reported the administration of a group II *Campylobacter* phage (CP220) to *C. coli* and *C. jejuni* colonized chickens (El-Shibiny *et al.*, 2009). The results showed that a 2 $\log_{10}$ CFU/g reduction in *Campylobacter* counts was observed when a single 10^7 or 10^9 PFU dose of CP220 was administered to *C. jejuni* or *C. coli* colonized chickens, respectively. After this treatment, only 2% of the recovered *Campylobacter* displayed resistance to CP220.

Recently, Carvalho *et al.* (2010) tested a phage cocktail composed by three group II *Campylobacter* phages (Figure 3) in chicks that were previously challenged with *C. coli* strain A11 or *C. jejuni* strain 2140 (Carvalho *et al.*, 2010a). Again, colonization models were established before phage therapy experiments were performed in order to access the effective reduction in *Campylobacter* numbers. In order to determine the optimum dose of *Campylobacter* that should be given to birds, the animals were challenged with three different concentrations (10^4, 10^6 or 5.5×10^7 CFU) of an overnight culture of *C. jejuni* and sampling points were obtained during seven days. The results obtained revealed that the dose of *Campylobacter* appeared to have little effect on the outcome of subsequent colonization and that the mean level of colonization was 2.4×10^6 CFU/g, which is within the range of the infection levels found in commercial broilers (Rosenquist *et al.*, 2006). Seven days post-infection, a single dose of a phage cocktail was administered to chicks by two different routes: oral gavage and incorporated into their feed. Sampling points were taken for seven days after phage administration and showed that the phage cocktail was able to reduce by approximately 2 $\log_{10}$ CFU/g the titre of both *C. coli* and *C. jejuni* in faeces of colonized chickens. This reduction persisted throughout the experimental period and none of the pathogens regained their former numbers. The administration in feed led to an earlier and more sustainable reduction of *Campylobacter* than administration by oral gavage. The phage titers from faecal samples of the chicks infected with *Campylobacter* remained approximately constant throughout the experimental period showing that phages delivered to chicks (either by oral gavage or in feed) were able to replicate and therefore able to reduce the *Campylobacter* populations.

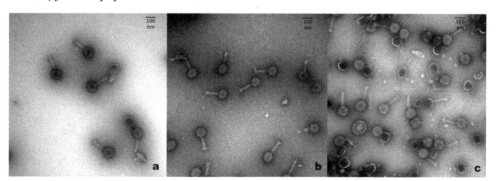

Fig. 3. TEM observation of the three *Campylobacter* phages, belonging to the family *Myoviridae*, which make up the cocktail used in the *in vivo* experiments by Carvalho *et al.* (2010): a) phiCcoIBB12, b) phiCcoIBB35, c) phiCcoIBB37 (Carvalho *et al.*, 2010a)

The appearance of phage resistant mutants has always been seen as a major drawback of phage therapy. Nevertheless some of the above mentioned studies found that phage resistance can be associated with a reduced colonization potential in the chicken intestine, suggesting that there is a fitness cost to phage resistance in which phage-resistant *Campylobacter* revert to a sensitive phenotype when re-colonizing the chicken intestine in the absence of phage predation. It was also suggested that genomic instability of *C. jejuni* in the avian gut can be seen as a mechanism to temporarily survive phage predation and later competition for resources (Carvalho *et al.*, 2010a; Loc Carrillo *et al.*, 2005; Scott *et al.*, 2007a; Scott *et al.*, 2007b). Conversely, recent studies by Carvalho *et al.* (2010) reported that *Campylobacter* strains resistant to phage infection were recovered from phage-treated chickens at a frequency of 13% and that these resistant phenotypes did not exhibit a reduced ability to colonize the chicken guts and did not revert to sensitive types (Carvalho *et al.*, 2010a).

4.2.2 *Salmonella*

The importance and impact of *Salmonella* has elicited several studies on the use of phages on the control of this pathogen. Although poultry represents the most commonly used models, studies also exist in pigs.

In 1991, Berchieri and colleagues used a chicken model to assess the potential of phages to control *Salmonella* Typhimurium in chickens. In the established model, oral infection with 10^9 CFU of *Salmonella* was fatal in 53% of the chickens. Oral administration of phage at high concentration (greater than 10^{10} PFU/ml), soon after *Salmonella* infection, was able to significantly reduce mortality (from 60% to 3%) as well as the number of pathogens in the alimentary tract. The high dose required for the reduction of *Salmonella* suggests that control by the phage was made by "lysis from without" or by a single cycle of replication. Although pathogen numbers were reduced shortly after phage administration, they increased soon after. Moreover, the phage was only present in the intestine when *Salmonella* was found in numbers above 10^6 CFU/ml (Berchieri *et al.*, 1991). Therefore, an efficient application of the phage required that administration should be right after infection and even so bacterial control was possible only for a short period. This may be attributed to the ability of *Salmonella* to internalize cells thus escaping from phages. As with the *Campylobacter* studies (see above) phage resistant bacteria were also found to present rough morphology and to display less virulence than the wild strain. More than one phage was used in this study and one of these, although lytic *in vitro*, was not effective *in vivo* showing the importance of *in vivo* trials.

Fiorentin *et al.* (2005) used a cocktail of three phages to control *S*. Enteritidis PT4 in broilers. The phage dose administered was also very high (10^{11} PFUs of each phage). In this study phage was administered only seven days post infection and a 3.5 $\log_{10}$ CFU reduction per gram of caecal content was recorded five days later. This level remained for 25 days. It seems here that time of administration was not critical but phage dose was. Although a high dose and a cocktail of phages were used with the purpose of reducing the possibility of phage resistant bacteria emergence, the presence of such cells was not assessed (Fiorentin *et al.*, 2005).

When a similar dose (10^{11} PFUs) was given orally two days after infection, Atterbury *et al.* (2007) were able to achieve a reduction of the caecal colonization of both *S*. Enteritidis and *S*.

Typhimurium in broiler chickens by up to 4.2 $\log_{10}$ CFU/g. Lower doses were ineffective in achieving similar results. A third phage tested in this study against *S*. Hadar did not reduce bacterial counts despite its strong lytic activity *in vitro* against that strain. The emergence of resistant bacteria was here also a reality with the number of phage resistant *Salmonella* increasing with the PFUs administered. This subpopulation of resistant bacteria was able to recolonize the broiler chickens within 72 hours after phage treatment. Interestingly, when these resistant bacteria were used to colonize a new group of chickens, they reverted to their phage-sensitive phenotype (Atterbury *et al.*, 2007).

Borie *et al.* (2008) pretreated ten day old broiler chicks, by coarse spray or drinking water containing a cocktail of three phages 24 hours before administering roughly 10^6 CFUs of *Salmonella* Enteritidis (calculated multiplicity of infection (MOI) of 10^3). After ten days of infection phages were recovered from the intestinal and other organs. A significant reduction in *Salmonella* Enteritidis was obtained at that time for both routes of administration with a reduction of more than 1 $\log_{10}$ CFU/ml in challenged bacterial numbers. This study demonstrates not only that phages are able to reduce *Salmonella* bacterial loads in broiler chickens but also that aerosol spray and drinking water are conceivable routes of administration in the application of a phage product which will surely facilitate application and establishment of phage biocontrol in an industrial environment (Borie *et al.*, 2008). In a later study these authors, using seven days old chicks were able to replicate the aerosol results. Moreover, they showed that phage treatment coupled with competitive exclusion resulted in even better *Salmonella* reductions than each of the treatments alone (Borie *et al.*, 2009).

Toro *et al.* (2005) and coworkers also tested the association of phage therapy and competitive exclusion in the treatment of chickens infected experimentally with *Salmonella*. Phage treatment was given orally and included a cocktail of three different phages with different host ranges. In all treated groups, with phage or competitive exclusion alone or together, a decrease in the *Salmonella* counts was observed with a reduction to marginal levels in the ileum and a six fold reduction in the caeca in the case of the group treated with the phage cocktail. Moreover, there was a marginally improved weight gain in the treated animals. Although both approaches were able to reduce the *Salmonella* counts, unlike the previous study (Borie *et al.*, 2009) a synergistic effect was not observed (Toro *et al.*, 2005).

In 2007 Andreatti Filho and colleagues reported on the isolation and testing of two different phage cocktails (one with 4 phages and another with 45 phages) in the control of *Salmonella* *in vivo* and *in vitro*. *In vitro* test of these two preparations at concentrations of 10^5 to 10^9 PFU/ml in simulated crop environment resulted in a 1.5 or 5 $\log_{10}$ reduction of *Salmonella* Enteritidis, respectively for the 4-phage and 45-phage cocktails in two hours after treatment. Although the 4-phage cocktail did not produce a reduction at six hours post-treatment, the 45-phage cocktail was able to reduce bacterial counts by 6 $\log_{10}$. This study clearly shows the advantage of a cocktail with a large number of phages, probably enabling complementary host range between phages and broader action in bacteria. Phage cocktails were administered at 10^8 PFU by oral gavage to day-of-hatch chicks infected with 9×10^3 CFU of *Salmonella* Enteritidis. These showed significant reduction of *Salmonella* recovered from caecal tonsils after 24 hours of treatment but no difference was observed at 48 hours when compared with the control group. Another experiment combined the use of the 45-phage cocktail with a commercial probiotic (and controls with each alone) to treat day-of-hatch chicks achieving significant reductions in *Salmonella* recovery from caecal tonsils at 24 hours but no additive or synergistic effect was observed when combining both approaches. Once again, phage therapy was only efficient during a short period and no long term protection was observed (Andreatti Filho *et al.*, 2007).

Recently, Wall *et al.* (2010) tested the efficacy of a cocktail to treat *S.* Typhimurium experimentally infected pigs shortly before processing in order to reduce carcass contamination. Administration of the phage at the time of infection resulted in a 2-3 $\log_{10}$ reduction of *Salmonella* colonization (Wall *et al.*, 2010).

O'Flynn *et al.*, (2006) isolated two broad host range *Salmonella* phages (st104a and st104b) which were partially resistant to the porcine gastric juice (pH 2.5). *In vitro* tests showed a reduction of more than 2 $\log_{10}$ in *Salmonella* numbers in just one hour demonstrating their potential in controlling *Salmonella* in pigs by oral administration. Even so, as seen in other studies, for an efficient assessment of their potential as therapeutic agents, *in vivo* tests are needed (O'Flynn *et al.*, 2006).

Target pathogen	Livestock	Results	References
Campylobacter	chicken	Reduction of up to 3 $\log_{10}$ CFU/g in caecal samples	(Wagenaar *et al.*, 2005)
	chicken	Reduction of more than 5 $\log_{10}$ CFU in caecal samples	(Loc Carrillo *et al.*, 2005)
	chicken	Reduction of up to 2 $\log_{10}$ CFU/g in caecal samples	(El-Shibiny *et al.*, 2009)
	chicken	Reduction of up to 2 $\log_{10}$ CFU/g (faster reduction when administrated in feed)	(Carvalho *et al.*, 2010a)
Salmonella	chicken	Reduction from 60% to 3% in chicken mortality	(Berchieri *et al.*, 1991)
	chicken	Reduction of up to 3.5 $\log_{10}$ CFU/g of caecal content	(Fiorentin *et al.*, 2005)
	chicken	Reduction of up to 4.2 $\log_{10}$ CFU of the caecal colonization	(Atterbury *et al.*, 2007)
	chicken	Reduction of up to 1.63 $\log_{10}$ CFU/ml in *Salmonella* recovery from the intestine.	(Borie *et al.*, 2008; Borie *et al.*, 2009)
	chicken	Reduction of up to marginal levels in the ileum and six fold in the caeca in the *Salmonella* counts	(Toro *et al.*, 2005)
	chicken	Reduction of up to 6 $\log_{10}$ in simulated crop environment / 55% in recovery from caecal tonsils of chicks	(Andreatti Filho *et al.*, 2007)
	pig	Reduction of up to 3 $\log_{10}$ reduction of *Salmonella* colonization	(Wall *et al.*, 2010)
	pig	Reduction of up to 2 $\log_{10}$ in *Salmonella* numbers (*in vitro* test)	(O'Flynn *et al.*, 2006)

Table 1. Studies on phage biocontrol in livestock

4.3 Phage biocontrol in food material

Another way that phages may be used to improve food safety is to apply them directly onto raw food products. The practical applicability of this approach may be compromised by the minimum density of host cells that are suggested to be required for phage replication (Payne *et al.*, 2000; Payne & Jansen, 2001). Nevertheless it was demonstrated that phages can be effective biocontrol agents when the population of host cells is as low as 46 CFU/cm^2 (Greer, 1988). These contradictory results may be a consequence of differences in phage/host combinations, in the matrix used, in the presence of non-host decoys or even in the models applied. Therefore, the efficacy of phage-based biocontrol should be determined on a case-by-case basis (Atterbury, 2009).

The application of phages as biocontrol agents has been investigated in a variety of food matrices. Nevertheless most studies focus in poultry products, with *Campylobacter* and *Salmonella* being the most frequently targeted zoonotic pathogens.

4.3.1 *Campylobacter*

Most studies on the use of phages as biocontrol are devoted to pre-harvesting studies. In fact, there are only few studies reporting the efficacy of *Campylobacter* phages in foodstuff.

Atterbury *et al.* (2003) studied the survival of *C. jejuni* and phages on chicken skin at temperatures of 4°C or -20°C. A dose of 10^6 CFU *C. jejuni* NCTC 12662 PT14 was inoculated on the surface of chicken skin and then samples were stored at these two temperatures for a period of one hour to ten days (Atterbury *et al.*, 2003a). The results showed that there was a reduction of 1 or 2 log$_{10}$ CFU on *Campylobacter* counts for skin stored at 4°C or -20°C, respectively. This reduction was lower than the value normally reported for *Campylobacter* cells stored at -20°C (3 log$_{10}$ CFU) (Chan *et al.*, 2001), which indicates that chicken skin may have had a protective effect. Phage φ2 at 10^7 PFU was added to chicken skin samples and was show to survive for ten days at 4°C and at -20 °C. In order to access whether this phage would attach non-specifically to non-susceptible *Campylobacter* strains at 4°C, this phage was added to *C. coli* NCTC 12667 at a MOI of 10 for ten days. In fact, *Campylobacter* counts and phage titer did not fall during the experimental period which indicates that a nonspecific adsorption did not occur. Conversely when the same phage was added to its susceptible host, under the same conditions described above, the *Campylobacter* numbers were reduced by 0.8 log$_{10}$ CFU/ml (Atterbury *et al.*, 2003a).

In another experiment performed by the same authors, phage φ2 and *C. jejuni* NCTC 12662 PT14 were added to chicken skin samples at different MOI ranging from 10^{-3} to 10^5 with samples stored at 4°C and -20°C for five days. The results again showed a significant reduction in *Campylobacter* counts when the virus were administered at the highest MOI: a 1.1 log$_{10}$ to 1.3 log$_{10}$ CFU reduction in samples stored at 4°C and a 2.3 log$_{10}$ to 2.5 log$_{10}$ CFU reduction for frozen samples. Overall, the treatment that showed the best reduction in *Campylobacter* counts in chicken skin was the one in which high phage titers were applied following by storage of samples at freezing temperatures. Nevertheless, in all the treatments phages were not able to replicate. Furthermore, *Campylobacter* strains recovered after phage treatments were shown to be identical to the inoculated strains and did not display resistance to this phage (Atterbury *et al.*, 2003a).

Goode *et al.* (2003) also demonstrated the reduction in *Campylobacter* levels, after the application of a lytic phage to chicken skin, experimentally contaminated with *C. jejuni*. Chicken portions were initially inoculated with 10^4 CFU/cm^2 of *C. jejuni* strain C222 and then half of the chicken pieces were treated with *C. jejuni* typing phage 12673 at an approximate density of 10^6 PFU/cm^2. The samples were incubated at 4°C and swabs were taken after 24h. The results obtained show that *Campylobacter* counts were reduced by 1.3 $\log_{10}$ CFU comparing with the control. However, there was even a reduction of approximately 1 $\log_{10}$ CFU in the *Campylobacter* numbers from the non-phage-treated chicken portions, which meant that *Campylobacter* did not survive well on exposed chicken surfaces at 4°C (Goode *et al.*, 2003).

In 2008 Bigwood *et al.* (2008) reported on an investigation into the use of phage Cj6 against *C. jejuni* inoculated on cooked and raw meat and incubated for 24h at two different temperatures (5°C and 24°C) in order to simulate refrigerated and room temperature storage. Experiments were performed using different conditions: low (<100 cells/cm^2) or high (10^4 cells/cm^2) host densities and low (10) or high (10^4) MOIs. When the experimental conditions were set for 5°C, significant differences were obtained for samples inoculated with high MOI and high host density. A reduction of 2.4 $\log_{10}$ CFU/cm^2 and 1.5 $\log_{10}$ CFU/cm^2 was obtained on cooked and raw meat, respectively. The titer of phages inoculated was also reduced by 10%, after 24h of incubation on cooked meat. When samples were stored at 24°C, at high MOI and low host density, the reduction in *Campylobacter* counts was not significant. However when samples were inoculated at high MOI and high host density, reduction of 2.8 $\log_{10}$ CFU/cm^2 (after 6h) and 2.2 $\log_{10}$ CFU/cm^2 (after 24h) were obtained on cooked and raw meat, respectively. Nevertheless, *Campylobacter* counts were reduced, even in the untreated samples, which may be explained by their sensitivity to the experimental temperatures. Therefore the results did not allow an accurate assessment of the effective reduction by the phage treatment (Bigwood *et al.*, 2008).

Overall the studies suggest that high MOI values are more effective in the control of *Campylobacter* in foods. From the studies outlined above it is evident that *Campylobacter* are not able to grow and multiply under the conditions found on refrigerated raw meat, which renders unlikely phage replication. Nevertheless even if the phage cannot replicate at that temperatures, when reaching the human intestine, bacteria increases its metabolic activity and phages may eventually attach and lyse the target bacteria, leading to a control effect (Goode *et al.*, 2003).

4.3.2 *Salmonella*

Different approaches have been used to assess the effectiveness of phages to control pathogens in foodstuff. *Salmonella* control in poultry products has been constantly highlighted. Goode *et al.* (2003) also used phages to treat chicken skin experimentally contaminated with *Salmonella* Enteritidis (10^3 CFU/cm^2). Different MOIs were tested and for a MOI=1 the phages were able to replicate increasing their number with the consequent reduction of bacteria by less than 1 $\log_{10}$ in 24 hours. For higher MOIs (10^2 and 10^3) the levels of recovered *Salmonella* were reduced by roughly 2 $\log_{10}$ over 48h. Using even a higher MOI (10^5) to treat a more realistic *Salmonella* contamination level (<10^2 CFUs) no recoverable *Salmonella* was obtained, resulting in the total elimination of the pathogen (Goode *et al.*, 2003).

The use of phages to reduce contamination of pig skin was studied by Hooton *et al.* (2011). In their study on pig skin artificially contaminated with *Salmonella* Typhimurium U288 (the most prevalent serovar found in pigs) at levels of 10^3 CFU, a cocktail of four phages which included Felix 01 was employed at a temperature of 4°C. Although the application of the phage cocktail at MOI of 1 or less produce little or no reduction, the use of MOI above 10 led to a reduction of the pathogen below detectable levels (Hooton *et al.*, 2011).

The effectiveness of phages to control *Salmonella* in carcasses of broiler chickens and turkeys was tested by Higgins *et al.* (2005). In this study, 10^6 PFU of a phage applied to carcasses was deemed to be inefficient in the removal of *S.* Enteritidis at levels below 10^3. Instead, application of $\geq 10^8$ PFU resulted in a marked reduction in the numbers of carcasses with recoverable *Salmonella*. Higgins *et al.* also used a cocktail composed of 72 different *Salmonella* phages to treat naturally contaminated turkey carcasses. The results are promising showing that the cocktail effectively reduced *Salmonella* recovery from the contaminated carcasses. These studies suggest that a high concentration of phage, preferably a cocktail of different phages, should be used to efficiently treat carcasses contaminated with *Salmonella* (Higgins *et al.*, 2005).

Control of *Salmonella* through the action of phages was also tested in raw and cooked beef by Bigwood *et al.* (2008). The samples were inoculated with *Salmonella* at low or high densities (respectively $<10^2$ or 10^4 CFU/cm²). Afterwards, phages were added at a MOI of 10^1 or 10^4 and samples incubated at different temperatures to simulate refrigerated and room temperature storage (respectively 5°C and 24°C). Phages were able to reduce *Salmonella* counts in up to 2.3 $\log_{10}$ CFU/cm² for samples incubated at 5°C and in more than 5.9 $\log_{10}$ CFU/cm² for samples incubated at 24°C when compared to controls (samples inoculated with *Salmonella* without phage). These results were obtained for both high densities of *Salmonella* (10^4 CFU/cm²) and phages applied at a high MOI (10^4) with samples incubated for 24h. The reductions were even higher after eight days of incubation. For low *Salmonella* densities the reductions were not significant for the majority of the samples. Interestingly, recovered *Salmonella* cells were found to still be sensitive to phage infection (Bigwood *et al.*, 2008).

The well-known broad host range *Salmonella* phage Felix 01 was used by Whichard *et al.* (2003) to treat chicken sausages contaminated with *S.* Typhimurium DT104 (300 CFU) and a reduction of up to 2.1 $\log_{10}$ in the *Salmonella* levels was obtained (Whichard *et al.*, 2003).

Control of *Salmonella* in foodstuff has not been restricted only to meat and derivates. A study carried by Leverentz and colleagues examined the efficiency of phage to control *Salmonella* on fresh-cut fruit, a rapidly developing industry. Treatment with a phage mixture was able to reduce the numbers of *Salmonella* by nearly 3.5 $\log_{10}$ CFU/g in honeydew melon slices stored at 5 °C and 10°C, temperatures at which *Salmonella* can survive or increase up to 2 $\log_{10}$ respectively within a week. At 20°C, where *Salmonella* loads can increase up to 5 $\log_{10}$, the decrease of *Salmonella* in slices was of 2.5 $\log_{10}$. The reductions obtained were greater than the maximal amount achieved using chemical sanitizers. Although this marked reduction in melons, in apple slices this did not happen. The reason may rely in the lower pH of apples that does not enables phage survival since it was not possible to reisolate phages in the apple 48 hours post treatment while in the melons the phage concentration was stable during this time period (Leverentz *et al.*, 2001a).

Pao *et al.* (2004) used phages to treat sprouting seeds where it was found that *Salmonella* grows during soaking. Treatment with phages could reduce the numbers of *Salmonella* in 1.37 $\log_{10}$ and 1.5 $\log_{10}$ in mustard seeds and in the soaking water of broccoli seeds by using one or two phages respectively (Pao *et al.*, 2004).

Cheese contamination by *Salmonella* was also subject of study on possible biocontrol with phages. Modi *et al.* (2001) following standard procedures made cheddar cheese from raw and pasteurized milk (an important vehicle of *Salmonella* transmission to humans) (European Food Safety Authority, 2011). The cheese was inoculated with 10^4 CFU/ml of S. Enteritidis and 10^8 PFU/ml of the *Salmonella* phage SJ2 (Modi *et al.*, 2001). A decrease in *Salmonella* counts by 1 to 2 $\log_{10}$ was observed in raw and pasteurized milk cheeses while in the controls (cheeses made from milks inoculated with only *Salmonella* and no phage) an increase of 1 $\log_{10}$ was observed. After storage of the cheeses for 99 days at 8°C *Salmonella* was present in the controls at a final concentration of 10^3 CFU/g. In the phage treated cheeses only 50 CFU/g were present in the ones from raw milk and no *Salmonella* was recovered from pasteurized milk cheeses after 89 days.

Composting is a complex process used not only to obtain a nutrient-rich substrate but also to significantly reduce pathogen contamination. Even so, improperly composting may result in *Salmonella* survival and thus constitute a vehicle of *Salmonella* transmission to animals and humans. Heringa *et al.* (2010) used a cocktail of five phages to treat a dairy manure compost inoculated with S. Typhimurium and observed a 2 $\log_{10}$ reduction within four hours and 3 $\log_{10}$ reduction within 34 hours compared to the controls (Heringa *et al.*, 2010).

Reduction of *Salmonella* through the action of phages was also investigated in wastewater. Turki *et al.* (2011) isolated three different phages and tested their ability to reduce *Salmonella* in TSB medium at two temperatures (30°C and 37°C). Phages were applied at three different MOI (10^0, 10^2 and 10^4) alone or as a cocktail of two or three phages. The three phage cocktail was able to reduce all *Salmonella* cultures at both temperatures when using a high MOI. Although, addition of individual or combination of two phages led to the emergence of phage resistant bacteria. Even so, the use of two phages presented better results than the use of an individual phage. The most important result was the eradication of *Salmonella* from the samples when the three phage cocktail was inoculated in raw wastewater (Turki *et al.*, 2011).

4.4 Considerations

The above studies showed inconsistencies in the ability of phages to act as biocontrol agents of *Campylobacter* and *Salmonella* in livestock. Even so, it seems that the proof of principle, that phages are able to reduce the number of these pathogens (at least in a short period after treatment), has been established.

At a first glance, *in vivo* biocontrol of foodborne pathogens made "on-farm" seems to be a good approach since, theoretically, the problem would be treated on its origin. Ideally, phages would be applied through the use of a sole administration of a low dose of phage. The virulent phage should then amplify at the expense of the target bacteria by repeated cycles of replication leading to the host eradication. The increasing number of the progeny phage would remain in the system for some period of time acting prophylactically in a possible subsequent infection. Although, as shown by several studies, eradication of the target bacteria is an extremely unlikely event in part due to the coexistence of a phage-

Target pathogen	Foodstuff	Results	References
Campylobacter	chicken skin	Reduction of up to 2 log_{10} on frozen samples	(Atterbury *et al.*, 2003a)
	chicken skin	Reduction of up to 2.31 log_{10} CFU/g at 4°C.	(Goode *et al.*, 2003)
	raw and cooked meat	Reduction of up to 2.8 log_{10} CFU on cooked meat	(Bigwood *et al.*, 2008)
Salmonella	chicken skin	Reduction of up to no recoverable *Salmonella*	(Goode *et al.*, 2003)
	pig skin	Reduction of up to below detectable levels at 4°C	(Hooton *et al.*, 2011)
	chickens and turkeys carcasses	Reduction of up to 93% reduction in *Salmonella* recovery	(Higgins *et al.*, 2005)
	raw and cooked meat	Reduction of more than 5.9 log_{10} CFU on raw meat	(Bigwood *et al.*, 2008)
	chicken sausages	Reduction of up to 2.1 log_{10} in the *Salmonella* levels	(Whichard *et al.*, 2003)
	fresh-cut fruit	Reduction of up to 3.5 log_{10} in *Salmonella* numbers in honeydew melon slices stored at 5 °C	(Leverentz *et al.*, 2001b)
	sprouting seeds	Reduction of up to 1.5 log_{10} in *Salmonella* numbers	(Pao *et al.*, 2004)
	cheese	Reduction of up to no recoverable *Salmonella* in pasteurized milk cheeses	(Modi *et al.*, 2001)
	dairy manure compost	Reduction of up to 3 log_{10}	(Heringa *et al.*, 2010)

Table 2. Studies on phage biocontrol in foodstuff

resistant bacterial subpopulation. Eradication is even difficult *in vivo* since the number of *Salmonella* and *Campylobacter* is usually higher in broiler chicken intestines than in the carcasses and derived products. Moreover, the colonization of animals in herds or flocks spreads exponentially and infection may be through direct contact with pens and holding facilities which were used before by infected animals. Therefore, the on-farm treatment may lead to the emergence of phage resistant strains with their consequent spread to all animals in the farm. In addition, repeatedly use of phages may induce production of antibodies that will afterwards neutralize the phages, diminishing their effectiveness (Johnson *et al.*, 2008). The emergence of phage resistant bacteria is a major concern in phage biocontrol and long term studies on the resistance and resistant bacteria should be performed.

The consistent reduction of the target bacteria shortly after phage administration, and the need to avoid the emergence of phage-resistant strains, suggests that phages should be applied shortly before slaughter. Indeed, in the period preceding slaughter, the animal is an epidemiological endpoint and the phages, as well the bacteria, will be removed from the contaminated source. This will prevent the emergence, spread and establishment of phage-resistant strains and will not have an impact on the farm microbial balance. Moreover this approach constitutes a realistic application of phages enabling the administration of a single dose. This will result in a low number of pathogens during food processing with the consequent decrease of cross-contamination, contributing thus to the consumer safety.

Even so, some considerations should be taken into account for an efficient application of phages. Phages can be administered by different routes that will have impact in the efficacy of the phage action but also economic and practical implications. Obviously, the most practical, and probably economical, route of phage application is through food and drinking water, at least for large scale treatments as is the treatment of *Salmonella* and *Campylobacter* in the poultry industry. These routes of administrations will require further studies on the dose (volume and concentration) that will be incorporated into food or added to water as well the need for a way to protect phages from the low pH of the gastrointestinal tract that was shown to often inactivate the majority of phages (Smith *et al.*, 1987a). Protection may be enabled through the simultaneously administration of an antacid or through phage proper encapsulation as tested by Carvalho *et al.* (2010) with a *Campylobacter* phage, phiCcoIBB35 (Carvalho *et al.*, 2010a), and Ma *et al.* (2008) with *Salmonella* phage Felix 01 (Ma *et al.*, 2008), respectively. Alternatively, higher phage concentrations and/or phage mutants resistant to low pH may be used to increase the efficacy of phage treatment. Intramuscular inoculation seems only feasible to treat a low number and/or animals which represent an added value. Another important consideration is the need for *in vivo* studies for each phage in an appropriate model able to mimic the system in which the particular phage is to be administered since *in vitro* behaviour usually does not reflects the phage behaviour *in vivo*, in part due to the immune system response. This fact has also been often observed in the above studies.

In the case where phage therapy immediately before slaughter is not possible due to inefficiency or impracticability, treatment of meat and foodstuff is also a possible approach. When phages are applied in foodstuff (post-slaughter), as in the case of a pre-slaughter phage treatment, the emergence of phage-resistant is also prevented since the phages and bacterial populations will be removed from the contamination source. Some studies, as described above, have addressed the possibility of using phages in the control of foodborne pathogens in foodstuff although, the majority of the *Salmonella* studies were carried at a temperature that is the optimum for the target bacteria to grow. This may lead to erroneous phage efficacy assessment because phage behaviour is highly dependent on the bacterial physiology which in turn also depends on the temperature. Consequently, a reliable study should be carried at the same temperature and other conditions at which the foodstuff is prepared, processed and/or stored. Commercial storage conditions of carcasses, meat and other foodstuff often include a refrigerated temperature of 4°C in order to halt bacterial growth. Experiments performed at this temperature with *Campylobacter* and *Salmonella* have shown that a relative high dose (high MOI) of phage is needed probably due to the reduced host growth that will prevent phage replication. Even so, this is not much different from what was observed at higher temperatures or even in studies in livestock where, as it is shown here, a high MOI seems to be a requirement to reduce pathogenic loads. The presented studies have shown that little or no reduction was observed for MOIs of 1 or less suggesting that the therapeutic effect of the phages is passive, without taking advantage of their replicating and self-amplification ability. This passive effect of phages is known as "lysis from without" in which the adsorption and attachment of many phages to the bacterial surface of a single cell results in lysis without phage replication. The passive effect is corroborated by the results of Bigwood *et al.* (2008) who found that phages did not increase in numbers after reducing the *Salmonella* counts. This passive effect performed by the phages is not impaired at temperatures below the minimum for bacterial host growth as seen in these experiments performed at 4°C and 5°C reinforcing phage application in foodstuff at storage conditions (Atterbury *et al.*, 2003a; Bigwood *et al.*, 2008). Moreover, this passive effect seems to be less specific than the active one (where phages replicate inside their host) as has been observed by Goode *et al.* (2003) where *Salmonella* strains resistant to phage through restriction were also eliminated by passive effect of phages as long as the phages were able to attach (Goode *et al.*, 2003). Consequently, treatment with high MOIs, although economically not so attractive due to the need of a higher dose, are able to reduce the emergence of phage-resistant strains.

Another requirement of phage biocontrol is the need for broad host range phages and/or the use of a cocktail. Higher reductions in pathogenic bacteria were consistently obtained in livestock and foodstuff when a cocktail of phages with complementary host ranges and target receptors were used. The advantages of using cocktails and broad host range phages will be discussed below.

Risk analysis modelling have shown that a reduction of 2 log_{10} in faeces of the slaughtered animal or in chicken carcasses can reduce the risk to consumers in 75% or 30 times respectively, in the incidence of campylobacteriosis associated with chickens consumption (Havelaar *et al.*, 2007; Rosenquist *et al.*, 2003). This 2 log_{10} reduction was shown in these studies to be a possible, practical and realistic goal.

It can thus be concluded that (by using broad host range phages or cocktails of phages with a broad lytic spectrum, applied at high MOI, complemented with *in vivo* studies) both approaches (phage biocontrol in livestock at pre slaughter and in foodstuff) can be used in order to decrease the number of pathogenic bacteria in the food chain and consequently to reduce the incidence of foodborne diseases caused by *Campylobacter* and *Salmonella*. Although the studies only address the use of these two strategies separately, and because their effectiveness relies in the passive ability of phages to lyse the cells (and thus do not depends on the host concentration and physiological state), it can be argued that both approaches can be used together in order to achieve higher reductions of the pathogenic bacterial loads.

5. Prerequisites for an acceptable phage product

The better understanding of phage biology, infectious process and host-range specificity has been tracing the path by which the problems with early phage research may be partially or totally solved. Therefore the knowledge acquired so far coupled with the awareness of the mistakes committed, should be used to develop phage preparations which should meet several criteria whether they are intended to be applied in foodstuff or more importantly in animals and humans. Newly isolated phages should always be characterized in detail and biocontrol should not be attempted before the biology and genomics of the therapeutic phage is well understood. Consequently, the phage host range, virulence, stability and interaction with both innate and active immune systems should be determined.

5.1 Lytic phages

In order to assure safety of phage therapy and eliminate potential risks of failure or even complications, it is critical that phages used in therapy are strictly lytic and without: transducing capabilities, gene sequences having significant homology with known major antibiotic resistance genes, genes for toxins and genes for other bacterial virulence factors (Carlton et al., 2005). Consequently, it is critical to avoid temperate phages. Reasons for this discrimination involve the fact that the latter phages will not kill all the target bacteria due to their ability to lysogenize them and probably more important, there is a high risk of possessing genes that can render the bacteria more virulent (Los et al., 2010). This can happen because certain temperate phages, such as those associated with Staphylococcus, E. coli, Salmonella, Corynebacterium and Clostridium actually carry virulence genes. In addition, during the transition between the lysogenic cycle and the lytic cycle, the excision of the prophage DNA may be accompanied with small pieces of the bacterial genome thus producing a specialized transducing phage. In addition, certain viruses such as Salmonella phage P22 are generalized transducers that are capable of randomly packaging any part of the host genome. Transducing phages will then transfer the host DNA fragments to newly infected bacteria producing changes in the bacterial genomes through recombination or reintegration. This may produce undesired phenotypic changes in their hosts such as resistance to antibiotics, restriction systems and increased bacterial virulence such as: bacterial adhesion, colonization, invasion and spread through human tissues, resistance to immune defences and exotoxin production (e.g. cholera toxin encoded by Vibrio cholerae-phage CTX) (Los et al., 2010; Waldor & Mekalanos, 1996). Strictly lytic phages usually do not pose these risks but should be tested for transducing (Waddell et al., 2009).

5.2 Cocktails and broad-host-range phages

Within a given bacterial pathogen different mutants may exist with different susceptibility for a given phage. Although therapy should always match the phage with the target bacteria there are situations where treatment is urgently need, turning that approach impossible. Moreover, bacterial pathogens may mutate during the treatment time period and become resistant to the phage. This risk is real in part due to the narrow host range of phages. A way to circumvent these problems is by using a cocktail of phages targeting different receptors in the pathogen cell and with cross-resistant lytic activity. Consequently, if a bacterium is, or stays, resistant to one phage it is likely that it would not be resistant to a second phage which is immediately available in the cocktail and that targets a different receptor. This approach has been successfully used by Smith et al. (1987) in their E. coli diarrhoeal work (Smith et al., 1987b).

An alternative to the phage cocktails is the use of broad host range phages that can be isolated through the use of common ubiquitous receptors in the target bacteria as in the case of phages that recognize TolC. In Salmonella this outer membrane protein is involved in the adhesion and invasion of host intestinal epithelial cells (Ricci & Piddock, 2010). Broad host range phages are polyvalent phages capable of infecting across bacterial species or genera and thus able to infect the majority of the target bacteria. These phages present a huge advantage as therapeutic agents and from the biotechnological point of view they are much more attractive because a single phage is far easier to characterize and get approved by the regulatory authorities. The existence of such polyvalent phages in nature is rare and only a few of them are known. Examples include *Salmonella* phage Felix O1 which infects most *Salmonella* serovars (Kallings, 1967) and *Salmonella* phage PVP-SE1 which presents a lytic spectrum even broader than that of Felix 01 (Santos *et al.*, 2010). Nevertheless these phages may present a drawback if they are able to target other non-pathogenic bacteria causing dysbiosis. Therefore the ideal phage for use as biocontrol agent should have a broad host range among the target pathogen without infecting the commensal flora. Examples of these phages include two virulent coliphages which were able to lyse a high percentage of pathogenic *E. coli* strains of various serotypes whilst showed low lytic ability to lyse non-pathogenic *E. coli* strains (Viscardi *et al.*, 2008). These broad host range phages may also be used, not only as alternatives to cocktails, but also to design new cocktails with broader host ranges and consequently more efficient. Furthermore, phage cocktails and broad host range phages will also combat and prevent the emergence of phage resistant strains. Another possibility would be to engineer phages in order to target numerous receptors. Although very interesting and promising from a scientific point of view, this approach would hardly get acceptance from the food authorities (Kropinski, 2006).

5.3 Genome sequencing

A full knowledge of phage genome sequences is important to ensure that the phage does not carry obvious virulence factors, resistance or lysogeny genes. The identification of gene homologies requires detailed bioinformatics analysis. The latter is essential to evaluate possible complications that might arise during phage therapy. It was suggested that data from phage genome sequences could be used to establish a bank of "safe" therapeutic phages increasing the availability, safety and efficacy of phage therapy (Petty *et al.*, 2007).

Phages that break down the bacterial DNA to recycle bacterial host genome for their own DNA synthesis should be selected for therapy since this will hamper coexistence between phage and host. Such phages usually encode enzymes involved in nucleotide metabolism and the corresponding genes can be identified through sequence analysis (Carlton *et al.*, 2005). Even so, due to the usually high number of genes in the phage genome with unknown function, it is never possible to assure at 100% that no virulence factors, resistance or lysogeny genes exist. Correspondingly, the genome sequencing analysis enables to reject the use of phages for which genes were found to code for virulence, resistance or lysogeny but that does not present any experimental biologic evidence of it, that is, which never have shown to lysogenize a bacterium or increase the bacterium virulence and/or resistance.

5.4 Models for host-phage interaction

Phage therapy experimental design is not straightforward due to the self-replicating nature of phages. This means that for each phage, pharmacokinetic information is required which can be achieved by the determination of the phage infection parameters and by the use of a reliable population dynamic model able to predict the phage-bacteria behaviour. Understanding these dynamics will help the transference of *in vitro* results to *in vivo* predictions, explaining why a phage that replicates extremely well in the target bacteria *in vitro* fails to do it *in vivo*. It is expected that mathematical models may help to design experimental studies of population dynamics by identifying and evaluating the relative contribution of phage and bacteria in the course and outcome of an infection (Levin & Bull, 2004). Therefore, *in vivo* assessment of the phage, in a suitable animal model, should always be accomplished.

5.5 Phage production in a large-scale and storage

The increasing interest in phages as therapeutic or biocontrol agents and the intention of commercially distribute a phage or a phage based product demands a large scale production that is not compatible with the conventional double-agar overlay method. Consequently, production of phages will certainly make use of bioreactors and a control and optimization of phage production will play an important role. Good manufacturing practice requirements demands for the development of methods that ensure phage preparations highly purified, free of bacteria, toxins, pyrogenic substances and other harmful components. Although in the majority of animal studies phages were administered as crude lysates without adverse effects for the animal, the removal of endotoxins, exotoxins and cell debris is very important for the safety of the phage product and also for an easier acceptance by the consumers. An option would be the propagation of phages in a non-pathogenic or in a non-toxin producer host (Santos *et al.*, 2010). The storage should also be validated and suitable for the particular phage in order to assure that the preparations contain viable phage particles able to infect the target bacteria. Moreover, stability and pH control of the preparation is important as shown in the past by the rising problems observed when these facts were neglected (Merabishvili *et al.*, 2009).

6. Commercialization of phage-based products

Despite the good scientific results and the economic viability of phage products an important issue that cannot be forgotten is the public acceptance which can constitute a serious obstacle to the introduction of phages in food. It is likely that consumers will feel an antipathy when knowing that live viruses are being added in their foods and that will be ingested by them. First of all, phages are viruses that only infect bacteria and not eukaryotic cells as the human cells. Moreover they are very specific to the target bacteria avoiding this way dysbiosis. It is also important to note that the use of a virus to combat a pathogen is not so strange since some vaccinations are carried out using live, albeit attenuated, eukaryotic viruses. Since phages have been identified in drinking water, and foods such a yoghurt and salami they are generally considered safe (Rohwer, 2003). Also, this means not only that phages are already inside our body but also that they are constantly being ingested. Different phages, applied at different doses, using different routes of administration in humans during the long history of phage therapy in the Eastern Europe did not produced serious complications (Sulakvelidze *et al.*, 2001). Moreover, in a carefully controlled double-blind study involving ingestion of phage T4 by volunteers, no side effects were reported (Bruttin & Brussow, 2005). These facts show that phages are nontoxic to animals and plants and apparently innocuous from a clinical standpoint. Along these lines, it can be conclude that the introduction of phages in human food-chain through the usage of phages as biocontrol agents in living animals or carcasses can be considered safe and may be seen as a valuable alternative to the use of antibiotics in animal production.

With respect to regulations, the introduction of a biocontrol phage product in animals and foodstuff may not be as stringent as its introduction in human therapy. The way for its introduction has been recently (2006) paved by the approval by the American Food and Drug Administration (FDA) of a mixed *Listeria*-phage preparation, ListShield (www.intralytix.com) to be used as a food additive in poultry derivatives and ready-to-eat meat. Another product based on a *Listeria*-phage, Listex P-100 (www.microsfoodsafety.com), has received the status "generally recognized as safe" (GRAS) to be used in all food products in 2007. Other phage-based products to control *E. coli* and *Salmonella* exist to be used at pre-slaughter (www.omnilytics.com and www.intralytix.com). The approval of such products proves the safety of phages and anticipates the development and introduction of new phage based products to be applied not only in foodstuff but also, at a long term, in animals and humans.

7. Conclusions

The emergence of multidrug-resistant bacteria has opened a second window for phage biocontrol. The recent work reviewed here shows that it has been established the proof-of-principle and evidences are more than enough to state that phage biocontrol, if well-conceived, is very effective in the treatment and prophylaxis of many problematic infectious diseases. Of particular interest is the potential that phage biocontrol has demonstrated against the global problematic multidrug-resistant bacteria. While the results of phage based products efficacy are very promising some consideration need to be taken into account. Efficient phage biocontrol requires the use of broad host range phages and administration of cocktails of phages with complementary host ranges and target receptors (showing thus cross-resistant lytic activity) in order to circumvent the emergence of phage resistant bacteria. Also, *in vivo* studies with suitable models should always be performed to assess the efficacy of the phage based product. From an economical and practical point of view, the best route of administration at an industrial scale is obviously through food and drinking water. For a successful application of phages it is important to understand the epidemiology of the pathogen against which the phage preparation is to be used in order to identify the critical intervention points in the processing cycle where phage application would be most beneficial. On the other hand, consistent pathogenic bacteria reduction is only achieved short after phage administration suggesting that phages should be applied in livestock shortly before slaughter and/or post-slaughter in carcasses and foodstuff. Besides reducing significantly the bacterial loads in the food chain with the consequent reduction in foodborne incidences it will impair the emergence of phage resistant bacteria.

Finally, although the consumers may be reluctant to the introduction of phages in the food chain, they have already shown to be safe for the environment, animals and humans with high efficiency in the reduction of foodborne pathogens. Moreover, some phage products are already commercially available and thus the way for the introduction of new phage based products is now open.

Overall, it can be concluded that phages can and should be used, not only as alternative, but also as substitutes of antibiotics and chemical antibacterials, in the control of foodborne pathogens in livestock and foodstuff.

8. References

Aarestrup, F. M., Wegener, H. C., & Collignon, P. (2008). Resistance in bacteria of the food chain: epidemiology and control strategies. *Expert review of anti-infective therapy*, Vol. 6, No. 5, pp. 733-750.

Aarestrup, F. M., Seyfarth, A. M., Emborg, H. D., Pedersen, K., Hendriksen, R. S., & Bager, F. (2001). Effect of abolishment of the use of antimicrobial agents for growth promotion on occurrence of antimicrobial resistance in fecal enterococci from food animals in Denmark. *Antimicrobial Agents and Chemotherapy*, Vol. 45, No. 7, pp. 2054-2059.

Ackermann, H. W. (2001). Frequency of morphological phage descriptions in the year 2000. Brief review. *Archives of Virology*, Vol. 146, No. 5, pp. 843-857.

Ackermann, H. W. (2007). *Salmonella* Phages Examined in the Electron Microscope, In: *Methods in Molecular Biology*, pp. 213-234.

Ackermann, H. W. (2009). Phage Classification and Characterization, In: *Bacteriophages - Methods and Protocols*, A. M. Kropinski & M. Clookie, pp. 127-140, Humana Press.

Allen, V. M., Bull, S. A., Corry, J. E. L., Domingue, G., Jorgensen, F., Frost, J. A., Whyte, R., Gonzalez, A., Elviss, N., & Humphrey, T. J. (2007). *Campylobacter* spp. contamination of chicken carcasses during processing in relation to flock colonisation. *International Journal of Food Microbiology*, Vol. 113, No. 1, pp. 54-61.

Andreatti Filho, R. L., Higgins, J. P., Higgins, S. E., Gaona, G., Wolfenden, A. D., Tellez, G., & Hargis, B. M. (2007). Ability of bacteriophages isolated from different sources to reduce *Salmonella enterica* serovar enteritidis in vitro and in vivo. *Poultry science*, Vol. 86, No. 9, pp. 1904-1909.

Antunes, P., Reu, C., Sousa, J. C., Peixe, L., & Pestana, N. (2003). Incidence of *Salmonella* from poultry products and their susceptibility to antimicrobial agents. *International Journal of Food Microbiology*, Vol. 82, No. 2, pp. 97-103.

Atterbury, R. J., Connerton, P. L., Dodd, C. E., Rees, C. E., & Connerton, I. F. (2003a). Application of host-specific bacteriophages to the surface of chicken skin leads to a reduction in recovery of *Campylobacter jejuni*. *Applied and Environmental Microbiology*, Vol. 69, No. 10, pp. 6302-6306.

Atterbury, R. J., Connerton, P. L., Dodd, C. E., Rees, C. E., & Connerton, I. F. (2003b). Isolation and characterization of *Campylobacter* bacteriophages from retail poultry. *Applied and Environmental Microbiology*, Vol. 69, No. 8, pp. 4511-4518.

Atterbury, R. J., Dillon, E., Swift, C., Connerton, P. L., Frost, J. A., Dodd, C. E., Rees, C. E., & Connerton, I. F. (2005). Correlation of *Campylobacter* bacteriophage with reduced presence of hosts in broiler chicken ceca. *Applied and Environmental Microbiology*, Vol. 71, No. 8, pp. 4885-4887.

Atterbury, R. J., Van Bergen, M. A., Ortiz, F., Lovell, M. A., Harris, J. A., De Boer, A., Wagenaar, J. A., Allen, V. M., & Barrow, P. A. (2007). Bacteriophage therapy to reduce *Salmonella* colonization of broiler chickens. *Applied and Environmental Microbiology*, Vol. 73, No. 14, pp. 4543-4549.

Atterbury, R. J. (2009). Bacteriophage biocontrol in animals and meat products. *Microbial Biotechnology*, Vol. 2, No. 6, pp. 601-612.

Bell, C. & Kyriakides, A. (2002). *Salmonella*, In: *Foodborne Pathogens: Hazards, Risk Analysis and Control*, C. W. Blackburn & P. J. McClure, pp. 307-335, CRC Press.

Berchieri, A. J., Lovell, M. A., & Barrow, P. A. (1991). The activity in the chicken alimentary tract of bacteriophages lytic for *Salmonella typhimurium*. *Research in Microbiology*, Vol. 142, No. 5, pp. 541-549.

Berndtson, E., Tivemo, M., & Engvall, A. (1992). Distribution and numbers of *Campylobacter* in newly slaughtered broiler chickens and hens. *International Journal of Food Microbiology*, Vol. 15, No. 1-2, pp. 45-50.

Bigwood, T. & Hudson, J. a. (2009). Campylobacters and bacteriophages in the surface waters of Canterbury (New Zealand). *Letters in applied microbiology*, Vol. 48, No. 3, pp. 343-348.

Bigwood, T., Hudson, J. a., Billington, C., Carey-Smith, G. V., & Heinemann, J. a. (2008). Phage inactivation of foodborne pathogens on cooked and raw meat. *Food Microbiology*, Vol. 25, No. 2, pp. 400-406.

Borie, C., Albala, I., Sànchez, P., Sánchez, M. L., Ramírez, S., Navarro, C., Morales, M. A., Retamales, J., & Robeson, J. (2008). Bacteriophage Treatment Reduces *Salmonella* Colonization of Infected Chickens. *Avian Diseases*, Vol. 52, No. 1, pp. 64-67.

Borie, C., Sanchez, M. L., Navarro, C., Ramirez, S., Morales, M. A., Retamales, J., & Robeson, J. (2009). Aerosol spray treatment with bacteriophages and competitive exclusion reduces *Salmonella enteritidis* infection in chickens. *Avian Diseases*, Vol. 53, No. 2, pp. 250-254.

Brenner, F. W., Villar, R. G., Angulo, F. J., Tauxe, R., & Swaminathan, B. (2000). *Salmonella* nomenclature. *Journal of Clinical Microbiology*, Vol. 38, No. 7, pp. 2465-2467.

Bruttin, A. & Brussow, H. (2005). Human volunteers receiving *Escherichia coli* phage T4 orally: a safety test of phage therapy. *Antimicrobial Agents and Chemotherapy*, Vol. 49, No. 7, pp. 2874-2878.

Bryan, F. L. & Doyle, M. P. (1995). Health risks and consequences of *Salmonella* and *Campylobacter jejuni* in raw poultry. *Journal of Food Protection*, Vol. 58, No. 3, pp. 326-344.

Butzler, J. P. (2004). *Campylobacter*, from obscurity to celebrity. *Clinical microbiology and infection*, Vol. 10, No. 10, pp. 868-876.

Carlton, R. M., Noordman, W. H., Biswas, B., de Meester, E. D., & Loessner, M. J. (2005). Bacteriophage P100 for control of *Listeria monocytogenes* in foods: genome sequence, bioinformatic analyses, oral toxicity study, and application. *Regulatory Toxicology and Pharmacology*, Vol. 43, No. 3, pp. 301-312.

Carvalho, C., Gannon, B., Halfhide, D., Santos, S., Hayes, C., Roe, J., & Azeredo, J. (2010a). The *in vivo* efficacy of two administration routes of a phage cocktail to reduce numbers of *Campylobacter coli* and *Campylobacter jejuni* in chickens. *BMC Microbiology*, Vol. 10, pp. 232-.

Carvalho, C., Susano, M., Fernandes, E., Santos, S., Gannon, B., Nicolau, A., Gibbs, P., Teixeira, P., & Azeredo, J. (2010b). Method for bacteriophage isolation against target *Campylobacter* strains. *Letters in applied microbiology*, Vol. 50, No. 2, pp. 192-197.

Carvalho, C. M., Kropinski, A. M., Lingohr, E. J., Santos, S. B., King, J., & Azeredo, J. (2011). The genome and proteome of a *Campylobacter coli* bacteriophage vB_CcoM-IBB_35 reveal unusual features. *Virology Journal*, Vol. *In Press*.

Castanon, J. I. (2007). History of the use of antibiotic as growth promoters in European poultry feeds. *Poultry science*, Vol. 86, No. 11, pp. 2466-2471.

Chan, K. F., Le, T. H., Kanenaka, R. Y., & Kathariou, S. (2001). Survival of clinical and poultry-derived isolates of Campylobacter jejuni at a low temperature (4 degrees C). *Applied and Environmental Microbiology*, Vol. 67, No. 9, pp. 4186-4191.

Chen, H. C. & Stern, N. J. (2001). Competitive exclusion of heterologous *Campylobacter* spp. in chicks. *Applied and Environmental Microbiology*, Vol. 67, No. 2, pp. 848-851.

Connerton, I. F., Connerton, P. L., , B. P., , S. B. S., & Atterbury, R. J. (2008). Bacteriophage therapy and *Campylobacter*, In: *Campylobacter*, I. Nachamkin & M. J. Blaser, pp. 679-693, American Society for Microbiology, Washington, D.C..

Connerton, P. L., Loc Carrillo, C. M., Swift, C., Dillon, E., Scott, A., Rees, C. E., Dodd, C. E., Frost, J., & Connerton, I. F. (2004). Longitudinal study of *Campylobacter jejuni* bacteriophages and their hosts from broiler chickens. *Applied and Environmental Microbiology*, Vol. 70, No. 7, pp. 3877-3883.

Connerton, P. L., Timms, A. R., & Connerton, I. F. (2011). *Campylobacter* bacteriophages and bacteriophage therapy. *Journal of Applied Microbiology*, Vol. 111, No. 2, pp. 255-265.

Corry, J. E. & Atabay, H. I. (2001). Poultry as a source of *Campylobacter* and related organisms. *Symposium series (Society for Applied Microbiology)*, Vol. 30, pp. 96S-114S.

Dabrowska, K., Switala-Jelen, K., Opolski, A., Weber-Dabrowska, B., & Gorski, A. (2005). Bacteriophage penetration in vertebrates. *Journal of Applied Microbiology*, Vol. 98, No. 1, pp. 7-13.

Doyle, M. P. & Erickson, M. C. (2006). Reducing the carriage of foodborne pathogens in livestock and poultry. *Poultry science*, Vol. 85, No. 6, pp. 960-973.

El-Shibiny, A., Scott, A., Timms, A., Metawea, Y., Connerton, P., & Connerton, I. (2009). Application of a group II *Campylobacter* bacteriophage to reduce strains of *Campylobacter jejuni* and *Campylobacter coli* colonizing broiler chickens. *Journal of Food Protection*, Vol. 72, No. 4, pp. 733-740.

El-Shibiny, A., Connerton, P. L., & Connerton, I. F. (2005). Enumeration and diversity of campylobacters and bacteriophages isolated during the rearing cycles of free-range and organic chickens. *Applied and Environmental Microbiology*, Vol. 71, No. 3, pp. 1259-1266.

Emborg, H. D., Andersen, J. S., Seyfarth, A. M., Andersen, S. R., Boel, J., & Wegener, H. C. (2003). Relations between the occurrence of resistance to antimicrobial growth promoters among *Enterococcus faecium* isolated from broilers and broiler meat. *International Journal of Food Microbiology*, Vol. 84, No. 3, pp. 273-284.

European Food Safety Authority (2006). The Community Summary Report on Trends and Sources of Zoonoses, Zoonotic Agents, Antimicrobial Resistance and Foodborne Outbreaks in the European Union in 2005. *The EFSA Journal*, Vol. 94, pp. 2-288.

European Food Safety Authority (2009). The Community Summary Report on Trends and Sources of Zoonoses and Zoonotic Agents in the European Union in 2007. *The EFSA Journal*, Vol. 223, pp. 1-320.

European Food Safety Authority (2010). The Community Summary Report on Trends and Sources of Zoonoses, Zoonotic Agents and Food-borne Outbreaks in the European Union in 2008. *EFSA Journal*, Vol. 8, pp. 1-410.

European Food Safety Authority (2011). The European Union Summary Report on Trends and Sources of Zoonoses, Zoonotic Agents and Food-borne Outbreaks in 2009. *The EFSA Journal*, Vol. 9, pp. 1-378.

European Food Safety Authority & European Centre for Disease Prevention and Control (2011). European Union summary report on antimicrobial resistance in zoonotic and indicator bacteria from animals and food in the European Union in 2009. *EFSA Journal*, Vol. 9, pp. 1-321.

Fiorentin, L., Vieira, N. D., & Barioni, W. (2005). Oral treatment with bacteriophages reduces the concentration of *Salmonella* Enteritidis PT4 in caecal contents of broilers. *Avian pathology*, Vol. 34, No. 3, pp. 258-263.

Friedman, C. R., Neimann, J., , W. H. C., & Tauxe, R. V. (2000). Epidemiology of *Campylobacter jejuni* infections in the United States and other industrialzed nations, In: *Campylobacter*, I. Nachamkin & M. J. Blaser, pp. 121-138, American Society for Microbiology, Washington, D.C..

Frost, J. A., Kramer, J. M., & Gillanders, S. a. (1999). Phage typing of *Campylobacter jejuni* and *Campylobacter coli* and its use as an adjunct to serotyping. *Epidemiology and infection*, Vol. 123, No. 1, pp. 47-55.

García, P., Martínez, B., Obeso, J. M., & Rodríguez, a. (2008). Bacteriophages and their application in food safety. *Letters in applied microbiology*, Vol. 47, No. 6, pp. 479-485.

Goode, D., Allen, V. M., & Barrow, P. A. (2003). Reduction of Experimental *Salmonella* and *Campylobacter* Contamination of Chicken Skin by Application of Lytic Bacteriophages. *Applied and Environmental Microbiology*, Vol. 69, No. 8, pp. 5032-5036.

Goodridge, L. & Abedon, S. T. (2003). Bacteriophage biocontrol and bioprocessing: Application of phage therapy to industry. *Society for Industrial Microbiology News*, Vol. 53, No. 6, pp. 254-262.

Gorman, R., Bloomfield, S., & Adley, C. C. (2002). A study of cross-contamination of food-borne pathogens in the domestic kitchen in the Republic of Ireland. *International Journal of Food Microbiology*, Vol. 76, No. 1-2, pp. 143-150.

Greer, G. G. (1988). Effects of Phage Concentration, Bacterial Density, and Temperature on Phage Control of Beef Spoilage. *Journal of Food Science*, Vol. 53, No. 4, pp. 1226-1227.

Guttman, B., Raya, R., & Kutter, E. (2005). Basic Phage Biology, In: *Bacteriophages. Biology and applications.*, E. Kutter & A. Sulakvelidze, CRC Press.

Gyles, C. L. (2008). Antimicrobial resistance in selected bacteria from poultry. *Animal Health Research Reviews*, Vol. 9, No. 2, pp. 149-158.

Hagens, S. & Loessner, M. J. (2010). Bacteriophage for biocontrol of foodborne pathogens: calculations and considerations. *Current Pharmaceutical Biotechnology*, Vol. 11, No. 1, pp. 58-68.

Hammerl, J. A., Jackel, C., Reetz, J., Beck, S., Alter, T., Lurz, R., Barretto, C., Brussow, H., & Hertwig, S. (2011). *Campylobacter jejuni* group III phage CP81 contains many T4-like genes without belonging to the T4-type phage group: implications for the evolution of T4 phages. *Journal of Virology*, Vol. 85, No. 17, pp. 8597-8605.

Hansen, V. M., Rosenquist, H., Baggesen, D. L., Brown, S., & Christensen, B. B. (2007). Characterization of *Campylobacter* phages including analysis of host range by selected *Campylobacter* Penner serotypes. *BMC Microbiology*, Vol. 7, pp. 90-.

Hansson, I., Ederoth, M., Andersson, L., Vågsholm, I., & Olsson Engvall, E. (2005). Transmission of *Campylobacter* spp. to chickens during transport to slaughter. *Journal of Applied Microbiology*, Vol. 99, No. 5, pp. 1149-1157.

Havelaar, A. H., Mangen, M. J., de Koeijer, A. A., Bogaardt, M. J., Evers, E. G., Jacobs-Reitsma, W. F., van, P. W., Wagenaar, J. A., de Wit, G. A., van der, Z. H., & Nauta, M. J. (2007). Effectiveness and efficiency of controlling *Campylobacter* on broiler chicken meat. *Risk Analysis*, Vol. 27, No. 4, pp. 831-844.

Heringa, S. D., Kim, J., Jiang, X., Doyle, M. P., & Erickson, M. C. (2010). Use of a mixture of bacteriophages for biological control of *Salmonella enterica* strains in compost. *Applied and Environmental Microbiology*, Vol. 76, No. 15, pp. 5327-5332.

Higgins, J. P., Higgins, S. E., Guenther, K. L., Huff, W., Donoghue, A. M., Donoghue, D. J., & Hargis, B. M. (2005). Use of a specific bacteriophage treatment to reduce *Salmonella* in poultry products. *Poultry science*, Vol. 84, No. 7, pp. 1141-1145.

Hooton, S. P., Atterbury, R. J., & Connerton, I. F. (2011). Application of a bacteriophage cocktail to reduce *Salmonella* Typhimurium U288 contamination on pig skin. *International Journal of Food Microbiology*, Vol. 151, No. 2, pp. 157-163.

Jacobs-Reitsma, W. (2000). *Campylobacter* in the food supply, In: *Campylobacter*, I. Nachamkin & M. Blaser, ASM Press, Washington DC.

Joerger, R. D. (2001). Alternatives to Antibiotics : Bacteriocins , Antimicrobial Peptides and Bacteriophages. *Poultry science*, Vol. 82, No. 4, pp. 640-647.

Johannessen, G. S., Johnsen, G., Okland, M., Cudjoe, K. S., & Hofshagen, M. (2007). Enumeration of thermotolerant *Campylobacter* spp. from poultry carcasses at the end of the slaughter-line. *Letters in applied microbiology*, Vol. 44, No. 1, pp. 92-97.

Johnson, R. P., Gyles, C. L., Huff, W. E., Ojha, S., Huff, G. R., Rath, N. C., & Donoghue, a. M. (2008). Bacteriophages for prophylaxis and therapy in cattle, poultry and pigs. *Animal Health Research Reviews*, Vol. 9, No. 2, pp. 201-215.

Kallings, L. O. (1967). Sensitivity of various salmonella strains to felix 0-1 phage. *Acta pathologica et microbiologica Scandinavica*, Vol. 70, No. 3, pp. 446-454.

Keener, K. M., Bashor, M. P., Curtis, P. A., Sheldon, B. W., & Kathariou, S. (2004). Comprehensive Review of *Campylobacter* and Poultry Processing. *Comprehensive Reviews In Food Science And Food Safety*, Vol. 3, No. 2, pp. 105-116.

Kropinski, A. M. (2006). Phage therapy - everything old is new again. *Canadian Journal of Infectious Diseases & Medical Microbiology*, Vol. 17, No. 5, pp. 297-306.

Kropinski, A. M., Arutyunov, D., Foss, M., Cunningham, A., Ding, W., Singh, A., Pavlov, A. R., Henry, M., Evoy, S., Kelly, J., & Szymanski, C. M. (2011). The genome and proteome of *Campylobacter jejuni* bacteriophage NCTC 12673. *Applied and Environmental Microbiology*, Vol. *In Press*.

Kropinski, A. M., Sulakvelidze, A., Konczy, P., & Poppe, C. (2007). *Salmonella* phages and prophages--genomics and practical aspects. *Methods in Molecular Biology*, Vol. 394, pp. 133-175.

Leverentz, B., Conway, W. S., Alavidze, Z., Janisiewicz, W. J., Fuchs, Y., Camp, M. J., Chighladze, E., & Sulakvelidze, A. (2001b). Examination of bacteriophage as a biocontrol method for salmonella on fresh-cut fruit: a model study. *Journal of Food Protection*, Vol. 64, No. 8, pp. 1116-1121.

Leverentz, B., Conway, W. S., Alavidze, Z., Janisiewicz, W. J., Fuchs, Y., Camp, M. J., Chighladze, E., & Sulakvelidze, A. (2001a). Examination of bacteriophage as a biocontrol method for salmonella on fresh-cut fruit: a model study. *Journal of Food Protection*, Vol. 64, No. 8, pp. 1116-1121.

Levin, B. R. & Bull, J. J. (2004). Population and evolutionary dynamics of phage therapy. *Nature Reviews Microbiology*, Vol. 2, No. 2, pp. 166-173.

Loc Carrillo, C., Atterbury, R. J., Connerton, P. L., Dillon, E., Scott, A., & Connerton, I. F. (2005). Bacteriophage Therapy To Reduce *Campylobacter jejuni* Colonization of

Broiler Chickens. *Applied and Environmental Microbiology*, Vol. 71, No. 11, pp. 6554-6563.

Los, M., Kuzio, J., McConnell, M., Kropinski, A., Wegrzyn, G., & Christie, G. (2010). Lysogenic conversion in bacteria of importance to the food industry, In: *Bacteriophages in the Detection and Control of Foodborne Pathogens*, P. Sabour & M. Griffiths, pp. 157-198, ASM Press, Washington, DC..

Luber, P., Brynestad, S., Topsch, D., Scherer, K., & Bartelt, E. (2006). Quantification of *Campylobacter* Species Cross-Contamination during Handling of Contaminated Fresh Chicken Parts in Kitchens. *Applied and Environmental Microbiology*, Vol. 72, No. 1, pp. 66-70.

Ma, Y., Pacan, J. C., Wang, Q., Xu, Y., Huang, X., Korenevsky, A., & Sabour, P. M. (2008). Microencapsulation of bacteriophage felix O1 into chitosan-alginate microspheres for oral delivery. *Applied and Environmental Microbiology*, Vol. 74, No. 15, pp. 4799-4805.

Marks, T. & Sharp, R. (2000). Bacteriophages and biotechnology: a review. *Journal of Chemical Technology and Biotechnology*, Vol. 75, No. 1, pp. 6-17.

Mathew, A. G., Cissell, R., & Liamthong, S. (2007). Antibiotic resistance in bacteria associated with food animals: a United States perspective of livestock production. *Foodborne Pathogens and Disease*, Vol. 4, No. 2, pp. 115-133.

Matsuzaki, S., Rashel, L., Uchiyama, J., Sakurai, S., Ujihara, T., Kuroda, M., Ikeuchi, M., Tani, T., Fujieda, M., Wakiguchi, H., & Imai, S. (2009). Bacteriophage therapy: a revitalized therapy against bacterial infectious diseases. *Journal of Infection and Chemotherapy*, Vol. 11, No. 5, pp. 211-219.

Mellon, M., Benbrook, C., & Benbrook, K. L. (2001). Hogging It: Estimates of Antimicrobial Abuse in Livestock,Union of Concerned Scientists, pp. pp. 1-67.

Merabishvili, M., Pirnay, J. P., Verbeken, G., Chanishvili, N., Tediashvili, M., Lashkhi, N., Glonti, T., Krylov, V., Mast, J., Van, P. L., Lavigne, R., Volckaert, G., Mattheus, W., Verween, G., De, C. P., Rose, T., Jennes, S., Zizi, M., De, V. D., & Vaneechoutte, M. (2009). Quality-controlled small-scale production of a well-defined bacteriophage cocktail for use in human clinical trials. *PLoS One.*, Vol. 4, No. 3, pp. e4944-.

Modi, R., Hirvi, Y., Hill, A., & Griffiths, M. W. (2001). Effect of phage on survival of *Salmonella enteritidis* during manufacture and storage of cheddar cheese made from raw and pasteurized milk. *Journal of Food Protection*, Vol. 64, No. 7, pp. 927-933.

Moran, A. P. & Upton, M. E. (1987). Factors affecting production of coccoid forms by *Campylobacter jejuni* on solid media during incubation. *Journal of Applied Microbiology*, Vol. 62, No. 6, pp. 527-537, 1365-2672 .

Nachamkin, I. (2002). Chronic effects of *Campylobacter* infection. *Microbes and infection*, Vol. 4, No. 4, pp. 399-403.

Newell, D. G. & Fearnley, C. (2003). Sources of *Campylobacter* Colonization in Broiler Chickens. *Applied and Environmental Microbiology*, Vol. 69, No. 8, pp. 4343-4351.

Nyachuba, D. G. (2010). Foodborne illness: is it on the rise? *Nutrition Reviews*, Vol. 68, No. 5, pp. 257-269.

O'Flynn, G., Coffey, A., Fitzgerald, G. F., & Ross, R. P. (2006). The newly isolated lytic bacteriophages st104a and st104b are highly virulent against *Salmonella enterica*. *Journal of Applied Microbiology*, Vol. 101, No. 1, pp. 251-259.

OzFoodNet Working Group (2009). Monitoring the incidence and causes of diseases potentially transmitted by food in Australia: annual report of the OzFoodNet Network, 2008. *Communicable diseases intelligence*, Vol. 33, No. 4, pp. 389-413.

Pao, S., Rolph, S. P., Westbrook, E. W., & Shen, H. (2004). Use of Bacteriophages to Control *Salmonella* in Experimentally Contaminated Sprout Seeds. *Journal of Food Science*, Vol. 69, No. 5, pp. M127-M130.

Payne, R. J., Phil, D., & Jansen, V. A. (2000). Phage therapy: the peculiar kinetics of self-replicating pharmaceuticals. *Clinical pharmacology and therapeutics*, Vol. 68, No. 3, pp. 225-230.

Payne, R. J. H. & Jansen, V. A. A. (2001). Understanding Bacteriophage Therapy as a Density-dependent Kinetic Process. *Journal of Theoretical Biology*, Vol. 208, pp. 37-48.

Petrov, V. M., Ratnayaka, S., Nolan, J. M., Miller, E. S., & Karam, J. D. (2010). Genomes of the T4-related bacteriophages as windows on microbial genome evolution. *Virology Journal*, Vol. 7, pp. 292-.

Petty, N. K., Evans, T. J., Fineran, P. C., & Salmond, G. P. C. (2007). Biotechnological exploitation of bacteriophage research. *Trends in Biotechnology*, Vol. 25, No. 1, pp. 7-15.

Rabinovitch, A., Aviram, I., & Zaritsky, A. (2003). Bacterial debris-an ecological mechanism for coexistence of bacteria and their viruses. *Journal of Theoretical Biology*, Vol. 224, No. 3, pp. 377-383.

Rabsch, W., Tschape, H., & Baumler, A. J. (2001). Non-typhoidal salmonellosis: emerging problems. *Microbes and infection*, Vol. 3, No. 3, pp. 237-247.

Raetz, C. R. & Whitfield, C. (2002). Lipopolysaccharide endotoxins. *Annual Review of Biochemistry*, Vol. 71, pp. 635-700.

Ricci, V. & Piddock, L. J. (2010). Exploiting the role of TolC in pathogenicity: identification of a bacteriophage for eradication of *Salmonella* serovars from poultry. *Applied and Environmental Microbiology*, Vol. 76, No. 5, pp. 1704-1706.

Robinson, D. A. (1981). Infective dose of *Campylobacter jejuni* in milk. *British Medical Journal*, Vol. 282, No. 6276, pp. 1584-.

Rohwer, F. (2003). Global phage diversity. *Cell*, Vol. 113, No. 2, pp. 141-.

Rosenquist, H., Sommer, H. M., Nielsen, N. L., & Christensen, B. B. (2006). The effect of slaughter operations on the contamination of chicken carcasses with thermotolerant *Campylobacter*. *International Journal of Food Microbiology*, Vol. 108, No. 2, pp. 226-232.

Rosenquist, H., Nielsen, N. L., Sommer, H. M., Nørrung, B., & Christensen, B. B. (2003). Quantitative risk assessment of human campylobacteriosis associated with thermophilic *Campylobacter* species in chickens. *International Journal of Food Microbiology*, Vol. 83, No. 1, pp. 87-103.

Sails, a. D., Wareing, D. R., Bolton, F. J., Fox, a. J., & Curry, a. (1998). Characterisation of 16 *Campylobacter jejuni* and *C. coli* typing bacteriophages. *Journal of medical microbiology*, Vol. 47, No. 2, pp. 123-128.

Santos, S. B., Fernandes, E., Carvalho, C. M., Sillankorva, S., Krylov, V. N., Pleteneva, E. A., Shaburova, O. V., Nicolau, A., Ferreira, E. C., & Azeredo, J. (2010). Selection and characterization of a multivalent *Salmonella* phage and its production in a nonpathogenic *Escherichia coli* strain. *Applied and Environmental Microbiology*, Vol. 76, No. 21, pp. 7338-7342.

Santos, S. B., Kropinski, A. M., Ceyssens, P. J., Ackermann, H. W., Villegas, A., Lavigne, R., Krylov, V. N., Carvalho, C. M., Ferreira, E. C., & Azeredo, J. (2011). Genomic and Proteomic Characterization of the Broad-Host-Range *Salmonella* Phage PVP-SE1: Creation of a New Phage Genus. *Journal of Virology*, Vol. 85, No. 21, pp. 11265-11273.

Sapkota, A. R., Lefferts, L. Y., McKenzie, S., & Walker, P. (2007). What do we feed to food-production animals? A review of animal feed ingredients and their potential impacts on human health. *Environmental Health Perspectives*, Vol. 115, No. 5, pp. 663-670.

Scott, A. E., Timms, A. R., Connerton, P. L., El-Shibiny, A., & Connerton, I. F. (2007a). Bacteriophage influence *Campylobacter jejuni* types populating broiler chickens. *Environmental microbiology*, Vol. 9, No. 9, pp. 2341-2353.

Scott, A. E., Timms, A. R., Connerton, P. L., {Loc Carrillo}, C., {Adzfa Radzum}, K., & Connerton, I. F. (2007b). Genome dynamics of *Campylobacter jejuni* in response to bacteriophage predation. *PLoS pathogens*, Vol. 3, No. 8, pp. e119-.

Sharp, R. J., Ahmad, S. I., Munster, A., Dowsett, B., & Atkinson, T. (1986). The isolation and characterization of bacteriophages infecting obligately thermophilic strains of *Bacillus*. *Journal of General Microbiology*, Vol. 132, No. 6, pp. 1709-1722.

Sillankorva, S., Pleteneva, E., Shaburova, O., Santos, S., Carvalho, C., Azeredo, J., & Krylov, V. (2010). *Salmonella* Enteritidis bacteriophage candidates for phage therapy of poultry. *Journal of Applied Microbiology*, Vol. 108, No. 4, pp. 1175-1186.

Smith, D. L., Dushoff, J., & Morris, J. G. (2005). Agricultural antibiotics and human health. *PLoS Medicine*, Vol. 2, No. 8, pp. e232-.

Smith, D. L., Harris, A. D., Johnson, J. a., Silbergeld, E. K., & Morris, J. G. (2002). Animal antibiotic use has an early but important impact on the emergence of antibiotic resistance in human commensal bacteria. *Proceedings of the National Academy of Sciences of the United States of America*, Vol. 99, No. 9, pp. 6434-6439.

Smith, H. W., Huggins, M. B., & Shaw, K. M. (1987a). Factors Influencing the Survival and Multiplication of Bacteriophages in Calves and in Their Environment. *Journal of General Microbiology*, Vol. 133, No. 5, pp. 1127-1135.

Smith, H. W., Huggins, M. B., & Shaw, K. M. (1987b). The control of experimental *Escherichia coli* diarrhoea in calves by means of bacteriophages. *Journal of General Microbiology*, Vol. 133, No. 5, pp. 1111-1126.

Sulakvelidze, A., Alavidze, Z., & Morris, J. G. (2001). Bacteriophage therapy. *Antimicrobial Agents and Chemotherapy*, Vol. 45, No. 3, pp. 649-659.

Swann, M. M. (1969). Joint Committee on the use of Antibiotics in Animal Husbandry and Veterinary Medicine, Her Majesty's Stationery Office, London, United Kingdom.

Tacconelli, E., De, A. G., Cataldo, M. A., Pozzi, E., & Cauda, R. (2008). Does antibiotic exposure increase the risk of methicillin-resistant *Staphylococcus aureus* (MRSA) isolation? A systematic review and meta-analysis. *Journal of Antimicrobial Chemotherapy*, Vol. 61, No. 1, pp. 26-38.

Timms, A. R., Cambray-Young, J., Scott, A. E., Petty, N. K., Connerton, P. L., Clarke, L., Seeger, K., Quail, M., Cummings, N., Maskell, D. J., Thomson, N. R., & Connerton, I. F. (2010). Evidence for a lineage of virulent bacteriophages that target *Campylobacter*. *BMC genomics*, Vol. 11, pp. 214-.

Toro, H., Price, S. B., McKee, A. S., Hoerr, F. J., Krehling, J., Perdue, M., & Bauermeister, L. (2005). Use of bacteriophages in combination with competitive exclusion to reduce *Salmonella* from infected chickens. *Avian Diseases*, Vol. 49, No. 1, pp. 118-124.

Tsuei, A. C., Carey-Smith, G. V., Hudson, J. A., Billington, C., & Heinemann, J. A. (2007). Prevalence and numbers of coliphages and *Campylobacter jejuni* bacteriophages in New Zealand foods. *International Journal of Food Microbiology*, Vol. 116, No. 1, pp. 121-125.

Turki, Y., Ouzari, H., Mehri, I., Ammar, A. B., & Hassen, A. (2011). Evaluation of a cocktail of three bacteriophages for the biocontrol of *Salmonella* of wastewater. *Food Research International*, No. doi:10.1016/j.foodres.2011.05.041.

Uzzau, S., Brown, D. J., Wallis, T., Rubino, S., Leori, G., Bernard, S., Casadesus, J., Platt, D. J., & Olsen, J. E. (2000). Host adapted serotypes of *Salmonella enterica*. *Epidemiology and infection*, Vol. 125, No. 2, pp. 229-255.

van de Giessen, A. W., Bouwknegt, M., Dam-Deisz, W. D., van, P. W., Wannet, W. J., & Visser, G. (2006). Surveillance of *Salmonella* spp. and *Campylobacter* spp. in poultry production flocks in The Netherlands. *Epidemiology and infection*, Vol. 134, No. 6, pp. 1266-1275.

Velge, P., Cloeckaert, A., & Barrow, P. (2005). Emergence of *Salmonella* epidemics: the problems related to *Salmonella enterica* serotype Enteritidis and multiple antibiotic resistance in other major serotypes. *Veterinary Research*, Vol. 36, No. 3, pp. 267-288.

Viscardi, M., Perugini, A. G., Auriemma, C., Capuano, F., Morabito, S., Kim, K. P., Loessner, M. J., & Iovane, G. (2008). Isolation and characterisation of two novel coliphages with high potential to control antibiotic-resistant pathogenic *Escherichia coli* (EHEC and EPEC). *International Journal of Antimicrobial Agents*, Vol. 31, No. 2, pp. 152-157.

Waddell, T. E., Franklin, K., Mazzocco, A., Kropinski, A. M., & Johnson, R. P. (2009). Generalized transduction by lytic bacteriophages. *Methods Mol.Biol.*, Vol. 501, pp. 293-303.

Wagenaar, J. A., {Van Bergen}, M. A., Mueller, M. A., Wassenaar, T. M., & Carlton, R. M. (2005). Phage therapy reduces *Campylobacter jejuni* colonization in broilers. *Veterinary Microbiology*, Vol. 109, No. 3-4, pp. 275-283.

Waldor, M. K. & Mekalanos, J. J. (1996). Lysogenic conversion by a filamentous phage encoding cholera toxin. *Science*, Vol. 272, No. 5270, pp. 1910-1914.

Wall, S. K., Zhang, J., Rostagno, M. H., & Ebner, P. D. (2010). Phage therapy to reduce preprocessing *Salmonella* infections in market-weight swine. *Applied and Environmental Microbiology*, Vol. 76, No. 1, pp. 48-53.

Whichard, J. M., Sriranganathan, N., & Pierson, F. W. (2003). Suppression of *Salmonella* growth by wild-type and large-plaque variants of bacteriophage Felix O1 in liquid culture and on chicken frankfurters. *Journal of Food Protection*, Vol. 66, No. 2, pp. 220-225.

Wierup, M. (2001). The Swedish experience of the 1986 year ban of antimicrobial growth promoters, with special reference to animal health, disease prevention, productivity, and usage of antimicrobials. *Microbial Drug Resistance*, Vol. 7, No. 2, pp. 183-190.

World Health Organization (2002). The use of antimicrobials outside human medicine: information from the World Health Organization on the health consequences. *Journal of Environmental Health*, Vol. 64, No. 9, pp. 66, 62-.

12

A Phage-Guided Route to Discovery of Bioactive Rare Actinomycetes

D.I. Kurtböke
University of the Sunshine Coast, Faculty of Science, Health, Education and Engineering
Maroochydore DC,
Australia

1. Introduction

The discovery, development and exploitation of antibiotics was one of the most significant advances in medicine in 20th century and in a golden era lasting from 1940s to late 1960s, antibiotic research provided mankind with a wide range of structurally diverse and effective agents to treat microbial infections (Table 1) (McDevitt and Rosenberg, 2001; Hopwood, 2007). However, antibiotic resistance has developed steadily as new agents have been introduced and there has been a dramatic increase in the occurrence of resistant organisms in both community and hospital settings for the past 10-15 years. In particular, pathogens such as *Staphylococcus aureus* and *Streptococcus pneumoniae* and *Enterococcus faecalis* capable of resulting in severe and fatal infections have become increasingly resistant to multiple antibiotics. In hospital and community environments, Methicillin-resistant *S. aureus* (MRSA) and Vancomycin resistant enterococci (VRE) have become persistent pathogens. Other multiple drug resistant organisms currently include *Mycobacterium tuberculosis* and *Pseudomonas,* and related species in the hospital environment. Last line of antibiotics such as vancomycin might also become ineffective against super-bugs such as vancomycin-intermediate-resistant *S. aureus* isolates. New classes of antibiotics with a new mode of action (e.g. Linezolid™) are necessary to combat existing and emerging infectious diseases deriving from multiple drug resistant agents (McDevitt and Rosenberg, 2001; Hopwood, 2007).

An extreme example for yet to be faced outbreaks has been the recent occurrence of multi-drug resistant enterohaemorrhagic *E. coli* in Germany claiming the lives of many (Chattaway *et al.*, 2011). Interestingly, this strain acquired virulence genes from another group of diarrhoeagenic *E. coli,* the enteroaggregative *E. coli* (EAEC), which is the most common bacterial cause of diarrhoea. This event once more stressed the importance of powerful diagnostic systems to detect all diarrheagenic *E. coli* as part of routine surveillance systems, which would thus contribute to the mapping of the global distribution of EAEC (Chattaway *et al.*, 2011; Mellmann *et al.*, 2011).

For more than a century, infectious diseases have been controlled by vaccination and the administration of antibiotics (Muzzi *et al.*, 2007). In spite of the technical progress of the past century, innovation in both fields came exclusively from traditional approaches, and antibiotics have been identified by screening natural compounds for their ability to kill

DISCOVERY YEARS	NAME OF ANTIBIOTICS
1940-1950	Streptomycin Streptothricin Actinomycin
1950-1960	Neomycin Chlorotetracycline Candicidin Chloramphenicol Spiramycin Tetracycline Erythromycin Oxytetracyline Nystatin Kanamycin
1960-1970	Mitomycin Novobiocin Amphotericin Vancomycin Virginiamycin Gentamicin Tylosin Pristinamycin Polyoxin Rifamycin Bleomycin
1970-1980	Monensin Adriamycin Avoparcin Kasugamycin Fosfomycin Bialaphos Lincomycin Teicoplanin
1980-1990	Thienamycin Rapamycin Avermectin Nikkomycin
1990-2000	Spinosyn Tacrolimus

Table 1. Antibiotics since discovery of Streptomycin (adapted from Hopwood, 2007).

bacteria grown *in vitro*. Furthermore, by improving existing drugs such as glycylcylines and fluoroquinolones deriving from tetracyclines and quinolones, pharmaceutical industries aimed to stay "one step ahead" of resistant microorganisms. Although such an approach has been effective, it is becoming increasingly difficult to meet the needs of the community and to provide sufficient coverage for all emerging infectious agents (McDevitt and Rosenberg, 2001; Muzzi *et al.*, 2007).

To keep pace with microbial resistance, objective and target-directed strategies are needed to discover and develop new classes of antibiotics. In the light of the global threat outlined above, this chapter will overview emerging novel strategies with particular emphasis on bacteriophages as tools in the search for new and potent therapeutic agents from actinomycetes.

2. Genomics based approaches to drug discovery

Since early 2000s, information from completed genome sequences and genomic based technologies has been a driving force in antibiotic discovery resulting in new target identification of pathogens as well as in the enhancement of action studies of antimicrobial compounds. Exploitation of high-throughput automated DNA sequencing capabilities and genome sequences of microbial pathogens advanced rapidly producing full genome sequence results (e.g. *Enterococcus faecium* genome) (Amber, 2000; McDevitt and Rosenberg, 2001). In the past, antimicrobial studies were conducted on model microorganisms such as *E. coli* and *Bacillus subtilis*, however, with the new advances, research has become possible by directly using pathogens such as *Staphylococcus aureus* and *Streptococcus pneumoniae* (McDevitt and Rosenberg, 2001; Payne *et al.*, 2007). These developments have led to a shift in the discovery of novel vaccines and antimicrobials from the traditional empirical approach to a novel knowledge-based approach.

In conventional drug discovery, whole-cell screening approaches are adapted. This approach first identifies an antimicrobial compound and later seeks to establish the cellular target of that compound, and the vast majority of antibiotics that are currently used have this mode of action (e.g. targeting a limited number of proteins involved in critical cellular functions) (McDevitt and Rosenberg, 2001; Mills, 2003; Ricke *et al.*, 2006). Whereas in the modern era of genome-driven and target-based approach a target gene is selected and its spectrum is identified. After it is validated for its role, cloned and sequenced, its corresponding protein product is expressed in an optimized expression system (e.g. *Pichia pastoris*, Baculovirus or *E. coli*). The target protein is then purified and screened against a large and diverse collection of low-molecular weight compounds in order to identify target inhibitors to investigate their potency, mechanism of inhibition and enzyme spectrum and selectivity (McDevitt and Rosenberg, 2001; Mills, 2003).

Increasing knowledge of bacterial diversity based on genomics and pangenomics now suggests that the way forward should be to focus discovery strategies on the identification of targets that are essential for the formation and persistence of an *in vivo* infection or in the expression of virulence factors (Muzzi *et al.*, 2007).

Sequencing the entire genome of pathogens has revealed all of their open reading frames (ORFS), which can be utilized as selected targets in drug discovery. As summarized by McDevitt and Rosenberg (2001), there are several key criteria to be considered in target selection: "(1) the target should be present in a required spectrum of organisms; (2) it should be either absent in humans or, if present, it should be significantly different to allow confidence that selective inhibitors of the bacterial target over a human counterpart can be developed; (3) it should be essential for bacterial growth or viability under the conditions of the infection; (4) it should be expressed and relevant to the infection process; and (5) some aspects of its function should be understood to allow the relevant assays and high throughput screens to be developed".

One example was the use of peptide deformylase (PDF) as a protein target to facilitate discovery of a broad spectrum antibacterial drug (Mills, 2003). The protein is encoded by the *def* gene which is present in all pathogenic bacteria, but does not share a functionally equivalent gene in mammalian cells, which is one of the most sought after characteristics related to a drug candidate (McDevitt and Rosenberg, 2001; Yuan *et al.*, 2001; Mills, 2003). This example was a good proof-of-principle illustration of the genomics-driven, target-based approach; starting with a conserved gene and leading to an antimicrobial compound (Clements *et al.*, 2001; Mills, 2003).

Genome sequencing studies can also be utilized from the angle of drug producer microorganisms (Kurtböke, 2012). An example is the genome sequencing of *Salinispora tropica*, which has revealed its complex secondary metabolome as a rich source of drug-like molecules. Such a discovery has been a powerful interplay between genomic analysis and traditional natural product isolation studies (Udwary *et al.*, 2007). Other examples that reveal the superior ability of actinobacteria to produce potent bioactive compounds facilitating discovery of novel bioactive compounds include genome sequences of *Streptomyces coelicolor* A3(2) (Bentley *et al.*, 2002) and *S. avermitilis* (Ikeda *et al.*, 2003).

3. Bacteriophages in chemotherapy

Therapeutic use of bacteriophages for the prevention and treatment of bacterial diseases, has been targeted since the discovery of phages in 1917 **by** Félix d'Hérelle. Following his discovery, he first attempted to use these against dysentery and since then, bacteriophages have been used to treat human infections as an alternative or a complement to antibiotic therapy (Hermoso *et al.*, 2007). Particularly, from 1920s to 1950s, phage therapy has exploded and centres in the US, France and Georgia were established (Kütter and Sulakvelidze, 2005; Hermoso *et al.*, 2007; Chanishvili, 2009), however, there have been limitations to antibacterial phage therapy that hamper its application as an antibiotic alternative. These have been summarized recently by Hermoso *et al.* (2007) as follows: (i) phages generally have narrow host range and only strongly lytic phage against bacterial strain infecting the patient, should be given to the patient; (ii) phages may not always remain lytic under the physiological conditions and bacteria can become resistant to phages after infection; (iii) phage preparations should be free of bacteria and their toxic components to meet clinical safety requirements, but sterilization of phage preparations could inactivate the phages; (iv) phages can be inactivated by a neutralizing antibody, and there is some risk of promoting allergic reactions to them; (v) the pharmacokinetics of phage treatments are more complicated than those of chemical drugs because of their self-replicating nature; (vi) phages might endow bacteria with toxic or antibacterial resistance genes.

Due to the above-listed limitations of bacteriophage therapy, bacteriophages might have more value as tools in drug discovery such as for target discovery and validation, assay development and compound design (Brown, 2004; Projan, 2004), and some of these exploitations are discussed below.

3.1 Bacterial virulence and injection mechanisms of bacteriophages

Efficient host infection relies on bacterial virulence factors being localized outside the producing cell where they are identically placed to interact with host defences and subvert host cells for the pathogen's benefit. Pathogenic bacteria have thus developed powerful molecular strategies to deliver their virulence factors across the bacterial cell envelope as well as powerful mechanisms to adverse host cell plasma membrane (Cambronne and Roy, 2006; Filloux, 2009; 2011; Russell *et al.*, 2011).

In Gram-negative bacteria, the cell envelopes have two hydrophobic inner and outer membranes with a hydrophilic space in between. The secreted hydrophobic molecules of proteins, enzymes or toxins have to travel through the hydrophobic environment of the membranes in an aqueous channel, or another type of conduit, that spans the cell envelope. These paths to the external medium are built by assembling macromolecular complexes, called secretion machines (Filloux, 2009; 2011) and they are distinguishable by the number and characteristics of the components such as types I, II and V secretion systems and they play important roles in the virulence of pathogens (Filloux *et al.*, 2008; Leiman *et al.*, 2009; Pukatzki *et al.* 2009; Bönemann *et al.*, 2010; Schwarz *et al.*, 2010).

In type VI secretion systems (T6SS) of Gram-negative bacteria the lack of an outer membrane channel for the T6SS might suggest an alternative delivery strategy such as local puncturing of the cell envelope to avoid cell lysis whilst allowing transient assembly of the secretion machine (Filloux, 2009; 2011). Filloux (2011) points out that the structural proteins of the T6SS are very similar to those that make up the injection machinery found in bacteriophages. Bacteriophages inject their DNA into bacterial cytosol and use the bacterium as a phage factory to replicate phage DNA. Bacterial cell envelope is perforated by bacteriophage puncturing device and its DNA is injected into bacterial cell via a tail tube. T6SS seems to use the same mechanism used by bacteriophages to inject their DNA into bacteria in which some components like the T6SS-specific exoproteins might have a similar tail-spike puncturing device of the T4 phage and might create a channel across the bacterial envelope which resembles the phage tail tube. T6SS translocation mechanism operate from the inside to the outside of the bacterial cell, and might be a mirror image of the phage translocation mechanism, which operates from outside to the inside of the bacterial cell Filloux (2011). Therefore, a sound understanding of bacteriophage injection and bacterial secretion systems might bring new insights to the development of effective therapeutic agents.

3.2 Bacteriophage-guided route to biodiscovery

Bacteriophages have evolved multiple strategies to interfere with bacterial growth. As a result, improved understanding of the bacteriophage-host interactions can also bring a new perspective to drug discovery (Young *et al.*, 2000; Brown, 2004; Projan, 2004; Parisien *et al.*, 2008). Examples include successful use of phage encoded lytic enzymes to destroy bacterial targets (Fishetti *et al.*, 2003) and use of lysostaphin to achieve sterilization in an endocarditis model (Climo *et al.*, 1998). Furthermore, in a novel approach, Liu *et al.* (2004) applied information deriving from phage genome to target discovery of gene products that inhibit pathogenic bacterium such as *Staphylococcus aureus*. They uncovered strategies used by bacteriophage to disable bacteria for design of a method, which uses key phage proteins to identify and validate vulnerable targets and exploit them in the identification of new antimicrobials.

Polysaccharide-specific phages were also suggested to treat encapsulated pathogenic bacteria since exolysaccharide production in bacteria involves biofilm formation and acts as a barrier to the penetration of therapeutic agents. Phages that can polymerize these substances and/or kill the bacteria may potentially be useful for control of bacteria forming biofilms on medical devices (Hermoso *et al.*, 2007). Protein antibiotics, which are the gene products of some small phages that do not produce endolysins, have also been shown to inhibit cell wall synthesis (Bernhardt *et al.*, 2002). Genetic engineering of bacteriophages to carry toxic genes or proteins to produce cell death without lysis and hence avoiding the release of unwanted endotoxins has also been suggested (Westwater *et al.*, 2003). Furthermore, Hagens *et al.* (2006) proposed a bacteriophage-based strategy to reduce effective doses of antibiotics during treatment for resensitization of antibiotic resistant pathogen via the presence of phage *in vivo*. In addition, it has been reported that phage host-cell lysis proteins, encoded by holins and amidases and elaborated late in the infection cycle, maintain their potent antibacterial activity when administered from outside cell (Loeffler *et al.*, 2001; Schuch *et al.*, 2002).

3.3 From bacteriophage genomics to drug discovery

Over evolutionary time, bacteriophages have developed unique proteins that arrest critical cellular processes to commit bacterial host metabolism to phage reproduction (Liu *et al.*, 2004). Bacterial key metabolic processes can be shut off via inactivation of critical cellular proteins with these unique bacteriophage proteins, and host metabolism can be directed into the production of progeny phages. As an example; phages of *E. coli*, host physiology shuttoff is typically performed early during the phage lytic cycle by small phage-encoded proteins that target particularly vulnerable and accessible proteins involved in crucial host metabolic processes. Thus, Liu *et al.* (2004) using a high-throughput bacteriophage genomics strategy, exploited the concept of phage-mediated inhibition of bacterial growth to systematically identify antimicrobial phage-encoded polypeptides. They found that four proteins of the *Staphylococcus aureus* DNA replication machinery were targeted by a total of seven unrelated phage polypeptides leading to a superior approach to currently available antibiotics which only target topoisomerases. In some cases, sequence-unrelated polypeptides from different phages were found to target the same proteins in *S. aureus*, and such susceptibility might have uses in antimicrobial drug discovery.

All these developments including increased understanding of the mechanism of injection, beginning with adsorption to the host and ending with complete delivery of genomic material (McPartland and Rothman-Denes, 2009) are now paving the way towards recruitment of phages in the search for new antibiotics with previously unknown antibacterial mechanisms.

3.4 From endolysins to enzybiotics

Phages have different methods of progeny release from bacterial cells: filamentous phages are ejected from bacterial cell walls without destroying the host cell, whereas non-filamentous phages induce lysis through lytic enzymes. Phage lytic enzymes are highly evolved murein hydrolases to quickly destroy the cell wall of the host bacterium to release the progeny. Lysis is a result of abrupt damage to the bacterial cell wall by means of specific proteins and as stated by Hermoso *et al.* (2007) it can be completed in two different ways: (i) inhibition of peptidoglycan synthesis by a single protein or (ii) enzymatic cleavage of peptidoglycan by lysins or holin-lysin system.

Tailed phages achieve correctly-timed lysis by the consequtive use of endolysins and holins. Holins are small hydrophobic proteins that are encoded by the phage and inserted into cytoplasmic membrane to form membrane lesions or holes for endolysin passage. Whereas endolysins are phage-coded enzymes that break down bacterial peptidoglycan at the terminal stage of the phage reproduction cycle (Moak and Molineux, 2004; Loessner, 2005; Hermoso et al., 2007). Target specificity in endolysin studies reveal differences such as bifunctional enzyme of *Streptococcus agalactiae* phage with glycosidase and endopeptidase activities or muramidase activity of *Lactobacillus helveticus* phage (Loessner, 2005; Hermoso et al., 2007). However, most enzymes like amidases from phage that infect Gram-positive bacteria feature narrow lysis ranges, which can be genus-specific (*Streptomyces aureofaciens*) and even species-specific (*Clostridium perfringens*) (Loessner, 2005). Other examples include narrow specificity of endolysins only targeting *Clavibacter michiganensis* subspecies without affecting other bacteria in soil including closely related *Clavibacter* species (Wittmann et al., 2010).

Due to increasing antibiotic resistance, phage-derived lytic enzymes are now being exploited to control infections. In antibiotic resistant Gram-positive bacteria, it has been reported that even small quantities of purified recombinant lysin added externally lead to immediate lysis resulting in log-fold of death of the bacterial cells found on the mucosal surfaces and infected tissues. They have been suggested to make ideal antiinfectives due to lysin specificity for the pathogen that does not disturb the normal flora, the low chance of bacterial resistance towards lysins, and their ability to kill colonizing pathogens on mucosal surfaces illustrating a previously unavailable capacity (Hermoso et al., 2007, Fenton et al., 2010; Fishetti, 2010). These enzymes are suggested to particularly be useful to control antibiotic resistant Gram-positive pathogens. In this group of bacteria, lysins can make direct contact with their cell wall carbohydrates and peptidoglycan externally making them suitable candidates in clinical applications (Loessner, 2005; Hermoso et al., 2007).

Another example is *Mycobacterium*, phylogenetically related to Gram-positive bacteria but its cell envelope has a double-membrane structure similar to Gram-negative bacteria. Cell envelopes of mycobacteria contain peptidoglycan-arabinogalactan-mycolic acid complex (Sutcliffe, 2010). Mycobacteriophages must not only degrade the peptidoglycan layer but must also circumvent a mycolic acid-rich outer membrane covalently attached to the arabinogalactan-peptidoglycan complex. They utilize two lytic enzymes to produce lysis: (i) Lysin A that hydrolyzes peptidoglycan, and (ii) Lysin B, a novel mycolylarabinogalactan esterase, that cleaves the mycolylarabinogalactan bond to release free mycolic acids (Payne et al., 2009) and the study of phage ejection mechanisms in this group of bacteria might lead to the discovery of novel lytic systems and thus new antimicrobial agents.

Effective antimicrobial activity against Gram-positive bacterial pathogens including *Streptococcus pneumoniae* and *Bacillus anthracis* by exogenously applied phage-encoded endolysins has already been demonstrated. This approach has however, proved ineffective against Gram-negative bacteria since the outer membrane blocks access to the peptidoglycan targets (Fishetti, 2008). Due to their mycolic acid, rich outer membrane mycobacteria are likely to be similarly intractable to exogenously added endolysins. In order to overcome this resistance, a novel approach has been proposed by Payne et al. (2009) to render mycobacterial pathogens such as *M. tuberculosis* susceptible to endolysin treatment through co-treatment with LysA and LysB proteins.

In-depth understanding of the host-phage interaction and the full lytic-system is required to design effective biocontrol strategies using bacteriophage lysins. In this search, another rich source for mycobacterial phages might be the activated sludge systems where fascinating suborder, family, genus and species-specific host-phage interactions occur (Thomas *et al.*, 2002). Recent genome sequencing of a *Tsukamurella* phage again isolated from an activated sludge system reveals a modular gene structure that shares some similarity with those of *Mycobacterium* phages (Petrovski *et al.*, 2011). Accordingly, phylum level perspective and understanding of bacterial cell wall envelope architecture (Sutcliffe, 2010) with particular emphasis on monoderm and diderm bacteria, and translation of this understanding to phage lytic activity will advance current knowledge and contribute towards design and application of new phage-derived therapeutics. *Actinobacteria*-specific proteins, mainly specific for the *Corynebacterium*, *Mycobacterium* and *Nocardia* subgroups, have also been reported (Venture *et al.*, 2007) and such specific proteins might have implications for the control of these pathogens. Mycetoma, a chronic granumatous infection persistent worldwide and endemic to tropical and subtropical regions, is another example (Linchon and Khachemoune, 2006) and among bacteria *Actinomadura* species reportedly cause the disease. However, in spite of trials in many different laboratories, phages specific to *Actinomadura* species were not reported until early 1990s (Long *et al.*, 1993; Kurtböke *et al.*, 1993b). Phages isolated towards different species of *Actinomadura* from organic mulches used in avocado plantations revealed that they belonged to *Siphoviridae* group of phages (Kurtböke *et al.*, 1993b). Further studies on the *Actinodamura* phage and host-cell-wall interactions might shed light on the development of effective treatment strategies deriving from phage lytic activity on the pathogenic host.

Furthermore, metagenomics sequencing studies of uncultured viral populations have provided new insights into bacteriophage ecology. The cloning of phage lytic enzymes from uncultured viral DNA, and observations into colony lysis following exposure to inducing agent, revealed the value of viral metagenomes as potential sources of recombinant proteins with biotechnological value (Schmitz *et al.*, 2010). Functional screens of viral metagenomes will inevitably provide a large source of recombinant proteins which might subsequently be used to treat infections resulting from difficult to control pathogens.

3.5 Prophage genomics

Prophage genomics has increased our understanding of the phage-bacterium interaction at the genetic level. Data deriving from these studies has also revealed genetic rules that underlie the arms race between the host bacterium and the infecting virus (Wagner and Waldor, 2002; Canchaya *et al.*, 2003). Studies into non-pathogenic bacteria inhabiting varied but defined environments have also improved our understanding of the prophage contribution to the fitness increase of host bacterial cells. Even environmental and commensal bystander bacteria have been shown to be converted into toxin-producing ones via lysogenization (Chibani-Chennoufi *et al.*, 2004).

Prophage genomics studies will possibly lead to discovery of important genes for the ecological adaptation of bacterial commensals and symbionts (Canchaya *et al.*, 2003; Venture *et al.*, 2007). Moreover, prophage genomics studies will provide further information on the expression of many lysogenic conversion genes (Canchaya *et al.*, 2003) and all this information will then provide significant clues to be further exploited in drug discovery.

4. Natural products

Natural products have historically made significant contributions to the provision of new lead candidates in drug discovery programs (Newman and Cragg, 2004 a,b). Most characteristic features of the secondary metabolites are their incredible array of unique chemical structures and can be exploited as lead compounds, for chemical synthesis of new analogues or as templates, in the rational drug design studies. Their very frequent occurrence, versatile bioactivities and the rich structural and stereochemical attributes of natural products promote these compounds as valuable molecular scaffolds to explore their chemotherapeutic potential (Demain and Fang, 2000; Croteau *et al.*, 2000; Firn and Jones, 2002). However, to continue to be competitive with other drug discovery methods, natural product research needs to continually improve the speed of the screening, isolation, and structure elucidation processes, as well addressing the suitability of screens for natural product extracts and dealing with issues involved with large-scale compound supply (Butler, 2004). Current alternative strategies include exploitation opportunities for drug discovery arising from an understanding of the mode of action of existing antibiotics. In this way, biochemical pathways or processes (e.g. peptidoglycan synthesis, tRNA synthesis, transcription and DNA replication) inhibited by antibiotics already in clinical use may contain key functions that represent unexploited targets for further drug discovery. Since most of these antibiotics are of natural product origin they might provide further clues in the search for their alternatives (Chopra *et al.*, 2002).

4.1 Bioactive compounds from microbial resources

In industrial applications, microbial secondary metabolites are often defined as "low molecular mass products of secondary metabolism," which include antibiotics, pigments, toxins, effectors of ecological competition and symbiosis, pheromones, enzyme inhibitors, immunomodulating agents, receptor antagonists and agonists, pesticides, antitumor agents and growth promoters (Demain and Fang, 2000; Bérdy, 2005; Bull, 2004; 2007; 2010) (Table 2 and 3).

Amino sugars	Glycopeptides	Phenazines	Pyrrolines
Anthocyanins	Glycosides	Phenoxazinones	Pyrrolizines
Anthraquinones	Hydroxylamines	Phthaldehydes	Quinolines
Aziridines	Indole derivatives	Piperazines	Quinones
Benzoquinones	Lactones	Polyacetylenes	Salicylates
Coumarines	Macrolides	Polyenes	Terpenoids
Diazines	Naphthalenes	Polypeptides	Tetracyclines
Epoxides	Naphthoquinones	Pyrazines	Tetronic acids
Ergot alkoloids	Nitriles	Pyridines	Triazines
Flavonoids	Nucleosides	Pyrones	Tropolones
Furans	Oligopeptides	Pyrroles	Vanillin
Glutarimides	Perylenes	Pyrrolidones	Zeaxanthin

Table 2. Examples of classes of organic compounds deriving from microbial secondary metabolites (adapted from Demain, 1981 and reproduced from Kurtböke, 2010a)

ACTH-like	Complement inhibition	Hemolytic	Leukemogenic
Anabolic	Convulsant	Hemostatic	Motility inhibition
Analeptic	Dermonecrotic	Herbicidal	Nephrotoxic
Anesthetic	Diabetogenic	Hormone releasing	Paralytic
Anorectic	Diuretic	Hypersensitizing	Parasympathomimetic
Anticoagulant	Edematous	Hypochloresterolemic	Photosensitizing
Antidepressive	Emetic	Hypoglycemic	Relaxant (smooth muscle)
Antihelminthic	Enzyme inhibitory	Hypolipidemic	Sedative
Anti-infective	Erythematous	Hypotensive	Serotonin antagonist
Anti-inflammatory	Estrogenic	Immunostimulating	Spasmolytic
Anti-parasitic	Coagulative (blood)	Hallucinogenic	Telecidal
Anti-spasmodic	Fertility enhancing	Inflammatory	Ulcerative
Carcinogenesis inhibition	Complement inhibition	Hemolytic	Vasodilatory
Coagulative (blood)	Hallucinogenic	Insecticidal	Anti-viral

Table 3. Pharmacological activities of microbial secondary metabolites (adapted from Demain, 1983 and reproduced from Kurtböke, 2010a)

However, existing antibiotics have mode of actions directed at a narrow spectrum of targets, principally cell wall, DNA and protein biosynthesis and so far multidrug resistance among bacterial pathogens has been largely due to a limited repertoire of antibacterial drugs that eradicate bacteria using a narrow range of mechanisms (Brown, 2004; Baltz, 2005; 2006a,b; 2008). Novel structural attributes are also required and only one new class of antibiotics has reached the clinic since 2001 (Ford et al., 2001). Currently, many novel microorganisms are being isolated from extreme biological niches, revealing their own chemical defence mechanisms. These naturally occurring organisms, together with recombinant organisms generated using combinatorial genetics and the availability of new chimeric metabolic pathways, might deliver an abundance of new compounds (Payne et al., 2007; Goodfellow 2010).

As advocates of natural product screening to search for novel antibacterial leads, Payne et al. (2007) adapted an alternative innovative approach with the belief that leads were not going to come from screening, but from alternative approaches. They reconsidered known antibacterial molecules to see whether they could improve their antibacterial and developmental properties and along these lines, they modified the pleuromutilin core structure in ways to bring three derivatives into clinical development. They also found lead molecules by screening a small, discrete library of compounds for antibacterial activity, which resulted in the discovery of a novel compound class capable of inhibiting bacterial DNA replication, and reached the developmental stage.

In the light of the above-mentioned advances, revisiting natural products with target-directed strategies might again provide us with novel and potent therapeutic agents.

4.2 Actinomycetes and drug discovery

Among the bacteria, the members of the order *Actinomycetales* have proved to be a particularly rich source of secondary metabolites with extensive industrial applications (Table 4).

Source	Bioactive secondary metabolites			
	Antibiotics		Bioactive metabolites	Total bioactive metabolites
	Total	With other activity	No antibiotic activity	Antibiotics plus other active compounds	
Bacteria	2900	780	900	1680	3800
Actinomycetes	8700	2400	1400	3800	10100
Fungi	4900	2300	3700	6000	8600
Total	16500	5500	6000	11500	22500

Table 4. Bioactive compounds of microbial origin (adapted from Bérdy, 2005 and reproduced from Kurtböke, 2010b).

In particular, the capacity of the members of the genus *Streptomyces* to produce commercially significant compounds, especially antibiotics, remains unsurpassed, possibly because of the extra-large DNA complement of these bacteria (Goodfellow and Williams, 1986; Kurtböke, 2010a; 2012). Members from this genus are even predicted to be the producers of many novel yet to be discovered bioactive compounds (Watve *et al.*, 2001). As a result, selective isolation of previously undetected bioactive actinomycetes is one of the major targets of industrial microbiologists in the search for novel therapeutic agents (Bull *et al.*, 2000; Bull and Stach, 2007; Goodfellow, 2010; Goodfellow and Fiedler, 2010; Kurtböke, 2003; 2010a).

The range of versatility of actinomycete metabolites is enormous and yields significant economic returns, yet, biodiscovery from these sources depends on the

i.   detection and recovery of bioactive actinomycete fraction from previously unexplored environmental sources,
ii.   effective assessment of their metabolites in defined targets (Goodfellow, 2010; Kurtböke, 2003; 2010a).

4.3 Bacteriophage-guided route to detection of rare actinomycetes

Chemical diversity of bioactive compounds, particularly from those rare and "yet to be discovered" actinomycetes is promising, however, detection of bioactive actinomycete taxa requires in-depth understanding of their true diversity and eco-physiology through which target-directed isolation strategies can be implemented (Bull *et al.*, 2000; Bull, 2003; Kurtböke, 2012).

Isolation of bioactive rare actinomycete taxa requires highly specialised isolation techniques (Lazzarini *et al.*, 2000; Kurtböke, 2003; Goodfellow 2010), and those employed range from the use of antibiotics to chemotaxis chambers, and excessive heat treatments (Hayakawa, 2003; Terekhova, 2003; Okazaki, 2003; Goodfellow 2010). In this context, bacteriophages have also proved to be useful tools in different applications, such as naturally-present indicators of under-represented or rare actinobacterial taxa in environmental samples; or as tools for deselecting unwanted taxa on the isolation plates in the process of target specific search for rare actinomycete taxa (Kurtböke, 2003; 2009; 2010b; 2011).

4.3.1 Actinophages as naturally-present indicators of rare actinomycetes in environmental samples

Presently, more than 50 rare actinomycete taxa are reported to be the producers of 2500 bioactive compounds (Bérdy, 2005), including several clinically important antibiotics such as vancomycin, erythromycin, tobramycin, apramycin, and spinosyns. However, these actinomycetes are not commonly cultured from natural substrates. Vancomycin producer *Amycolatopsis* sp. or spinosyn producer *Saccharopolyspora* sp. were found to be 4% and 3% abundant (Baltz, 2005).

Bacteriophages indicate presence of their host bacteria in an environmental sample and increased phage titre to detectable levels reflects the growth of indigenous host cells, and failure to do so reflects their absence from that source (Goyal, 1987). High densities of phages were reported in soils with conditions favourable for the host proliferation (Reanney and Marsh, 1973; Goyal *et al.*, 1987). This ecological reality has been used to utilize bacteriophages as naturally-present indicators of under-represented or rare actinobacterial taxa in environmental samples (Williams *et al.*, 1993; Kurtböke, 2003; Kurtböke, 2005; 2007; 2010b; 2011). Examples include detection of indicator phages towards actinomycetes including members of the genera *Saccharopolyspora* and *Salinispora* species (Kurtböke, 2009).

4.3.2 Exploitation of phages as deselection agents of unwanted taxa on isolation plates to recover rare actinomycetes from environmental samples

Direct analysis of rRNA gene sequences and birth of metagenomic studies showed that the vast majority of microorganisms present in the environment had not been captured by culture-dependent methods (Handelsman, 2004). Current advances such as microarrays

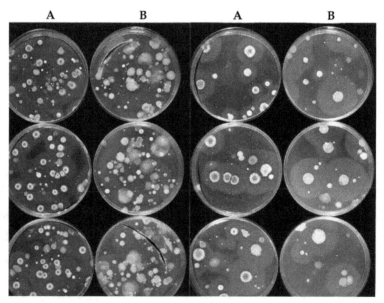

Fig. 1. Use of polyvalent streptomycete phage to reduce their numbers on isolation plates of a soil sample (A: without phage, B: with phage)

targeting the 16S rRNA gene of bacteria and archaea and the use of PhyloChips to identify specific members within a complex microbial community as well as targeting known functional gene markers to study functional gene diversity and activities of microorganisms in specific environment reveal true microbial diversity (Andersen *et al.*, 2010). Functional gene arrays (GeoChips) have also been used to analyse microbial communities, and provide linkages of microbial genes/populations to ecosystem processes and functions (Andersen *et al.*, 2010). Culturing representatives of these microorganisms with particular reference to previously explored environments such as those extreme and marine, has thus importance for biotechnological applications (Kennedy *et al.*, 2007; Joint *et al.*, 2010).

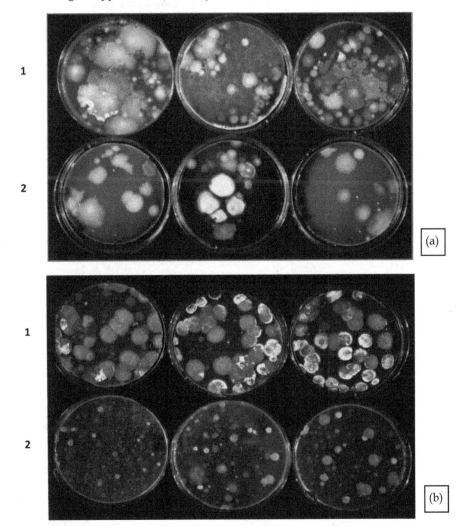

Fig. 2. Use of phage battery to reduce the numbers of (a) unwanted bacteria and (b) streptomycetes on ½ TSA plates to isolate rare actinomycetes. (1): Without phage, (2): With phage (reproduced from Kurtböke, 2010b)

Once information is generated on microbial diversity via above-listed molecular advances, phages can again be effective tools to remove unwanted taxa on the isolation plates in the process of target specific isolation of targeted taxa such as rare actinomycetes (Kurtböke *et al.*, 1992; Kurtböke, 2003; Kurtböke, 2011). Examples include removal of smearing bacterial contaminants (e.g. as *Bacillus* species) rendering isolation of rare actinomycetes difficult from heated material on the isolation plates via phage battery (Kurtböke *et al.*, 1993b; Kurtböke, 2003) (Figure 2 a,b).

Furthermore, layer by layer removal of unwanted soil taxa can also reveal bioactive fractions of the test sample under study (Kurtböke *et al.*, 2002; Kurtböke and French, 2007). This fact is illustrated in Figure 1 where removal of streptomycete fraction of the sample reveals the presence of other bacterial taxa which are obviously susceptible to the antibiotic activity of streptomycetes (Kurtböke *et al.*, 1992). This approach can particularly be useful in the detection of antimicrobial compound producing actinomycetes, even including novel streptomycetes in the samples, proved to be carrying antibiotic-resistant bacteria. It is a known fact that most studied environments can still yield novel members of bioactive genera (Williams *et al.*, 1984) and revisiting these environments via the aid of indicator phages might render new bioactive species.

It is important to note that in-depth understanding of each sample's natural characteristics and its microfloral diversity is required for successful application of phage battery as a tool for selective isolation. In every different sample, a new set of bacteriophages suitable for the nature of the sample, has to be used to remove layers only to be present in that sample. Accordingly, new sets of phages obtained against contaminating background will be required for complete reduction in the numbers of background bacteria in each different sample (Kurtböke *et al.*, 1992).

5. Conclusion

Bacteriophages can be powerful tools in the detection of bioactive actinomycetes and facilitate the discovery of novel bioactive compounds. They can offer more than we currently benefit from them if improved understanding of the host-phage ecology can be generated. Sound knowledge of microbial taxonomy is also a prerequisite for the effective use of bacteriophages in selective isolation procedures. Phage cross infectivity should also be interpreted carefully before they can be effectively exploited to select bioactive bacterial taxa (Kurtböke, 2011). In addition, current expansion of knowledge of phage and prophage genomics and phage infective mechanisms of host bacteria will provide a platform for the effective use of phages in biodiscovery.

Targeting host bacterial functional diversity, in which, certain metabolic activities might be triggered in a defined ecosystem following phage-mediated gene transfer might also offer clues for bioactivity (e.g. abolishment of rapamycin production as a consequence of phage insertion and its restoration upon the loss of the inserted phage by a second recombination (König *et al.*, 1997)). An evaluation of the role of host-phage interactions in antibiotic production as well as in rendering antibiotics ineffective via lysogenation or prophage exertion will also further complement therapeutic success, and all this provides enough reason for the value of phages to be reconsidered in the post-genomic era (Kurtböke, 2011).

Current expertise of host receptor recognition by phages and the specificity of phage-derived lytic enzymes also needs to be developed further as well as an in-depth understanding of the ecological and evolutionary reasons for monovalency and polyvalency (Kurtböke, 2011). Through such cumulative information, bacteriophages will gain increasing value as tools in drug discovery with their further use ranging from assay development to compound design (Brown, 2004; Projan, 2004).

6. In Memoriam

This chapter is dedicated to the memory of Professor Romano Locci (1937-2010), University of Milan and University of Udine, Italy.

7. References

Amber, D. (2000). DOE's microbial genomics program focuses on high-throughput draft sequencing. *The Scientist*, Vol. 14, p. 1.

Andersen, G.L.; He, Z.; DeSantis, T.Z.; Brodie, E.L. & Zhou, J. (2010). The use of microarrays in microbial ecology. *e-Scholarship*, Lawrence Berkeley National Laboratory, University of California, pp. 1-45.

Baltz, R.H. (2005) Antibiotic discovery from actinomycetes: Will a renaissance follow the decline and fall? *SIM News*, Vol., 55, No. 5, pp. 186-196.

Baltz, R.H. (2006a). Marcel Faber Roundtable: Is our antibiotic pipeline unproductive because of starvation, constipation or lack of inspiration. *Journal of Industrial Microbiology and Biotechnology*, Vol. 33, pp. 507-513.

Baltz, R.H. (2006b). Combinatorial biosynthesis of novel antibiotics and other secondary metabolites in actinomycetes. *SIM News*, Vol. 56, pp. 148-160.

Baltz, R.H. (2008). Renaissance in antibacterial discovery from actinomycetes. *Current Opinion in Pharmacology*, Vol. 8, pp. 557-563.

Bentley, S.D.; Chater, K.F.; Cerdeno-Tarraga, A.M.; Challis, G.L.; Thomson, N.R.; James, K.D.; Harris, D.E.; Quail, M.A.; Kieser, H.; Harper, D.; Bateman, A.; Brown, S.; Chandra, G.; Chen, C.W.; Collins, M.; Cronin, A.; Fraser, A.; Goble, A., Hidalgo, J.; Hornsby, T.; Howarth, S.; Huang, C.H.; Kieser, T.; Larke, L.; Murphy, L.; Oliver, O'Neil, K.S.; Rabbinowitsch, E.; Rajandream, M.A.; Rutherford, K.; Rutter, S.; Seeger, K.; Saunders, D.; Sharp, S.; Squares, R.; Squares, S.; Taylor, K.; Warren, T.; Wietzorrek, A.; Woodward, J.; Barrell, B.G.; Parkhill, J.; Hopwood, D.A. (2002). Complete genome sequence of model of actinomycete *Streptomyces coelicolor* A3(2). *Nature, Volume* 417, pp. 141-147.

Bérdy, J. (2005). Bioactive microbial metabolites. *Journal of Antibiotics*, Vol. 58, No. 1, pp. 1-26.

Bernhardt, T.G.; Wang, I-N.; Struck, D.K. & Young, R. (2002). Breaking free: "protein antibiotics" and phage lysis. *Research in Microbiology*, Vol. 153, pp. 493-501.

Bönemann, G.; Pietrosiuk, A. & Mogk, A. (2010). Tubules and donuts: a type VI secretion story. *Molecular Microbiology*, Vol. 76, No. 4, pp. 815–821.

Brown, E.D. (2004). Drugs against superbugs: private lessons from bacteriophages. *Trends in Biotechnology*, Vol. 22, No. 9, pp. 434-436.

Bull, A.T. (ed) (2004). *Microbial Diversity and Bioprospecting*. ASM Press, Washington.

Bull, A.T. (2007). Alice in Actinoland, and looking glass tales. *SIM News*, Vol. 57, No. 6, pp. 225-234.

Bull, A.T. (2010). Actinobacteria of the extremobiosphere. In: Horikoshi, K., Antranikian, G., Bull, A.T., Robb, F. & Stelter, K. (eds) *Extremophiles Handbook*. Springer-Verlag GmbH, Berlin.

Bull, A.T. & Stach, J.E.M. (2007). Marine actinobacteria: new opportunities for natural product search and discovery. *Trends in Microbiology*, Vol. 15, No. 11, pp. 491-499.

Bull, A.T.; Ward, A.C. & Goodfellow, M. (2000). Search and discovery strategies for biotechnology: the paradigm shift. *Microbiology and Molecular Biology Reviews*, Vol. 64, pp. 573-606.

Butler, M.S. (2004). The role of natural product chemistry in drug discovery. *Journal of Natural Products*, Vol. 67, pp. 2141-2153.

Cambronne, E.D. & Roy, C.R. (2006). Recognition and delivery of effector proteins into eukaryotic cells by bacterial secretion systems. *Traffic*, Vol. 7, pp. 929–939.

Canchaya, C.; Proux, C.; Fournous, G.; Bruttin, A. & Brüssow, H. (2003). Prophage genomics. *Microbiology and Molecular Biology Reviews*, Vol. 67, No. 2, , pp. 238-276.

Chanishvili, N. (2009). In: *A literature review of the practical application of bacteriophage research*, R. Sharp, (Ed.). Publication of the *Eliava Institute of Bacteriophage, Microbiology and Virology*, Tbilisi, Georgia.

Chattaway, M.A.; Dallman, T.; Okeke, I.N & Wain, J. (2011). Enteroaggregative *E. coli* O104 from an outbreak of HUS in Germany 2011, could it happen again? *The Journal of Infection in Developing Countries*, Vol. 5, No. 6, pp. 425-436.

Chibani-Chennoufi, S.; Bruttin, A.; Dillmann, M-L. & Brüssow, H. (2004). Phage–host interaction; an ecological perspective. *Journal of Bacteriology*, Vol. 186, No. 12, pp. 3677–3686.

Chopra, I.; Hesse, L. & O'Neill, A.J. (2002). Exploiting current understanding of antibiotic action for discovery of new drugs. *Journal of Applied Microbiology Symposium Supplement* 2002, Vol. 92, pp. 4S–15S.

Clements, J.M.; Beckett, R.P.; Brown, A.; Catlin, G.; Lobell, M.; Palan, S.; Thomas, W.; Whittaker, M.; Wood, S.; Salama, S.; Baker, P.J.; Rodgers, H.F.; Barynin, V.; Rice, D.W. & Hunter, M.G. (2001). Antibiotic activity and characterization of BB-3497, a novel peptide deformylase inhibitor. *Antimicrobial Agents and Chemotherapy*, Vol. 45, pp. 563–70.

Climo, M.W.; Patron, R.I.; Goldstein, B.P. & Archer, G.I. (1998). Lysostaphin treatment of experimental methicillin-resistant *Staphylococcus aureus* aortic valve endocarditis. *Antimicrobial Agents and Chemotherapy*, Vol. 42, pp. 1355-1360.

Crateu, R.; Kutchan, T.M. & Lewis, N.G. (2000). Natural Products (Secondary Metabolites). In: *Biochemistry & Molecular Biology of Plants*, Buchanan, B.; Gruissem, W. & Jones, R. (Eds.), 1250-1318. American Society of Plant Physiologists, ISBN: 10-0943088399.

Demain, A.L. (1981). *Industrial Microbiology. Science*, Vol. 214, pp. 987-995.

Demain, A.L. (1983). New applications of microbial products. *Science*, Vol. 219, pp. 709-714.

Demain, A.L. & Fang, A. (2000). The natural functions of secondary metabolites. *Advances in Biochemical Engineering and Biotechnology*, Vol. 69, pp. 1-39.

Fenton, M.; Ross, P.; McAuliffe, O.; Mahony, J.; & Coffey, A. (2010). Recombinant bacteriophage lysins as antibacterials. *Bioengineered Bugs*, Vol. 1, No. 1, pp. 9-16.

Filloux, A. (2009). The type VI secretion system: a tubular story. *The EMBO Journal*, Vol. 28, pp. 309-310.

Filloux, A. (2011). The bacterial type VI secretion system: on the bacteriophage trail. *Microbiology Today*, Vol. 38, No. 2, pp. 97-101.

Filloux, A.; Hachani, A. & Bleves, S. (2008). The bacterial type VI secretion machine: yet another player for protein transport across membranes. *Microbiology*, Vol. 154, pp. 1570-1583.

Firn, R. D. & Jones, C.G. (2003). Natural products-a simple model to explain chemical diversity. *Natural Product Reports*, Vol. 20, pp. 382–391.

Fischetti VA. (2003). Novel Method to Control Pathogenic Bacteria on Human Mucous Membranes. *Annals of New York Academy of Science*, Vol. 987, pp. 207-214.

Fischetti VA. (2008). Bacteriophage lysins as effective antibacterials. *Current Opinion in Microbiology*, Vol. 11, pp. 393–400.

Fishetti, V.A. (2010). Bacteriophage endolysins: a novel anti-infective to control Gram-positive pathogens. *International Journal of Medical Microbiology*, Vol. 300, pp. 357-362.

Ford C.W.; Zurenko, G.E. & Barbachyn M.R. (2001). The discovery of linezolid, the first oxazolidinone antibacterial agent. *Current Drug Targets and Infectious Disorders*, Vol.1, No. 2, pp. 181-199.

Goodfellow, M. (2010). Selective isolation of Actinobacteria. In: Section 1: Bull, A.T. & Davies, J.E. (section eds) *Isolation and Screening of Secondary Metabolites and Enzymes. Manual of Industrial Microbiology and Biotechnology* Baltz R.H.; Davies, J.; & Demain, A.L. (Eds), Washington: ASM Press, pp.13-27.

Goodfellow, M. & Williams, E. (1986). New strategies for the selective isolation of industrially important bacteria. *Biotechnology and Genetic Engineering Reviews*, Vol. 4, pp. 213-262.

Goyal, S.M.; Gerba, C.P. & Bitton, G. (1987). Phage Ecology. CRC Press, Boca Raton, Florida. OCLC 15654933, ISBN 0-471-82419-4.

Hagens, S.; Habel, A. & Bläsi, U. (2006). Augmentation of the antimicrobial efficacy of antibiotics by filamentous phage. *Microbial Drug Resistance*, Vol. 12; pp. 164-168.

Handelsman J. (2004). Soils-The Metagenomics Approach. In: *Microbial Diversity and Bioprospecting*, Bull, A.T., (Ed.), pp. 109-119. ASM Press, ISBN: 1-55581-267-8.

Hayakawa, M. (2003). Selective isolation of rare actinomycete genera using pretreatment techniques. In: Kurtböke, D.I. (Ed.) *Selective Isolation of Rare Actinomycetes*. Queensland Complete Printing Services, Nambour

Hermoso, J.A.; Garcia, J.L. & Garcia, P. (2007). Taking aim on bacterial pathogens: from phage therapy to ezybiotics. *Current Opinion in Microbiology*, Vol.10, pp. 461-472.

Hopwood, D.A. (2007) (Ed.). Streptomycetes in nature and medicine: The antibiotic makers. Oxford University Press. ISBN: 0-19-515066-X.

Ikeda, H.; Ishikawa, J.; Hanamoto, A.; Shinose, M.; Kikuchi, H.; 4, Shiba, T.; Sakaki, Y.; Hattori, M. & Ōmura, S. (2003). Complete genome sequence and comparative analysis of the industrial microorganism *Streptomyces avermitilis*. *Nature Biotechnology*, Volume 21, pp. 526-531.

Joint, I.; Mühling, M. & Querellous, J. (2010). Culturing marine bacteria–an essential prerequisite for biodiscovery. *Microbial Biotechnology*, Special Issue: *Marine Omics*, Giuliano, L.; Barbier, M. & Briand, F. (Eds.), Vol. 3, No. 5, pp. 564-575.

Kennedy, J., Marchesi, J.R. & Dobson, A.D.W. (2007). Metagenomic approaches to exploit the biotechnological potential of the microbial consortia of marine sponges. *Applied Microbiology and Biotechnology*, Vol. 75, pp. 11–20.

König, A.; Schwecke, T.; Molnár, I.; Böhm, G.A.; Lowden, P.A.S.; Staunton, J. & Leadley, P.F. (1997). The pipecolate-incorporating enzyme for biosynthesis of the immunosuppressant rapamycin. *European Journal of Biochemistry*, Vol. 247, pp. 526–534.

Kurtböke, D.I. (2003) Use of bacteriophages for the selective isolation of rare actinomycetes. In: *Selective Isolation of Rare Actinomycetes*, Kurtböke, D.I. (Ed.), Queensland Complete Printing Services, Nambour, Queensland, Australia. ISBN: 0646429108.

Kurtböke, D.I. (2005). Actinophages as indicators of actinomycete taxa in marine environments. *Antonie van Leeuwenhoek*, Vol. 87, pp. 19-28.

Kurtböke, D.I. (2009). Use of phage-battery to isolate industrially important rare actinomycetes. In: *Contemporary Trends in Bacteriophage Research*, Adams, H.T. (Ed.). NOVA Science Publishers, New York, ISBN-13: 978-1-60692-181-4.

Kurtböke, D.I. (2010a). Biodiscovery from microbial resources: Actinomycetes leading the way. *Microbiology Australia*, Vol. 31, No. 2, pp. 53-57.

Kurtböke, D.I. (2010b). Bacteriophages as tools in drug discovery programs. *Microbiology Australia*, Vol. 31, No. 2, pp. 67-70.

Kurtböke, D.I. (2011). Exploitation of phage battery in the search for bioactive actinomycetes. *Applied Microbiology and Biotechnology*, Vol. 89, pp. 931-937.

Kurtböke, D.I. (2012). Biodiscovery from rare actinomycetes: an eco-taxonomical perspective. *Applied Microbiology and Biotechnology, in press.* ((DOI) 10.1007/s00253-012-3898-2).

Kurtböke, D.I. & French, J.R.J. (2007). Use of phage battery to investigate the actinofloral layers of termite-gut microflora. *Journal of Applied Microbiology*, Vol. 103, No. 3, pp. 722-734.

Kurtböke, D.I.; Murphy, N.E. & Sivasithamparam, K. (1993a). Use of bacteriophage for the selective isolation of thermophilic actinomycetes from composted eucalyptus bark. *Canadian Journal of Microbiology*, Vol. 39, pp. 46-51.

Kurtböke, D.I.; Wilson, C.R. & Sivasithamparam, K. (1993b). Occurrence of *Actinomadura* phage in organic mulches used for avocado plantations in Western Australia. *Canadian Journal of Microbiology*, Vol. 39, pp. 389-394.

Kurtböke, D.I.; Chen, C-F. & Williams, S.T. (1992). Use of polyvalent phage for reduction of streptomycetes on soil dilution plates. *Journal of Applied Bacteriology*, Vol. 72, pp. 103-111.

Kütter, E. & Sulakvelidze, A. (2005). In: *Bacteriophages; Biology and Applications*, Kütter, E. & Sulakvelidze, A. (Eds.), pp. 1-4. CRC Press.

Lazzarini, A.; Cavaletti, L;, Toppo, G. & Marinelli, F. (2000). Rare genera of actinomycetes as potential producers of new antibiotics. *Antonie van Leeuwenhoek*, Vol. 78, pp. 388-405.

Leiman, P.G.; Bassler, M.; Ramagopal, U.A.; Bonanno, J.B.; Sauder, J.M.; Pukatzi, S.; Burley, S.K.; Almo, S.C. & Mekalanos, J.J. (2009). Type IV secretion apparatus and phage-tail -associated protein complexes share a common evolutionary origin. *Proceedings of National Academy of Science USA*, Vol. 106, pp. 4154-4159.

Loeffler, J.M.; Nelson, D. & Fishetti, V.A. (2001). Rapid killing of *Streptococcus pneumoniae* with a bacteriophage cell wall hydrolase. *Science*, Vol. 294, pp. 2170-2172.

Loessner, M.J. (2005). Bacteriophage endolysins-current state of research and applications. *Current Opinion in Microbiology*, Vol. 8, pp. 480-487.

Long, P.F.; Parekh, N.; Munro, J.C. & Williams, S.T (1993). Isolation of actinophage that attack some maduromycete actinomycetes. *FEMS Microbiology Letters*, Vol. 108, pp. 195- 200.

Linchon, V. & Khachemoune, A. (2006). Mycetoma: a review. *American Journal Clinical Dermatology*, Vol. 7, No. 5, pp. 315-21.

Liu, J.; Dehbi, M.; Moeck, G.; Arhin, F.; Bauda, P.; Bergeron, D.; Callejo, M.; Ferretti, V.; Ha, N.; Kwan, T.; McCarty, J.; Srikumar, R.; Williams, D.; Wu, L.J.; Gros, P.; Pelletier, J.

& DuBow, M. (2004). Antimicrobial drug discovery through bacteriophage genomics. *Nature Biotechnology*, Volume 22, No. 2, pp. 185-191.

McDevitt, D. & Rosenberg, M. (2001). Exploiting genomics to discover new antibiotics. *Trends in Microbiology*, Vol. 9, No. 12, pp. 611-616.

McPartland, J. & Rothman-Denes, L.B (2009). The tail sheath of bacteriophage N4 interacts with the *Escherichia coli* receptor. *Journal of Bacteriology*, 191(2): 525-532.

Mellmann, A.; Harmsenz, D.; Craig, C.A.; Zentz, E.B.; Leopold, S.R.; Rico, A; Prior, K.; Szczepanowski, R.; Ji, Y.; Zhang, W.; McLaughlin, S.F.; Henkhaus, J.K.; Leopold, B.; Bielaszewska, M.; Prager, R; Brzoska, P.M.; Moore, R.L.; Guenther, S.; Rothberg, J.M. & Karch, H. (2011). Prospective genomic characterization of the German Enterohemorrhagic *Escherichia coli* O104:H4 outbreak by rapid next generation sequencing technology. PLoS ONE, Vol. 6, No. 7, e22751.

Mills, S.D. (2003). The role of genomics in antimicrobial discovery. *Journal of Antimicrobial Chemotherapy*, Vol. 51, pp. 749-752.

Moak, M. & Molineux, I.J. (2004). Peptidoglycan hydrolytic activities associated with bacteriophage virions. *Molecular Microbiology*, Vol. 51, pp. 1169-1183.

Muzzi, A.; Masignani, V. & Rappuoli, R. (2007). The pan-genome: towards a knowledge-based discovery of novel targets for vaccines and antibacterials. *Drug Discovery Today*. Vol. 12, No. 11/12, pp. 429-439.

Newman, D.J. & Cragg, G.M. (2004). Advanced preclinical and clinical trials of natural products and related compounds from marine sources. *Current Medicinal Chemistry*, Vol. 11, pp. 1693-1713.

Newman, D.J. & Cragg, G.M. (2004). Marine natural products and related compounds in clinical and advanced preclinical trials. *Journal of Natural Products*, Vol. 67, pp. 1216-1238.

Okazaki, T. (2003). Studies on actinomycetes isolated from plant leaves. In: Kurtböke, D.I. (Ed.) *Selective Isolation of Rare Actinomycetes*. Queensland Complete Printing Services, Nambour.

Parisien, A.; Allain, B.; Zhang, J.; Mandeville, R. & Lan, C.O. (2008). Novel alternatives to antibiotics: Bacteriophages, bacterial cell wall hydrolases, and antimicrobial peptides. *Journal of Applied Microbiology*, Vol.104, pp. 1-13.

Payne, D.J.; Gwynn, M.N.; Holmes, D.J. & Pompliano, D.L. (2007). Drugs for bad bugs: confronting the challenges of antibacterial discovery. *Nature Reviews/Drug Discovery*, Vol.6, pp. 29-40.

Payne, K.; Sun, Q.; Sacchettini, J. & Hatfull, G.F. (2009). Mycobacteriophage Lysin B is a novel mycolylarabinogalactan esterase. *Molecular Microbiology*, Vol.73, No.3, pp.367-381.

Petrovski, S.; Seviour, R.J. & Tillett, D. (2011). Genome sequence and characterization of the *Tsukamurella* bacteriophage TPA2. *Applied and Environmental Microbiology*, Vol. 77, No. 4, pp. 1389-1398.

Projan, S. (2004). Phage-inspired antibiotics? *Nature Biotechnology*, Vol. 22; No. 2, pp. 167-168.

Pukatzki, S.; McAuley, S.B. & Miyata, S.T. (2009). The type VI secretion system: translocation of effectors and effector-domains. *Current Opinion in Microbiology*, Vol. 12, pp. 11-17.

Reanney D.C. & Marsh S.C.N. (1973). The ecology of viruses attacking *Bacillus stearothermophilus* in soil. *Soil Biology and Biochemistry*, Vol. 5, pp. 399-408.

Ricke, D.O.; Wang, S.; Cai, R. & Cohen, D. (2006). Genomic approaches to drug discovery. *Current Opinion in Chemical Biology*, Vol. 10, pp. 303-308.

Russell, A.B.; Hood, R.D.; Bui, N.K.; LeRoux, M.; Vollmer, W. & Mougous, J.D. (2011). Type VI secretion delivers bacteriolytic effectors to target cells. Nature, Vol. 475, pp. 343-349.

Schwarz, S.; Hood, R.D. & Mougous, J.D. (2010). What is type VI secretion doing in all those bugs? Trends in Microbiology, Vol. 18, No. 12, pp. 531-537.

Schuch, R.; Nelson, D. & Fishetti, V.A. (2002). A bacteriolytic agent that detects and kills Bacillus anthracis. Nature, Vol. 418, pp. 884-889.

Schmitz, J.E.; Schuch, R. & Fischetti, V.A. (2010). Identifying active phage lysins through functional viral metagenomics. Applied and Environmental Microbiology, Vol. 76, No. 21, pp. 7181-7187.

Sutcliffe, I. (2010). A phylum level perspective on bacterial cell envelope architecture. Trends in Microbiology, Volume 18, pp. 464-470.

Terekhova, L. (2003). Isolation of actinomycetes with the use of microwaves and electric pulses. In: Kurtböke, D.I. (ed) Selective isolation of rare actinomycetes. Queensland Complete Printing Services, Nambour.

Thomas, J.; Soddell, J. & Kurtböke, D.I. (2001). Fighting foam with phages? Water Science and Technology, Vol.46, No.1-2, pp. 511-8.

Udwary, D.W.; Zeigler, L., Asolkar, R.N.; Singan, V.; Lapidus, A.; Fenical, W.; Jensen, P.R. & Moore, B.S. (2007). Genome sequencing reveals complex secondary metabolome in the marine actinomycete Salinispora tropica. Proceedings of the National Academy of Sciences of the United States of America, Vol., 104, No. 25, pp. 10376-10381.

Ventura, M.; Canchaya, C.; Tauch, A.; Chandra, G.; Fitzgerald, G.F.; Chater, K.F. & van Sinderen, D. (2007). Genomics of Actinobacteria: tracing the evolutionary history of an ancient phylum. Microbiology and Molecular Biology Reviews, Vol. 71, No. 3, pp. 495-548.

Wagner, P.L. & Waldor. M.K. (2002). Bacteriophage control of bacterial virulence. Infection and Immunity, Vol. 70, pp. 3985-3993.

Watve, M.G.; Tickoo, R.; Jog, M.M. & Behole, B.D. (2001) How many antibiotics are produced by the genus Streptomyces? Archieves Microbiology, Vol. 176, pp. 386-390.

Westwater, C.; Kasman, L.M.; Schofield, D.A.; Werner, P.A.; Dolan, J.W.; Schmidt, M.G. & Norris, J.S. (2003). Use of genetically engineered phage to deliver antimicrobial agents to bacteria; an alternative therapy for treatment of bacterial infections. Antimicrobial Agents and Chemotherapy, Vol. 47, pp. 1301-1307.

Williams, S.T.; Vickers, J.C. & Goodfellow, M. (1984). New microbes from old habitats? In: Kelly, D.P., & Carr, N.G. (eds) The microbe 1984, II: prokaryotes and eukaryotes. Cambridge University Press, Cambridge, pp 219-256.

Williams, S.T., Locci, R., Beswick, A., Kurtböke, D.I., Kuznetsov, V.D., Le Monnier, F.J., Long, P.F., Maycroft, K.A., Palmit, R.A., Petrolini, B., Quaroni, S., Todd, J.I. & West, M. (1993). Detection and identification of novel actinomycetes. Research in Microbiology, Vol. 144, pp. 653-656.

Wittmann, J.; Eichenlaub, R. & Dreiseikelmann, B. (2010). The endolysins of bacteriophages CMP1 and CN77 are specific for the lysis of Clavibacter michiganensis strains. Microbiology, Vol. 156, pp. 2366-2373.

Young, I.; Wang, I. & Roof, W.D. (2000). Phages will out: strategies of host cell lysis. Trends in Microbiology, Vol. 8, pp. 120-128.

Yuan, Z.; Trias, J. & White, R. J. (2001). Deformylase as a novel antibacterial target. Drug Discovery Today, Vol. 6, pp. 954–61.

Permissions

The contributors of this book come from diverse backgrounds, making this book a truly international effort. This book will bring forth new frontiers with its revolutionizing research information and detailed analysis of the nascent developments around the world.

We would like to thank Ipek Kurtböke, for lending her expertise to make the book truly unique. She has played a crucial role in the development of this book. Without her invaluable contribution this book wouldn't have been possible. She has made vital efforts to compile up to date information on the varied aspects of this subject to make this book a valuable addition to the collection of many professionals and students.

This book was conceptualized with the vision of imparting up-to-date information and advanced data in this field. To ensure the same, a matchless editorial board was set up. Every individual on the board went through rigorous rounds of assessment to prove their worth. After which they invested a large part of their time researching and compiling the most relevant data for our readers. Conferences and sessions were held from time to time between the editorial board and the contributing authors to present the data in the most comprehensible form. The editorial team has worked tirelessly to provide valuable and valid information to help people across the globe.

Every chapter published in this book has been scrutinized by our experts. Their significance has been extensively debated. The topics covered herein carry significant findings which will fuel the growth of the discipline. They may even be implemented as practical applications or may be referred to as a beginning point for another development. Chapters in this book were first published by InTech; hereby published with permission under the Creative Commons Attribution License or equivalent.

The editorial board has been involved in producing this book since its inception. They have spent rigorous hours researching and exploring the diverse topics which have resulted in the successful publishing of this book. They have passed on their knowledge of decades through this book. To expedite this challenging task, the publisher supported the team at every step. A small team of assistant editors was also appointed to further simplify the editing procedure and attain best results for the readers.

Our editorial team has been hand-picked from every corner of the world. Their multi-ethnicity adds dynamic inputs to the discussions which result in innovative outcomes. These outcomes are then further discussed with the researchers and contributors who give their valuable feedback and opinion regarding the same. The feedback is then collaborated with the researches and they are edited in a comprehensive manner to aid the understanding of the subject.

Apart from the editorial board, the designing team has also invested a significant amount of their time in understanding the subject and creating the most relevant covers. They scrutinized every image to scout for the most suitable representation of the subject and create an appropriate cover for the book.

The publishing team has been involved in this book since its early stages. They were actively engaged in every process, be it collecting the data, connecting with the contributors or procuring relevant information. The team has been an ardent support to the editorial, designing and production team. Their endless efforts to recruit the best for this project, has resulted in the accomplishment of this book. They are a veteran in the field of academics and their pool of knowledge is as vast as their experience in printing. Their expertise and guidance has proved useful at every step. Their uncompromising quality standards have made this book an exceptional effort. Their encouragement from time to time has been an inspiration for everyone.

The publisher and the editorial board hope that this book will prove to be a valuable piece of knowledge for researchers, students, practitioners and scholars across the globe.

List of Contributors

E.V. Orlova
Institute for Structural and Molecular Biology, Department of Biological Sciences, Birkbeck College, UK

Philip Serwer
The University of Texas Health Science Center, San Antonio, USA

Luis Kameyama, Eva Martínez-Peñafiel, Omar Sepúlveda-Robles, Zaira Y. Flores-López and Rosa Ma. Bermúdez
Departamento de Genética y Biología Molecular, Centro de Investigación y de Estudios Avanzados del IPN, México

Leonor Fernández
Facultad de Química, Universidad Nacional Autónoma de México D. F., México

Francisco Martínez-Pérez
Laboratorio de Microbiología y Mutagénesis Ambiental, Escuela de Biología, Universidad Industrial de Santander, Bucaramanga, Colombia

Beatriz del Río, María Cruz Martín, Víctor Ladero, Noelia Martínez, Daniel M. Linares, María Fernández and Miguel A. Alvarez
Instituto de Productos Lácteos de Asturias, IPLA-CSIC, Spain

Toshirou Nagai
National Institute of Agrobiological Sciences, Japan

Maria M.F. Mesquita and Monica B. Emelko
Department of Civil and Environmental Engineering, University of Waterloo, Canada

Sanjay Chhibber and Seema Kumari
Department of Microbiology, Basic Medical Sciences Building, Panjab University, Chandigarh, India

L.R. Bielke, G. Tellez and B.M. Hargis
Department of Poultry Science, Division of Agriculture, University of Arkansas, USA

Takashi Yamada
Department of Molecular Biotechnology, Graduate School of Advanced Sciences of Matter, Hiroshima University, Higashi-Hiroshima, Japan

Bruce S. Seal
Poultry Microbiological Safety Research Unit, Agricultural Research Service, USDA, USA

Brian B. Oakley, Cesar A. Morales, Johnna K. Garrish and Mustafa Simmons
Poultry Microbiological Safety Research Unit, Agricultural Research Service, USDA, USA

Edward A. Svetoch and Nikolay V. Volozhantsev
State Research Center for Applied Microbiology & Biotechnology, Russian Federation

Gregory R. Siragusa
Danisco USA Inc., USA

Carla M. Carvalho, Sílvio B. Santos, Eugénio C. Ferreira and Joana Azeredo
IBB - Institute for Biotechnology and Bioengineering, Centre of Biological Engineering, Universidade do Minho, Portugal

Andrew M. Kropinski
Laboratory for Foodborne Zoonoses, Public Health Agency of Canada, Canada

D.I. Kurtböke
University of the Sunshine Coast, Faculty of Science, Health, Education and Engineering, Maroochydore DC, Australia

Printed in the USA
CPSIA information can be obtained
at www.ICGtesting.com
JSHW011443221024
72173JS00004B/918